WASTE AND WASTE MANAGEMENT

LANDFILLS

WASTE MANAGEMENT, REGIONAL PRACTICES AND ENVIRONMENTAL IMPACT

WASTE AND WASTE MANAGEMENT

Additional books in this series can be found on Nova's website under the Series tab.

Additional e-books in this series can be found on Nova's website under the e-books tab.

ENVIRONMENTAL REMEDIATION TECHNOLOGIES, REGULATIONS AND SAFETY

Additional books in this series can be found on Nova's website under the Series tab.

Additional e-books in this series can be found on Nova's website under the e-books tab.

WASTE AND WASTE MANAGEMENT

LANDFILLS

WASTE MANAGEMENT, REGIONAL PRACTICES AND ENVIRONMENTAL IMPACT

GUSTAVO B. C. CABRAL
AND
BEATRIZ A. E. BOTELHO
EDITORS

nova publishers
New York

For permission to use material from this book please contact us:
Telephone 631-231-7269; Fax 631-231-8175
Web Site: http://www.novapublishers.com

LIBRARY OF CONGRESS CATALOGING-IN-PUBLICATION DATA

Landfills : waste management, regional practices, and environmental impact / editors, Gustavo B.C. Cabral and Beatriz A.E. Botelho.
pages cm
Includes bibliographical references and index.
ISBN 978-1-62257-121-5 (hardcover)
1. Sanitary landfills. I. Cabral, Gustavo B. C. II. Botelho, Beatriz A. E.
TD795.7.L3486 2012
628.4'4564--dc23
2012020012

Published by Nova Science Publishers, Inc. † New York

CONTENTS

PREFACE

Although several countries have been introducing more stringent laws to reduce the amount of waste to be landfilled, in an attempt to maximize recycling and materials recovery, landfilling is still the most generalized practice for municipal solid waste treatment. In this book, the authors discuss waste management in landfills, regional practices and its environmental impact. Topics include the reduction of environmental impact of municipal landfill leachate during oxidative treatment; polymers recycling; management of electronic waste in the Basque Country; and toxicity of landfills by plant cytogenetic and mutagenic effects.

Chapter 1 – The aim of the authors' work was to evaluate the efficiency of oxidative processes for the treatment of municipal landfill leachates. Several oxidative treatment procedures were used for four selected landfill leachates, two of them were obtained in municipal landfills in Slovakia (LL1, LL2), while two of them originated from Slovenia (LL3, LL4). They all represent typical municipal landfill leachates of particular country, with high organic load and composition beyond effluent limits of the particular country. The first two landfill leachates (LL1 and LL2) were also treated biologically to compare the efficiency of their biological treatment and ozonation. Both raw and biologically treated leachates LL1 and LL2 were then used for ozonation experiments. Some ozonation experiments were also performed with the addition of Fenton reagents. An increase in biotreatability of ozonated leachate LL1 was investigated using kinetic testing with different types (adapted, non-adapted) of activated sludge. Slovenian landfill leachates LL3 and LL4 were treated only chemically. The efficiency of Fenton oxidation and ozonation was compared in terms of removal of organics as well as toxicity and biodegradability. The impact of the landfill leachates LL1 and LL2 to nitrification and removal of organics, and in the case of LL1 leachate biological denitrification was also evaluated in intermittently aerated completely mixed bioreactors enabling segregation of oxic and anoxic environments in the bioreactors. High efficiencies of ammonium nitrogen removal were observed for both of them. Intensive denitrification was also observed. On the other hand, high residual COD values were measured in the effluents of biological plant treating leachate LL1. This was the main reason to additionally investigate the availability of ozone and its combination with other processes, particularly with biological treatment processes, for the treatment of LL1 leachate. Post treatment of biologically pre-treated LL2 samples with regard to a decrease of ammonium nitrogen content resulted from the research. Approximately 3 to 39% reduction of COD in biologically pre-treated landfill leachate was observed after 10-12.5 h of ozonation. The maximum observed BOD_5 value, measured in ozonated biologically pre-treated landfill leachate by respirometric method, was 244 mg L^{-1}. Biodegradability expressed as BOD_5/COD ratio increased from zero up to a maximum value of 0.21, corresponding to the ozone consumption of 0.7 mg per mg of COD reduced. High efficiencies of ammonium nitrogen concentration decrease removal were observed for both landfill leachates. High denitrification efficiency was also observed. On the other hand, high residual values of COD were measured in the effluents of biological plant treating leachate LL1. This was the main reason to additionally investigate the availability of ozone treatment and its combination with other processes, particularly with biological treatment processes, for the treatment of LL1 leachate. Post treatment of biologically pre-treated LL2 samples with regard to decrease of ammonium nitrogen content resulted from the research. The efficiency of Fenton´s process and ozonation for the treatment of fresh (LL3) and mature (LL4) municipal landfill leachate from municipal waste disposal in Slovenia was also evaluated. Both samples had high organic load, high toxicity and low biodegradability. Fresh, more polluted landfill leachate (LL3) was more toxic than the mature one (LL4). Experimental results indicated that the Fenton's oxidative pretreatment of heavily polluted landfill leachate was more effective for mature landfill leachates. For this sample up to 88% of organics while in case of fresh leachate only 46% of COD could be removed. Zero or first order kinetics provided the best description of organics (in terms of COD and DOC) removal by the Fenton's oxidation for both landfill leachates. Higher reaction rates of the Fenton's oxidation were achieved with fresh leachate samples. The efficiency of initial organics removal during ozonation was about 70%, both in COD and TOC terms. In case of fresh LL3 leachate only 42% of COD were removed. The best fits of COD and DOC experimental data from ozonation treatment of fresh and mature LL4 leachates were obtained applying the combined kinetic model. Toxicity during ozonation was reduced more effectively than in the Fenton's oxidation. Similar to the Fenton reaction, ozonation was also assessed as appropriate for the removal of organics, while it did not improve the biotreatability potential of the leachates. Experiments indicated that an oxidative treatment applied as a single or post treatment step for landfill leachates should be evaluated carefully prior to its actual application. However, oxidative processes could be used as a valuable option when a combination of biological and chemical treatment is considered.

Chapter 2 - Although several countries have been introducing more stringent laws to reduce the amount of waste to be landfilled, in an attempt to maximize recycling and materials recovery, landfilling is still the most generalized practice for municipal solid waste (MSW) treatment. The generation of a high-polluted wastewater leaching from a sanitary landfill - the leachate - is inevitable and its treatment constitutes an important burden in landfill management, even after landfill closure. The selection and design of the management and treatment of a landfill leachate has to take into consideration factors such as leachate characteristics, effluent discharge alternatives, technological possibilities, final discharge requirements and economical aspects. Careful site management can increase the quality and reduce pollution potential of the formed leachate, however it does not allow total pollutant removal. The leachate complexity and variability prevents the formulation of general treatment recommendations of universal validity. Techniques successfully developed for one site might not necessarily being applicable elsewhere. In addition, treatment methods that work well at young landfills are likely to become progressively less effective with the tip's age. Therefore, the leachate treatment system must be flexible enough to produce the same effluent quality despite the sudden and large variations in the leachate strength. Currently, the treatment systems used for leachate treatment include biological, physico-chemical and a combination of these processes. The enforcement of environmental discharge rules and the ageing of landfill sites with more stabilized leachate in different countries have put pressure on operators to implement better processes for the reduction of pollutant. New treatment alternatives and conventional technology improvements, mainly specialized on tertiary treatment, have been developed in recent years. This work presents an overview of the most common processes used in landfill leachate treatment, their drawbacks and advantages. As the removal of nitrogenous compounds (mainly ammonium and nitrate) to values below the discharge limits has been one of the major problems in leachate treatment plants operating in Portugal, particular attention is focused on the Portuguese leachate treatment or pre-treatment plants. Furthermore, the results obtained by the authors' group with a lab-scale anoxic rotating biological contactor to remove the nitrate excess, from a real landfill leachate, collected in a Portuguese municipal landfill before being discharged to the municipal sewer, are presented. The results achieved so far point out this process as a promising alternative polishing treatment.

Chapter 3 - The accidental discharge of leachate from solid waste landfills often leads to groundwater contamination both by inorganic and organic compounds. A major contribution to groundwater pollution at solid waste landfills comes from chlorinated organic compounds such as Tetrachloroethylene (PCE) and Trichloroethylene (TCE). They are among the most dangerous pollutants, due to their high toxicity and persistence grade mainly originating from plastic material degradation due to rainfall water leaching. The environmental fate of these species strictly depends on their main physical and chemical properties; they are denser and less viscous than water, thus migrating from superficial water and contaminated soils to groundwater. Consequently, the remediation of polluted sites must be designed accordingly. Several remediation technologies have been developed for aquifers contaminated by organic and inorganic compounds; they range from ex-situ treatments, coupled with the *pump and treat* technique, to in situ treatments, performed with Permeable Reactive Barriers (PRB). In a PRB-based treatment, the barrier is commonly built with reactive materials whose hydraulic conductivity is higher than that of the surrounding soils, so that the contaminated groundwater, moving under natural hydraulic gradient, is forced to pass through the barrier without any external energy input. Nowadays, Pump and Treat (PT) technology is widely used for groundwater remediation. This technology has good remediation efficiency but has high operational costs. For this reason, there is a growing interest for PRB installations as an effective alternative to Pump and Treat remediation methods, thanks to its low operating and maintenance costs. The mechanism of action of a PRB depends on the reactive material chosen to build the barrier. Chlorinated organic compounds can be effectively removed from polluted water and wastewater also by adsorption, a process that combines good efficiency with a very simple process configuration. The pollutant is trapped into the barrier made of adsorbing material and does not precipitate. This particular PRB made of adsorptive material is named Permeable Adsorbing Barrier (PAB). The chapter, after a short description of the main remediation technologies, describes a specific procedure for the design of a PAB for the remediation of an aquifer contaminated by PCE and TCE. As case study, the design of a PAB at a solid waste landfill site in Giugliano in Campania (area North of Napoli, Italy) is presented. In this site, there are many solid waste landfills where, over the past 20 years, about eight million tons of urban and special wastes have been deposited and contamination by different compounds has been detected.

Chapter 4 - This chapter introduces readers to the processes involved in the creation of valley fills and how valley fills may affect the geochemistry of surface and underground waters. An ideal placement of mining waste in valley fills/landfills often requires a greater depth of prior knowledge on the chemical and physical properties of mining waste. Such knowledge allows for the development of appropriate methodologies and then procedures that can be designed to avoid landfill collapse and/or leaching of specific chemical substances. Solid mining waste or overburden containing substantial amounts of bioavailable and bio-accessible chemical species can be treated in-situ during placement in landfills to ameliorate subsequent release of certain targeted chemical species. Assessment of leaching characteristics using various laboratory analytical procedures is presented. The leaching processes are discussed with reference to myriad applications of a variety of instrumental methods of analysis. Emerging technologies on proactive and reactive control measures that include the sequestration and immobilization of dissolved minerals ions that encompass selenate, selenite, arsenate, and arsenite are also presented.

Chapter 5 - In the last decades sustainable waste management has become a high priority due to increased municipal waste production. In most countries landfilling of the waste still plays the most important role in waste management. As a result of stabilization processes, landfill leachate and biogas are formed. Heavily polluted landfill leachate must be collected and treated properly. Due to the complex matrix of the leachate there are no general recommendations of how to treat landfill leachate and usually a combination of physical, chemical and biological methods is used. Very promising are advanced oxidation processes (AOPs) with high efficiency for organics removal because of the formation of extremely reactive and non-selective hydroxyl radicals. Fenton's oxidation is one of these AOPs with high efficiency and low capital costs. Its main disadvantage is possible formation of different by-products with hazardous environmental impact. This is the reason that physico-chemical methods alone are not enough for reliable monitoring of treatment efficiency. They must be accompanied by different biotests to assure actual removal and decomposition of harmful pollutants prior to release of the leachate into the environment. Toxicity assessment should be performed using species belonging to different levels of aquatic ecosystem: decomposers (bacteria), producers (like duckweed *Lemna minor*) and consumers (like brine shrimp *Artemia salina* or water flea *Daphnia magna)* as well as species belonging to terrestrial ecosystem (like producers; white mustard *Sinapis alba* and white onion *Allium cepa)*. Toxicity of landfill leachates to plants, especially to higher plants, has not received much study. In this chapter an overview of different treatment methods and methods for ecotoxicity assessment for landfill leachates has been presented. In a case study municipal landfill leachate was treated by Fenton oxidation and its physico-chemical parameters, biodegradability and toxicity were determined prior to, and after treatment. Fenton oxidation was not successful for treatment of the investigated leachate. Maximal treatment efficiency was 58% in terms of COD removal at higher molar ration (1:10) and phytotoxicity decreased for just 30%. In some cases, formation of phytotoxic pollutants after Fenton oxidation was identified. Phytotoxicity tests were confirmed to be a necessary supplement for reliable evaluation of hazardous impact of landfill leachate. They are also important to confirm the efficiency of the selected treatment methods. Commonly applied toxicity tests with producers and consumers in the ecosystem do not assure enough data on the impact of hazardous landfill leachates to avoid long-term pollution problems.

Chapter 6 - Polymer materials with a range of excellent mechanical properties, low density, durability and low cost, are widely used in the daily needs of contemporary society, ranging from simple packaging to heavy construction. From another side, a continued increase in the use of plastics has led to an increasing amount of plastics ending up in the waste stream, causing in dangerous environmental problems. Recycling is the process for treatment and reusing these wastes. Generally, plastics are made from crude oil so that recycling of plastics helps to conserve this natural resource. This chapter presents a review of the recent works about the polymers recycling. Also it presents the last developments in the recycling of traditional polymers and their blends and composites using the mechanical and chemical methods.

Chapter 7 - Polyethylene (PE) is the highest volume thermoplastic consumed in the world, constituting the dominant component of plastics wastes. It is becoming the most recycled polymeric material. Polymer blends are becoming commercially important because they offer a property balance not found in either homopolymer. Furthermore, blending technology is used to enhance polymer waste recycling. This chapter is intended to focus on some recent scientific research in this field. Regenerated low density polyethylene (rLDPE) coming from various neat or dirty wastes sources (bags, ustabilized agricultural films, carbon black stabilized agricultural films, milk pouches) were used. They were obtained after milling, washing and extrusion in a conventional recycling plant. They were first characterized in terms of physico- chemical (density, melt flow index, water absorption, melting temperature and structure by Fourier transform infrared spectroscopy (FTIR) and mechanical properties (tensile properties and shore D hardness)). These rLDPE samples were recycled in the form of blends with ethylene- propylene-diene-monomer (EPDM) in the presence of peroxide, with polyvinyl chloride (PVC) in the presence of chlorinated polyethylene (CPE) as compatibizer, with polystyrene (PS) and polypropylene (PP) in presence of maleic anhydride functionalized elastomeric ethylene copolymer as comptiilizer. Blends of various compositions were realized and characterized physically (density, water absorption) and mechanically (tensile properties and shore D hardness). The experimental data showed an improvement of mechanical properties. Thus, from the point of view of practical applications, this kind of blending is an effective way of recycling LDPE wastes.

Chapter 8 - The success of a recycling solution as an alternative to disposal, in order to become economically feasible, depends mainly on two aspects: the responsibility of consumers to take back their appliances (reverse logistics) and the costs of handling and disassembly. The first issue is being dealt by several legislative initiatives, e.g.: European Directives on waste. For the second, a wide spectrum of technological solutions has been suggested and developed. This chapter will cover the first experiences of collecting the waste of electric and electronic appliances in the Basque Country, a region in Northern Spain. Enforcing laws, marketing strategies, related to logistics, etc, will be explained. The results over the first years of the experience applied in main Basque towns were positive, therefore the collection scheme was expanded to the whole region and other areas of Spain. Main outcomes regarding the economic impact, social awareness and influences over the environment will also be covered. The main problem concerning the disassembly of devices is their high diversity. A good technical approach should offer an intelligent combination of manual and automated operations, optimized logistics, flexible plant layout, etc. This chapter offers a review of the state of the art of industrial solutions, together with their shortcomings and future trends. In addition, some new optimization techniques will be explained, with the aim to increase the benefits of the recycling companies, promoting in this way the correct treatment of the electronic waste and avoiding the disposal in landfills. In order to solve the problem of a growing generation of electronic waste, the promotion of the reuse of devices and components is considered to be one of the most desirable solutions. The concept of reutilization of electronic equipment goes from the reuse of the complete device for a second life, to the alternative of making the most of its valuable components. In order to show specific examples of these solutions, a case study will be explained showing the recycling of TV sets and monitors. A local recycler of TV sets and monitors has been involved in the definition and planning of the recycling facility. This company has carried out a statistical study of the received TV sets and monitors, collecting technical data about the devices. Taking into account these records and other technical information, a recycling scheme was proposed in order to handle, separate, recycle and reuse them. The recycler defined some requirements to be fulfilled by the process in order to design the recycling facility properly. The components and materials obtained from the separation process have been reused in new devices (e.g. electronic components) and recycled as raw materials (e.g. glass) for new usages.

Chapter 9 - Every year, millions of tons of municipal solid waste are deposited in landfills. Upon rainfall, the production of leachates represents a serious environmental problem. Household waste dumped in landfills is mostly not correctly handled and treated. As a result, landfill leachates can pollute underground water. In this context, it is important to set up simple and accurate methods for monitoring the toxicity of landfill leachates. Many assays are available for this evaluation, employing plants, microbes, aquatic animals, mammals and human peripheral blood lymphocytes as test models. Among the bioassays used to detect toxicity and mutagenicity of environmental pollutants, plant systems have proved to be sensitive, cheap and effective alternatives, finding a wide range of applications as indicators of cytogenetic and mutagenic effects of contaminants. Some of these models evaluate mutations, e.g. the Stamen Hair Mutation (SHM) assay in *Tradescantia*, while others apply cytogenetic parameters to assess genomic modifications, e.g. the *Allium cepa* test. These species have been employed in studies of landfill leachate impact under natural and experimental conditions. The present chapter offers a review of these essays and their efficiency in evaluating cytogenotoxic and mutagenic effects of landfill leachates. Emphasis will be given to their applications, methodologies and endpoints.

In: Landfills
Editors: G. B. C. Cabral and B. A. E. Botelho

ISBN: 978-1-62257-121-5

Chapter 1

OXIDATIVE TREATMENT OF LANDFILL LEACHATE

Jan Derco*[*1], Andreja Žgajnar Gotvajn[2] and Angelika Mencáková[1]

[1]Slovak University of Technology,
Faculty of Chemical and Food Technology,
Institute of Chemical and Environmental Engineering,
Bratislava, Slovak Republic
[2]University of Ljubljana, Faculty of Chemistry
and Chemical Technology, Ljubljana, Slovenia

ABSTRACT

The aim of our work was to evaluate the efficiency of oxidative processes for the treatment of municipal landfill leachates. Several oxidative treatment procedures were used for four selected landfill leachates, two of them were obtained in municipal landfills in Slovakia (LL1, LL2), while two of them originated from Slovenia (LL3, LL4). They all represent typical municipal landfill leachates of particular country, with high organic load and composition beyond effluent limits of the particular country. The first two landfill leachates (LL1 and LL2) were also treated biologically to compare the efficiency of their biological treatment and ozonation. Both raw and biologically treated leachates LL1 and LL2 were then used for ozonation experiments. Some ozonation experiments were also performed with the addition of Fenton reagents. An increase in biotreatability of ozonated leachate LL1 was investigated using kinetic testing with different types (adapted, non-adapted) of activated sludge. Slovenian landfill leachates LL3 and LL4 were treated only chemically. The efficiency of Fenton oxidation and ozonation was compared in terms of removal of organics as well as toxicity and biodegradability. The impact of the landfill leachates LL1 and LL2 to nitrification and removal of organics, and in the case of LL1 leachate biological denitrification was also evaluated in intermittently aerated completely mixed bioreactors enabling segregation of oxic and anoxic environments in the bioreactors. High efficiencies of ammonium nitrogen removal were observed for both of them. Intensive denitrification was also observed. On the other hand, high residual COD values were measured in the effluents of biological plant treating

[*] E-mail: jan.derco@stuba.sk.

leachate LL1. This was the main reason to additionally investigate the availability of ozone and its combination with other processes, particularly with biological treatment processes, for the treatment of LL1 leachate. Post treatment of biologically pre-treated LL2 samples with regard to a decrease of ammonium nitrogen content resulted from the research. Approximately 3 to 39% reduction of COD in biologically pre-treated landfill leachate was observed after 10-12.5 h of ozonation. The maximum observed BOD_5 value, measured in ozonated biologically pre-treated landfill leachate by respirometric method, was 244 mg L^{-1}. Biodegradability expressed as BOD_5/COD ratio increased from zero up to a maximum value of 0.21, corresponding to the ozone consumption of 0.7 mg per mg of COD reduced. High efficiencies of ammonium nitrogen concentration decrease removal were observed for both landfill leachates. High denitrification efficiency was also observed. On the other hand, high residual values of COD were measured in the effluents of biological plant treating leachate LL1. This was the main reason to additionally investigate the availability of ozone treatment and its combination with other processes, particularly with biological treatment processes, for the treatment of LL1 leachate. Post treatment of biologically pre-treated LL2 samples with regard to decrease of ammonium nitrogen content resulted from the research.

The efficiency of Fenton′s process and ozonation for the treatment of fresh (LL3) and mature (LL4) municipal landfill leachate from municipal waste disposal in Slovenia was also evaluated. Both samples had high organic load, high toxicity and low biodegradability. Fresh, more polluted landfill leachate (LL3) was more toxic than the mature one (LL4). Experimental results indicated that the Fenton's oxidative pretreatment of heavily polluted landfill leachate was more effective for mature landfill leachates. For this sample up to 88% of organics while in case of fresh leachate only 46% of COD could be removed. Zero or first order kinetics provided the best description of organics (in terms of COD and DOC) removal by the Fenton's oxidation for both landfill leachates. Higher reaction rates of the Fenton's oxidation were achieved with fresh leachate samples. The efficiency of initial organics removal during ozonation was about 70%, both in COD and TOC terms. In case of fresh LL3 leachate only 42% of COD were removed. The best fits of COD and DOC experimental data from ozonation treatment of fresh and mature LL4 leachates were obtained applying the combined kinetic model. Toxicity during ozonation was reduced more effectively than in the Fenton's oxidation. Similar to the Fenton reaction, ozonation was also assessed as appropriate for the removal of organics, while it did not improve the biotreatability potential of the leachates.

Experiments indicated that an oxidative treatment applied as a single or post treatment step for landfill leachates should be evaluated carefully prior to its actual application. However, oxidative processes could be used as a valuable option when a combination of biological and chemical treatment is considered.

INTRODUCTION

Industrial and economic growth in many countries around the world has resulted in rapid increase in industrial and municipal solid waste generation (Neczaj et al., 2005). Many countries are still trying to implement separate collection and disposal of wastes based on Reduce-Recycle-Reuse principle, which should lead to zero waste management practice in the future. Currently the deposit on a landfill is still the most widely used method for municipal solid waste disposal. A major problem regarding waste disposal is lack of available landfill sites, as well as production of landfill leachates and biogas. Biogas is usually utilized for energy purposes of the site, while leachates could pose serious risk for nearby surface and underground waters (Klimiuk et. al., 2006).

Appropriate management can reduce the quantity and quality of the leachate, but it cannot be completely eliminated. Its composition is based on type and amount of waste deposited, as well as their maturity and construction of landfill site (Kurniawan et al., 2006). Concentrations of organic compounds, expressed as COD and BOD_5 in the leachate from young landfill are high, while leachates from stabilized landfills contain lower levels of organic matter (Klimiuk et. al., 2006). Several authors have reported that in the young landfills COD could vary from 5,000 to 60,000 mg L^{-1} while BOD_5 could start from 3,000 to even 40,000 mg L^{-1} (Robinson et al., 1983). Those values are significantly lower in leachates from mature landfills as a result of (a)biotic stabilization processes in the body of the landfill. This affects the attempts of uniform design of leachate treatment systems.

Traditionally, biological treatment is the most widely used treatment strategy for wastewaters, mostly because of its low operational costs and complete destruction of pollution; not only its transport into another phase of form (Renou et al., 2008). However, biological treatment is not always effective enough for toxic and low biodegradable wastewaters and in this case physico-chemical treatment can take place.

Usually the leachate treatment involves a combination of various biological and chemical methods. Conventional landfill leachate treatment can be classified into three major groups: i) recycling and combined treatment with domestic sewage, ii) biological treatment employing mainly aerobic biodegradation processes and iii) chemical and physical methods, such as chemical oxidation, adsorption, coagulation/flocculation, membrane techniques and air stripping (Renou et al., 2008).

Chemical oxidation processes are viable options for effective landfill leachate treatment, but they are relatively expensive for complete mineralization of landfill leachate pollutants because the oxidation intermediates formed during treatment tend to be more and more resistant to complete chemical degradation. However, the combination of oxidative processes and biological treatment has been investigated as one of the attractive potential alternatives. Fenton oxidation or ozonation in a pre-treatment process can convert persistent and non-biodegradable organic compounds into more biodegradable intermediates, which would be consequently treated by biological treatment process.

On the other hand, polishing of remaining organics in the effluent after biological treatment using one of the oxidation processes could also reduce environmental impact of landfill leachates. Several authors have been reported that Fenton's process can achieve 60-90% of COD removal of organics from landfill leachate (Hermosilla et al., 2009). Moreover, if leachates were pretreated by biological processes, the Fenton's oxidation added an additional 63% of COD removal (Lopez, 2004).

Landfill Leachate

The leachate results from the degradation of municipal (solid) waste together with the percolation of rain water. Their treatment is necessary in order to prevent contamination both to the surface (surface water receiver) and underground water. After collection, they can be treated on site or in a municipal WWTP. The most appropriate treatment has to be selected for each specific case, considering the state of the art in the area as well as local requirements.

The member countries of the European Union each year generate about 2,200 million tons of wastes, about 70% of which is landfilled. The necessity for landfills will always exist because certain materials cannot be recycled, composted or burned and because alternative options (sludge handling) such as recycling, composting and incineration will always produce residues. The sanitary landfill method for the ultimate disposal of solid waste material continues to be widely accepted and used also due to its economic advantages.

Comparative studies of the various possible means of eliminating solid urban waste (landfilling, incineration, composting, etc.) have shown that the cheapest, in terms of exploitation and capital costs, is landfilling (Renou et al., 2008). Leachate emissions from landfill sites are of concern, primarily due to their toxic impact when released untreated into the environment (Silva et al., 2004). The composition of landfill leachate is highly variable and depends on many factors such as refuse characteristics, hydrology of the site, climate, season, age of the site, height of refuse and moisture routing through the refuse. However, three types of leachate can basically be distinguished, depending on the landfill age (Kurniawan et al., 2004). Volatile fatty acids (VFA) are the main contributor to the high COD value of young leachate. COD/BOD ratio values range from 3.3 to 10.0 and VFA represent 10% to 30% of the total organic matter contained in intermediate leachate. The major components of the aged raw landfill leachate are large molecular refractory organic molecules; i.e., humic substances. This is due to the stabilization of most of its biodegradable fraction during the aging process (Wang at al., 2006).

Fulvic acids and humic substances represent the non-biodegradable COD and 76% to 90% of the total organic content in old leachate. COD/BOD ratio values are usually more than 10 (Table 1) for old landfill leachate (Bigot et al., 1995).

The second large group of pollutants is ammonium nitrogen, which is the dominant form of nitrogen impurities contained in leachates. The metal content of leachates from landfills receiving domestic wastes is usually low except for iron and it does not obstruct biological treatment.

Table 1. Landfill leachate characterization (Bigot et al., 1995)

	Young leachate	Intermediate leachate	Old leachate
Age [years]	< 5	5 – 10	> 10
pH [-]	< 6.5	6.5 – 7.5	> 7.5
DOC [g L^{-1}]	> 10	4 – 10	< 4
BOD_5/COD [-]	> 0.3	0.1 – 0.3	< 0.1
VFA [%]	70 – 90	10 – 30	0*
TKN [g L^{-1}]	/	0.1 – 2.0**	/

* No VFA, high molecular weight compounds.

** Not defined.

Biological and Advanced Oxidative Treatment of Landfill Leachates

The chemical oxidation process, including ozonation as well as Fenton oxidation, can either lead to a total degradation of chemical compounds to carbon dioxide or to transformation of highly hydrophobic organics to more polar easily biodegradable molecules

(Gulyás, 1997). Newly developed processes apply ozone for elimination of toxic compounds and for partial oxidation of resistant organic pollutants from wastewaters to overcome the expense of using large amounts of ozone to obtain their complete mineralization (Espulgas et al., 1996; Alvares et al., 2001). Hence, organics that are highly or completely resistant to conventional biological treatment can be totally mineralized using a combined chemical - biological system. This combination of processes is often applied for wastewater treatment, particularly for landfill leachate treatment.

Biological Processes

It is well known that the most common way to remove organic, nitrogen, or phosphorus impurities from wastewater are biological processes. One of the reasons is the possibility to achieve high elimination efficiency of these pollutants from wastewater. In addition to this there are relatively low capital and operational costs in comparison to physicochemical or chemical processes. Pollution is also completely destructed, not just transformed in another form. The wastewater treatment process is a highly complex system characterized by large disturbances, which are controllable with difficulties mainly due to mass and hydraulic load variations. Consequently, wastewater treatment plants (WWTPs) are practically never in a steady state. One of major objectives of biological wastewater treatment represents removal of organic matter. Domestic sewage contains approximately one-third suspended organics, one-third colloidal organics, and one-third soluble organics (Eckenfelder and Musterman, 1995).

The most often applied processes for this purpose are aerobic processes. Under an aerobic environment, i.e. the concentration of dissolved oxygen in a bioreactor is maintained above 2 mg L^{-1}, organic matter is used as a food source for microorganisms. The suspended organics are removed by being enmeshed in the biological activated sludge flocks (biomas). The colloidal organics are partially adsorbed and entrapped by the sludge flocks and a portion of the soluble organics is ′sorbed′ by the activated sludge flocks. Thus approximately up to 85% removal of the total COD (chemical oxygen demand) can be achieved after 10 to 15 min of activated sludge-wastewater contact. The remaining degradable soluble organic fraction undergoes a biological reaction in accordance with the kinetics of the system (Eckenfelder and Musterman, 1995). A portion of organic compounds (about 50% in terms of organic carbon) is oxidized to CO_2 and H_2O (catabolic process), and the rest is utilized for the synthesis of reserve material and incorporated into new cells (anabolic processes). Approximately 60% of energy content in wastewater organics is consumed for synthesis of new biomass and the rest (about 40%) represents reaction heat loss.

Polysacharides and lipids are the most frequent reserve substances. In the absence of organic matter, the cell tissue will be endogenously respired to gaseous end products and residuals in order to obtain energy for maintenance. In most biological treatment systems these three processes occur simultaneously. They can be represented for an aerobic process as follows (Pitter and Chudoba, 1990):

- Oxidation of organic substances (dissimilatory process):

$$C_xH_yO_z + (x + y/4 - z/2)O_2 \rightarrow xCO_2 + y/2H_2O \qquad (1)$$

- Synthesis of biomass (assimilatory process):

$$n(C_xH_yO_z) + nNH_3 + (x + y/4 - z/2 - 5)O_2 \rightarrow \quad (2)$$

- Endogenous respiration (auto-oxidation):

$$(C_5H_7NO_2)_n + n(x-5)CO_2 + n(y-4)/2H_2O \quad (3)$$

The formula $(C_5H_7NO_2)_n$ represents cell material. It is not a real chemical formula and represents a generalized value (formula), with regard to mutual ratios of the most important elements constituting cell tissue, obtained from experimental studies.

Nitrogen and phosphorus are required for biomass in order to perform synthesis, metabolism, and organics removal. This requirement can be approximately expressed as the following mass ratio: $BOD_5 : N : P = 100 : 5 : 1$. In general, obtaining this ratio is not the problem in the case of domestic and municipal wastewater. Deficiency of nutrients is usually the problem in the case of many industrial wastewaters and landfill leachates. In addition to these ′macronutrients′ trace levels of ′micronutrients′ (e.g. heavy metals) are required. These are usually present in sufficient quantities in raw wastewater.

The anaerobic microbial conversion of organic substrates to methane is a complex process involving a number of microbial populations. The first step in this process involves the enzyme mediated transformation of higher-weight molecular compounds into compounds suitable for use as a source of energy and cell carbon. The bacterial conversion of these products to lower-molecular-weight intermediate compounds occurs simultaneously. The conversion of the intermediate compounds into simpler end products, principally methane and carbon dioxide, is involved in the second step of the process.

The main portion of organics in the wastewater (about 95% in terms of organic carbon) is transformed to biogas, which can reach about 65 to 75% of methane content. The remaining organics in the wastewater is utilized for the synthesis of new cell material. Approximately 90% of energy content in wastewater organics are converted and conserved in biogas, about 5 to 7% of energy is consumed for synthesis of new biomass and the rest, about 3 to 5%, represent reaction heat loss. It is obvious that anaerobic processes of organics removal from wastewater are more sustainable when comparing to aerobic processes. On the other hand, it should be mentioned that specific removal rates of organics from wastewater by anaerobic activated sludge are significantly lower in comparison to aerobic activated sludge ones. Thus, anaerobic treatment of wastewater by activated sludge requires significantly higher values of HRT and consequently significantly greater reactor volumes in comparison to the aerobic activated sludge process.

The discovery of the effects of eutrophication upon lakes and seas has turned attention toward nutrients removal from wastewaters. Similarly to organic impurity, ammonium nitrogen can be transferred to nitrite or nitrate in an aerobic environment. Inorganic carbon, i.e. carbon dioxide, is utilized by chemoautotrophic nitrification bacteria (*Nitrosomonas, Nitrobacter*) in synthesis of new microorganisms. The energy-yielding two-step oxidation of ammonia to nitrate is generally represented as follows (Randall et al., 1992):

Nitrosomonas $$2NH_4^+ + 3O_2 \rightarrow 2NO_2^- + 4H^+ + 2H_2O \quad (4)$$

Nitrobacter $$2NO_2^- + O_2 \rightarrow 2NO_3^- \quad (5)$$

Total reaction $$NH_4^+ + 2O_2 \rightarrow NO_3^- + 2H^+ + H_2O \quad (6)$$

The removal of the products of nitrification can be accomplished in anoxic environment. In anoxic conditions, i.e. the level of dissolved oxygen is maintained near zero (less than 0.5 mg L^{-1}), the reduction of nitrite and nitrate occur. Nitrate nitrogen is utilized as the terminal electron acceptor, releasing nitrogen gas as the end product of the process of denitrification. Two models of nitrate reduction can occur in biological reactors: assimilating and dissimilating or denitrification (Randall et al, 1992). Assimilating nitrate reduction involves the reduction of nitrate to ammonia for use in cell synthesis. It occurs independently of oxygen concentration and when ammonium nitrogen is not available. On the other hand, dissimilating nitrate reduction or denitrification is coupled to the respiratory electron chain and involves the reduction of nitrate to nitrite to nitric oxide to nitrous oxide to nitrogen:

$$NO_3^- \rightarrow NO_2^- \rightarrow NO \rightarrow N_2O \rightarrow N_2 \quad (7)$$

Organic carbon source is necessary for the process of denitrification.

Thus the combination of aerobic and anoxic environment is necessary to remove organic and nitrogen pollutants from wastewater. On the other hand, the combination of anaerobic and aerobic biochemical environment is required in order to remove abundant phosphates from wastewater. The extension of biological processes from carbonaceous impurities removal to nitrogen and phosphorus removal has impacted the biological system configuration. Not only are the system configuration and its operation increased in complexity, but more stringent standards for effluent quality have to be satisfied. Thus, the system must be well designed, optimized and operated at its optimum in order to meet these effluent standards. Various biological system configurations for nutrient removal have been applied in practice. High removal efficiencies of nitrogen pollutants can be achieved in systems by simultaneous nitrification and denitrification processes.

Great flexibility in biochemical process control is typical for these systems. Aerobic and anoxic zones are well maintained, for example in an activated sludge reactor with long ditch channels, e.g. Carrousel systems. On the other hand, it is difficult to maintain different environmental conditions in a short ditch channel. The intermittent aeration method is available in these cases (Araki et al., 1990). The intermittent aeration method is widely used in the operation of small wastewater treatment plants (Nakajima and Kaneko, 1991). Intermittently operated continuously fed systems also seem to be suitable for the treatment of high-strength industrial wastewater (Bell and Hardcastle, 1984).

Advanced Oxidation Processes

When landfill leachate treatments have been discussed, growing interest has been focused on advanced oxidation processes (AOPs). They are mainly based on exploitation of the strong oxidation potential of hydroxyl radicals (HO·), which could achieve main goals: i) reduction of organics in terms of COD or DOC contents; and ii) at the same time more recalcitrant organics are transformed into more biodegradable ones.

In general, AOPs are defined as oxidation processes, which can generate hydroxyl radicals in sufficient quantity to affect organic pollution (Lopez, 2004). Processes could involve energy enhancement in terms of light involvement (UV photochemical methods) in combination with ozone, H_2O_2, ozone/H_2O_2, Fe^{2+}/H_2O_2 (photo-Fenton process) and TiO_2 (Lopez, 2004). There are also several non-photochemical methods available (Fe^{2+}/H_2O_2 (Fenton process), ozonation, combination of ozone and H_2O_2, ozone with different catalysts, etc.). State of the art of AOPs for wastewater treatment is presented in detail by Poyatos et al. (2010).

Ozone Oxidation

Ozonation represents one of the alternatives for the removal of biologically resistant and toxic compounds from wastewater. Ozone possesses oxidizing properties desirable for its use in wastewater treatment. It is a powerful oxidant capable of oxidative degradation of many organic compounds, it is readily available, soluble in water, and leaves more biodegradable by-products (Beltran-Heredia et al., 2000).

Likewise other chemical oxidation processes ozonation can either lead to a total degradation of chemical compounds to carbon dioxide or to transformation of highly hydrophobic organics to more polar biodegradable molecules (Gulyás, 1997). However, total mineralization by ozone is very expensive. Thus, new developed processes apply ozone only for elimination of toxic compounds and for partially oxidation of resistant organic pollutants from wastewaters (Espulgas et al., 1996; Alvares et al., 2001).

The ozonation process breaks molecules in smaller fragments, which contain a higher percentage of oxygen in their chemical structure in the form of hydroxyl, carboxyl or aldehyde functional groups that are more amenable to biodegradation (Painter, 1995). By this means ozonation processes can inactivate inhibitory compounds in wastewater treatment and improve the biodegradability of recalcitrant organics by altering their chemical structure. The extent of this partial oxidation needs only to be sufficient to facilitate subsequent biodegradation of the converted organic matter. Hence, organics that are highly or completely resistant to conventional biological treatment can be totally mineralized using a combined chemical - biological system. This combination of processes is very often applied for wastewater treatment, particularly for landfill leachate treatment.

Ozone may react with organic compounds in two ways: by direct reaction as molecular ozone or by indirect reaction through formation of secondary oxidants like free radical species (Hoigné, 1983; Arslan-Alaton, 2003; Sánchez-Polo et al., 2005). In practice, both mechanisms may occur depending on the chemical wastewater pollution. At low pH the predominant reaction mechanisms is the direct electrophilic attack by molecular ozone (Alvares et al., 2001). Molecular ozone is a rather selective oxidant. It can react directly with certain functional groups of organic compounds found in water and wastewaters, such as unsaturated and aromatic hydrocarbons with substitutes such as hydroxyl, methyl and amine groups giving rise to degradation products. Because of this high selectivity, many industrial wastewater treatment oxidations can be performed with molecular ozone. On the other hand, ozone decomposes in water to form •OH, which are stronger oxidizing agents than ozone itself, thus inducing so-called indirect ozonation. Indirect ozone oxidation is non-selective and is favored at alkaline pH conditions.

The presence of organic wastewater solutes which contain C-C double bonds (ligninsulfonic acid in the treated paper-mill effluent and humic acid in the oil reclaming model wastewater were shown to yield hydrogen peroxide by the reaction with ozone. Therefore, these wastewaters are efficiently ozonated even without the addition of hydrogen peroxide (Gulyas et al., 1995).

A major limitation of the ozonation process is the relatively high costs of ozone generation processes coupled with the very short half-life period of ozone. Thus, ozone must to be generated always at site. However, maximum concentration of the ozone produced in air or oxygen is approximately 4 to 8%, respectively, which coupled with the very low (5 to 10%) energy efficiency of the production and the requirement of an absolutely dry input of air or oxygen. Process efficiency is significantly dependent on the efficient gas liquid mass transfer, which is quite difficult to achieve due to the low solubility of ozone in aqueous solutions. Schematics of typical equipment used for ozonation and typical operating parameters are presented by Gogate and Pandit (2004).

There are other ways to use ozone in combination with other techniques, such as ultrasonic/UV radiation, hydrogen peroxide, or other hybrid methods (Gogate and Pandit, 2004; Rodrígues, 2009).

The attack by the OH radical, in the presence of oxygen, initiates a complex cascade of oxidative reactions leading to mineralization of the organic compound. The exact routes of these reactions are still not quite clear. For example, chlorinated organic compounds are first oxidized to intermediates, such as aldehydes and carboxylic acids, and finally to CO_2, H_2O, and the chloride ion. Nitrogen in organic compounds is usually oxidized to nitrate or free N_2, sulphur is oxidized to sulphate (Munter, 2001).

Ozonation in the presence of Fe^{2+} or other metals in solution or the heterogeneous behavior of solid catalysts is catalytic ozonation. Oxidation of Fe^{2+} consists of an electron transfer from the reduced metal to ozone, forming Fe^{3+} and the radical anion O_3^-, then the OH radical (Hart, 1983). Nowell and Hoigné (1987) maintain that the OH radicals are an intermediate in the reaction of Fe^{2+} with ozone and assume a mechanism involving an oxygen – transfer ozone to ferrous iron.

Volk et al (1997) studied effect of ozone, O_3/H_2O_2, and catalytic ozone processes. The oxidation system that generated the highest BDOC concentration was observed with O_3/H_2O_2, (0.35 H_2O_2 mg/mg O_3). Ozone mineralized 15%, O_3/H_2O_2 18% and catalytic ozone (10g of catalyst/1 L water) 24% of the initial DOC.

The addition of hydrogen peroxide to the ozone aqueous solution will enhance the O_3 decomposition with formation of OH radicals (Andreozzi, 1999). Hydroxyl radicals are strong and non-selective oxidants for organic pollutants in wastewater. The reaction of absorbed ozone with the hydrogen peroxide anion, HO_2^-, yields OH radicals. In aqueous solution ozone generates superoxide anions and HO_2 radicals by the reaction with hydroxide anions (Staehelin and Hoigné, 1982).

Another ozone molecule is transformed by superoxide anion radicals to O_3^- (Buehler et al., 1984). Taking up a proton O_3^- generates HO_3 radical, which are precursors of the hydroxyl radicals. If hydrogen peroxide is added to an aqueous solution of ozone the precursor O_3^- is produced by the reaction of ozone with the anion of H_2O_2. The oxidizing system O_3/H_2O_2 is a very efficient process for the elimination of biologically resistant organics. The optimum stoichiometric ratio is assumed to be 0.5 mol H_2O_2/mol O_3 (Gulyas, 1997), but it has to be determined experimentally for each case separately.

Fenton Oxidation

The Fenton process has been extensively studied and is one from the most commonly used advanced oxidation processes (AOPs). Analyses indicate that the Fenton process can be an effective and low-cost option for treatment of landfill leachates (Renou et al., 2008). Fenton oxidation is not efficient for oxidation of some organic compounds (e.g. chloroform, tetrachlorethane) that have been found from the leachate. However, organic compounds can be removed from the landfill leachate by both oxidation and coagulation processes, while COD removal efficiency by coagulation can be even higher that that by oxidation. The problem related to Fenton oxidation is production of high amounts of undesirable iron sludge (Robinson, 1983). In addition to conventional Fenton reagent, which is catalyzed by ferrous ion (Fe^{2+}), modified Fenton reagents relay also on catalysis applying ferric and zero-valet ion. They can be implemented alone or they can accompany other technologies for leachate treatment (Deng, 2006).

Production of OH radicals by Fenton reagent occurs by means of addition of H_2O_2 and Fe^{2+} salts. This reagent shows to be a very powerful oxidizing agent. The peroxide dose is important in order to obtain better degradation efficiency, while the iron concentration is important for the reaction kinetics. Other important operational parameters are reaction pH, aeration, coagulation pH, mode of reagent addition, temperature, and subsequent pH adjustment for final coagulation (Deng, 2007). One of the important advantages of this system is also the presence of reagents in the same phase to avoid mass transport problems. A total elimination of organic carbons requires large amount of oxidants and/or large residence time. The partial oxidation of toxic compounds enhances biodegradability. The recommended ratio of Fe/TOC is between 2 and 2.5 g g^{-1}, and 0.5 – 0.75 g H_2O_2/g of Fe^{2+}, but it can generally vary upon the composition of wastewater or leachate.

Fenton process follows four typical steps: i) oxidation, ii) neutralization, iii) coagulation/flocculation and, iv) solid-liquid separation (Deng, 2007). As a result, not only oxidation, but also coagulation plays an important role. The classical Fenton free radical mechanism in the absence of organic compounds involves several reactions, resulting in rapid formation of hydroxyl radicals (Reaction (8).

A number of different reactions also occur to enable cycling of iron between Fe^{2+} and Fe^{3+}, which form to play the role of catalyst. Their net reaction is actually the decomposition of hydrogen peroxide to water and oxygen (Deng, 2006).

$$Fe^{2+} + H_2O_2 \rightarrow Fe^{3+} + {}^{\cdot}OH + OH^- \tag{8}$$

Formed Fe^{3+} can be reduced back to Fe^{2+}. This reaction is much slower than their production from Fe^{2+} (Reaction (8)). Formed Fe^{3+} may also precipitate as oxyhydroxides in the case of pH increased (Deng, 2006). This especially occurs at the end of a treatment procedure, where the pH increase is required to stop the oxidation process. Depending upon the amounts of reagents used (Fe^{2+}/H_2O_2) and their ratio, red waste sludge is generated. It could also contain sorbed toxic and persistent organics and it must be treated properly before its final disposal.

Hydroxyl radicals could attack organics via different pathways: i) radical addition, ii) hydrogen abstraction, iii) electron transfer and iv) radical combination (Deng, 2006; Prousek,

1996). As a result, carbon-centered radicals (R˙) are formed and they rapidly and usually react reversibly with oxygen in water (Reactions (9) and (10)):

$$R^{\bullet} + O_2 \rightarrow R - H^{+} + HO_2^{\bullet} \qquad (9)$$

$$R^{\bullet} + O_2 \rightarrow R - OO^{\bullet} \rightarrow R - O^{\bullet} \qquad (10)$$

These R˙, R-OO˙ and R-O• radicals may couple or disproportionate to form relatively stable molecules or react further with iron ions (Deng, 2006; Pignatello et al., 2006). On the other hand they may continue to react with hydrogen radicals and oxygen proceeded by further decomposition or even complete mineralization (Prousek, 1996). However, it should be noted, that AOPs sometimes produce by-products, which may inhibit the biological processes when they are integrated into biological treatment systems. This is the reason that overall effluent toxicity and biodegradability has to be assessed prior to entrance of the oxidative pre-treated influent into the biological stage.

MATERIALS AND METHODS

Two oxidative treatment procedures were used for selected leachates. Four landfill leachates were investigated, two of them obtained in municipal landfills in Slovakia (LL1, LL2), while two of them originated in Slovenia (LL3, LL4). The first two landfill leachates (LL1 and LL2) were also treated biologically to compare the efficiency of the biological treatment and ozonation. Both raw and biologically treated leachates LL1 and LL2 were then used for ozonation experiments.

Some ozonation experiments were also performed with the addition of Fenton reagents. The increase in biotreatability of ozonated leachate LL1 was investigated using a kinetic test with different types of activated sludge. Slovenian landfill leachates LL3 and LL4 were only treated chemically. Efficiency of Fenton oxidation and ozonation was compared in terms of removal of organics as well as toxicity and biodegradability.

Treatment Procedures

Ozonation Experiments

Ozonation experiments were carried out in a lab-scale bubble ozonation column and jet loop ozonation reactor. Ozone was in both cases, i.e. LL1 and LL2 leachate collected in Slovakia as well as LL3 and LL4 leachates collected in Slovenia generated from pure oxygen. The experimental apparatus of *bubble ozonation column* is shown in Figure 1. The glass columns are 0.04 m diameter and 1.7 m height. The system was operated in the batch mode with regard to liquid. The column was filled with 1 litre of wastewater/landfill leachate. The second column was for excess ozone destruction and at the same time for evaluation of residual amount of ozone. The mixture of oxygen and ozone was injected at the bottom of the first/ozonation column through porous gas diffusers. Ozone is produced from pure oxygen by AZCO HTU, 500-DC with the maximum capacity 250 mg h^{-1} of ozone, Fischer Labour ozone generator with the maximum capacity 625 mg h^{-1}.

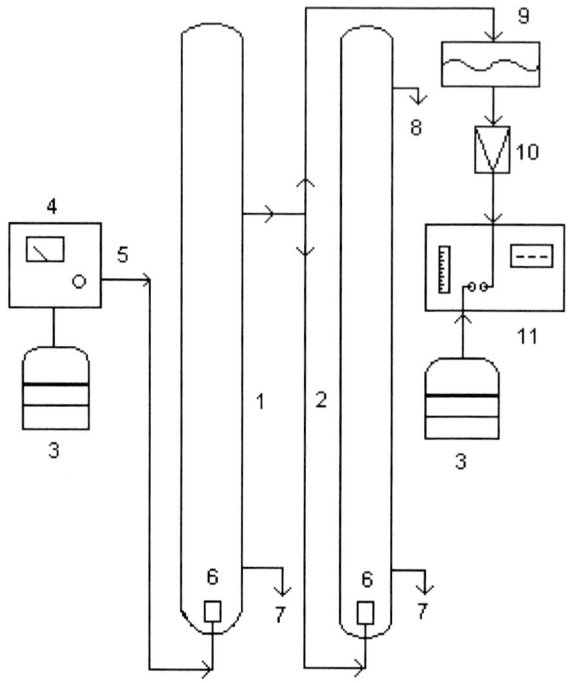

1 - ozonation column, 2 - column for destruction of residual ozone (KI solution), 3- oxygen gas holder, 4 - ozone generator, 5- mixture of oxygen and ozone, 6 - injection of ozone, 7- sampling, 8 – gas outlet, 9 – removal of moisture, 10 - glass fiber filter, 11 – ozone UV detector.

Figure 1. Schematic of experimental apparatus with ozonation bubble column.

The continuous flow of oxygen 100 L h^{-1} was applied for the generation of ozone when treating LL1 leachates. Ozonation trials were carried out at 100% of the ozone generator's power maximum. A scheme of the *jet loop ozonation equipment* is shown in Figure 2. The system was operated in the batch mode with regard to the wastewater samples. The samples were added into a jet loop ozonation reactor at the beginning of the trials.

The ozonation reactor was 0.06 m in diameter and 0.6 m in height. Effective volume of the reactor was 1.0 L. External circulation of the reaction mixture was maintained at 0.5 l^{-1} min^{-1} by a membrane pump. A diaphragm pulsation damper (SERA 721.1 Seybert & Rahier, Immenhausen, Germany) was used to minimize pulsation of external circulation. A Lifetech ozone generator with the maximum ozone production of 5 g h^{-1} was used. A mixture of O_3 and O_2 was injected into a wastewater sample through a Venturi ejector. At the same time, the ejector sucked the mixture of O_3 and O_2 from the reactor headspace. This, together with external circulation, should improve the efficiency of ozone utilization in the ozonation reactor. Lifetech ozone generator with the maximum capacity 5 g h^{-1} of ozone was used to generate ozone to supply this ozonation reactor. All ozonation reactors were operated in batch mode with regard to wastewater samples. The samples were added into reactors at the beginning of the trials. The outlet gas mixture from both ozonation reactors was conducted into a bubble column through a fine-bubble porous distribution element. The bubble column was 0.04 m in diameter and 1.7 m in height. The column was filled with a solution of potassium iodide. The excess ozone destruction was carried out in this column. Likewise to the ozonation reactor, effective volume of the bubble column was 1.0 L.

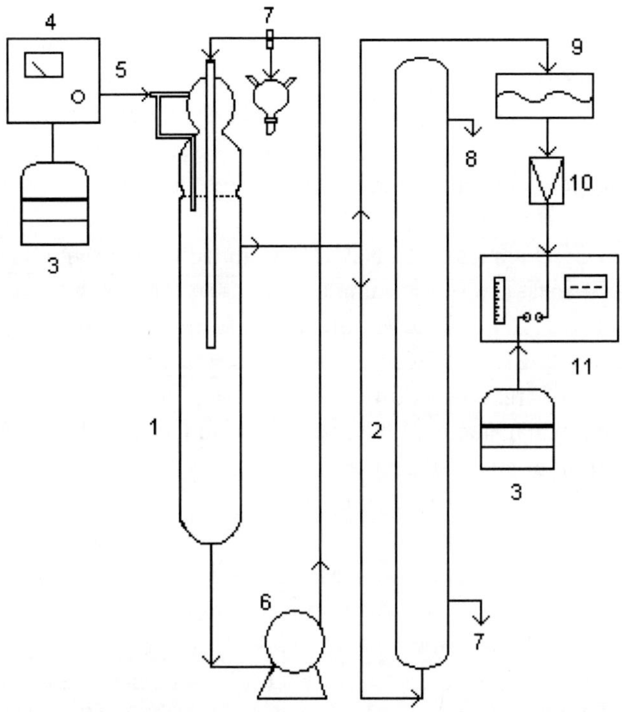

1 – ozonation jet loop reactor, 2 – destruction of excess ozone, 3 – ozone cylinder, 4 – ozone generator, 5- mixture of oxygen and ozone, 6 – external recirculation pump, 7 – sampling, 8 – gas outlet, 9 – moisture absorber, 10 - glass fiber filter, 11 – ozone UV detector.

Figure 2. Schematic of jet loop ozonation equipment.

The continuous flow of oxygen 10 L h^{-1} was applied for the generation of ozone for this ozonation reactor. Ozonation trials were carried out at 80% of the ozone generator's power maximum.

Ozonation in the Presence of Fenton Reagents

These experiments were carried out using non-diluted and non-filtered samples of landfill leachates LL1 and LL2. One liter of the relevant sample was added to each of four ozonation columns, which were arranged in parallel. The first and the third columns were operated as a reference, i.e. without the addition of any chemicals. $FeSO_4$, p.a. was added to the second and fourth columns to attain selected Fe^{2+} concentration (0.001 M). Ozonation trials were started after dosing H_2O_2 (30% w/v, p.a.) into the second and the fourth ozonation columns in order to achieve a concentration of 0.075 M. Ozonation assays were performed at laboratory temperature.

Fenton Oxidation

Experiments were performed using non-diluted and non-filtered samples of landfill leachates LL3 and LL4. pH was adjusted to 3.5 (± 0.2) before chemical oxidation experiments. A 250 mL sample was placed into 500 mL Erlenmeyer flask, which was submerged in a temperature controlled water bath to attain desired constant temperature (40 ± 2°C). $FeSO_4$, p.a. was added to attain selected Fe^{2+} concentration (0.3 M). Finally, the Fenton reaction was started with the addition of H_2O_2 (30% w/v, p.a.) to achieve a concentration of 4.0 M. The molar ratio 0.3 M Fe^{2+}/4.0 M H_2O_2 was selected on the basis of previous experiments (Žgajnar et al., 2005). The reaction was fast and exothermic, initial temperature was up to 40 - 45°C, so we conducted our experiments at approximately 40°C. The aqueous solution of Fenton reagent and diluted broth was stirred during the reaction period up to 35 minutes. Samples were redrawn at 5, 10, 20, 25 and 30 minutes and COD was determined immediately. Prior to COD analysis 1M NaOH was added to stop the oxidation at pH = 12 (± 0.2). To eliminate the excess H_2O_2, the sample was boiled for 10 minutes and then allowed to cool to room temperature. It was filtered afterwards to remove the formed ferric hydroxide and COD was determined. LL3 was also used for determination of COD behavior due to the pH regulation and thermal treatment of the sample. It was concluded, that the COD decrease was 5 ± 3% due to the management of the sample, while the rest of the COD decrease was a consequence of oxidation or coagulation processes.

Biological Treatment

Biological treatment of landfill leachates LL1 and LL2 was carried out in a lab-scale system, which consists of two completely mixed bioreactors in series (Figure 3). This arrangement of bioreactors offers high flexibility with regard to performance of aerobic or oxic/anoxic processes at different technological schemes (e.g. pre-denitrification, simultaneous denitrification and post-denitrification). Operational flexibility of this activated sludge system increases with involving an intermittently aeration operation mode as illustrated in Figure 4. This reaction system allows us to maintain different oxygen zones in the reactor (periodically alternated oxic and anoxic conditions) and consequently perform simultaneous nitrification and denitrification processes in one reactor.

Sedimentation is carried out in a separate sedimentation tank. Intermittent aeration can also be applied in a continuous flow activated sludge reactor. Intermittent aeration in such a reactor is used intensively for small WWTPs with regard to nitrogen removal (Araki et al, 1990; Nakajima and Kaneko, 1991; Heduit et al, 1990). An intermittently aerated completely mixed activated sludge reactor (CMASR, Figure 4) is convenient for lab-scale modelling of simultaneous nitrification and denitrification processes in highly recycled systems, such as the Carrousel reactor (Derco et al, 1994; 1999). Polyurethane biomass carriers (cubes of 2 cm edge) were also used with the aim of increasing biomass concentration in the CMASR reactor. The volume of the carrier corresponded to 10% of the active volume of the bioreactor. Semicontinuous cultivation (Pitter and Chudoba 1990) was operated for testing the impact of ozonated landfill leachate on activated sludge microorganism activity. Part of the biomass and cultivation medium was taken after a certain time from the reactor with the working volume, V, and the content was supplemented with a fresh substrate solution/wastewater.

When this procedure is repeated at a constant cultivation time and a constant amount of biomass and cultivation medium is taken for a selected time interval, the arrangement is called a semicontinuous system. Unlike a completely mixed bioreactor, e.g. an intermittently aerated CMASR, the microoganisms of activated sludge are cultivated at variable substrate concentration values within the operation cycle of the semicontinuous bioreactor due to the concentration gradient in the system.

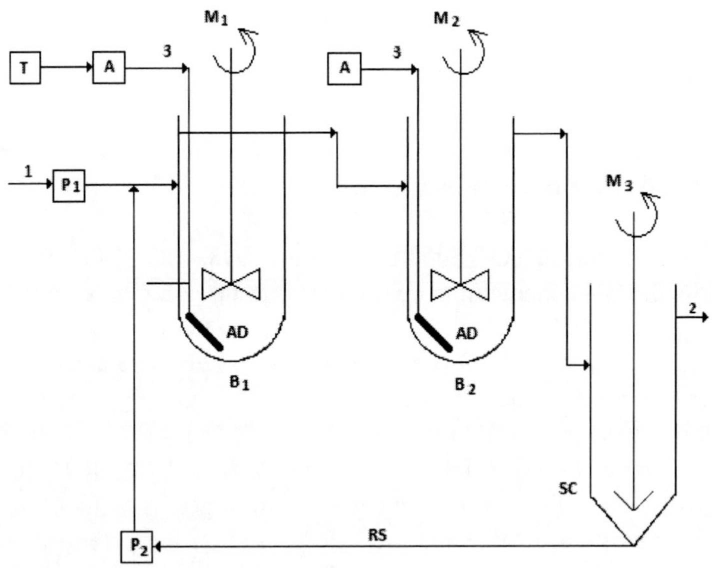

1 – influent, 2 - effluent, 3 – air supply, A – aeration, AD – air diffuser, B_1, B_2 – bioreactors. M_1, M_2 – mechanical agitation, P_1, P_2 pumps, R - return sludge,
SC – secondary clarifier, T – timer.

Figure 3. Layout of lab-scale biological treatment apparatus.

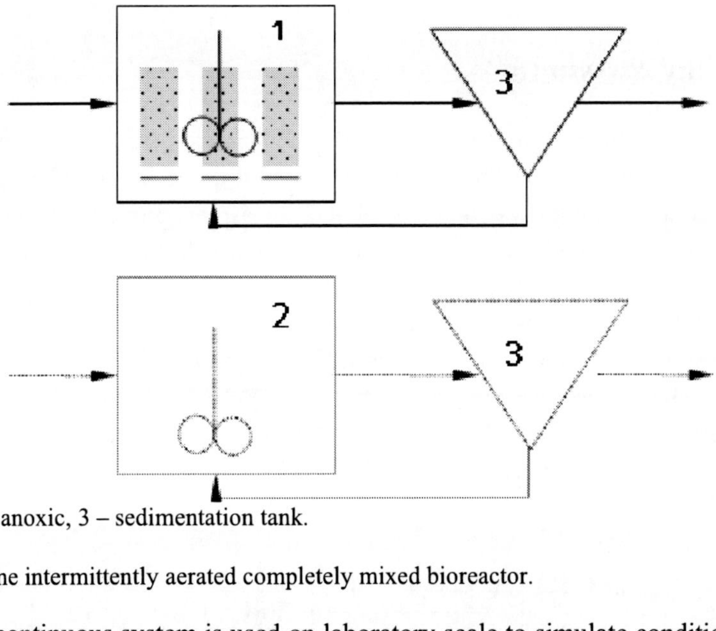

1 - aerobic, 2 – anoxic, 3 – sedimentation tank.

Figure 4. Scheme intermittently aerated completely mixed bioreactor.

The semicontinuous system is used on laboratory scale to simulate conditions of the plug flow continuous system. The semicontinuous system with parameters t_a = 23 h (aeration time) and t_s = 1 h (sedimentation time) is often used on laboratory scale when a fresh mixed culture of activated sludge is to be obtained daily for various purposes (e.g., for the study of kinetics, biodegradability testing of organic compounds, etc.). The realization of such a system is very simple and easy; no feeding device and no devices for separation and recirculation of the activated sludge are required.

Analytical Procedures

Analytical control of the raw leachates and monitoring of the treatment procedures included pH, BOD_5 (Biochemical Oxygen Demand (ISO 5813, 1983; ISO 5815, 2003)), COD (Chemical Oxygen Demand, (ISO 6060, 1999)), DOC (Dissolved Organic Carbon), IC (Inorganic Carbon) (Shimadzu TOC 5000A Analyzer, 1998, ISO 8245, 1999)) and nitrogen as organic (ISO5663, 1984), ammonium (ISO 7150/1, 1984) and nitrate (ISO 10304-1, 1992) determination.

In some cases concentration of nitrite and chloride was also evaluated (APHA, AWWA, WEF, 2005). The content of heavy metal concentration was measured by FAAS (ATI UNICAM / SOLAR 939) (Fe), CVAFS (P.S. ANALYTICAL/MERLIN PLUS SYSTEMS) (Hg) and ET AAS THERMO ELECTRON CORPORATION/SOLAAR M) (Cu, Pb, Cr, Cd). Variations and removal efficiencies of ammonium nitrogen, nitrate/nitrite nitrogen and organic pollution in terms of COD were evaluated based on measurements of influent and effluent concentration values in the bioreactors. The effectiveness of supplied ozone transfer was determined experimentally by iodometric method. The principle of iodometric method consists of oxidation of calcium iodide in neutral pH and sequential titration of released iodine with natrium tiosulphate after acidification of the solution (Gordon et al., 2000; APHA, AWWA, WEF, 2005).

Biodegradability Assessment

A number of methods used to determine biodegradability have been elaborated. Biodegradability improvement can be commonly assessed using BOD and BOD/COD ratio. The OxiTop® Control System was applied to BOD measurements (WTW, Germany). The BOD values were calculated using the negative pressure values from the OxiTop®-C measuring head. If oxygen is consumed in a closed vessel at a constant temperature, a negative pressure develops. If gas is released, an overpressure develops.

The OxiTop®-C measuring head measures and stores data on pressure developed during the respirometric test. The pressure values from the measuring heads are collected and processed by the OxiTop® OC110 controller. The BOD values are calculated using the values from the OxiTop®-C measuring head by the formula as follows (System OxiTop®, 1998):

$$\mathrm{BOD}\left[\mathrm{mg.L^{-1}}\right]=\frac{M(O_2)}{R\cdot T_m}\cdot\left(\frac{V_t-V_l}{V_l}+\alpha\frac{T_m}{T_0}\right)\cdot\Delta p(O_2) \tag{11}$$

where M (O_2) is the molecular weight of oxygen [mg mmol^{-1}], R is gas constant (83.144 mbar L.mol^{-1} K^{-1}), T_0 is reference temperature (273.15 K), T_m is measuring temperature [K], V_t bottle volume (nominal volume in 10^{-2} L), V_l is sample volume [10^{-2} L], α is Bunsen absorption coefficient (0.03103) and Δp (O_2) is a difference of the oxygen partial pressure [mbar].

BOD_5 values with this system were measured for the calculation of BOD/COD ratio values during ozonation trials. The parameter values of the BOD equation (Equation (16)) were also assessed from the evaluation of these measurements. We also used oxygen uptake rate as an alternative means of biodegradability measuring. Respirometric measurements (Pitter and Chudoba, 1990; Spanjer et al., 1998) were carried out in order to evaluate an effluence of the raw and ozonated samples on activated sludge activity. The method was developed by Čech et al. (1984a; 1984b) adapting and using a respirometer developed by Lamb et al. (1964). The scheme of the respirometer is shown in Figure 5. Prior to measuring, the activated sludge was aerated without substrate for at least one hour. The working concentration of biomass X depends on the heterotrophic endogenous respiration rate, i.e. on the history of activated sludge, which is usually about 1.0 g L^{-1}. In most cases, the aeration-tank-mixed liquor can be used without concentrating or diluting. Allylthiourea (1.0 mg L^{-1}) was added into a suspension in order to suppress nitrification during respirometric measurements (Čech et al., 1984a; 1984b). The suspension of microorganisms was then transferred into a respirometric cell (Figure 5) and aerated to increase the dissolved oxygen concentration to 8.0 to 8.5 mg L^{-1}. When these concentrations are reached, the aeration was stopped. A slow decrease of the oxygen concentration is due to the heterotrophic endogenous respiration of activated sludge microorganisms. During the endogenous respiration phase, microorganisms utilize oxygen at a constant endogenous specific respiration rate $r_{X,e}$ [mg g^{-1} d^{-1}] over a relatively long time period. At a selected time, a calculated small volume (1.0 to 5.0 mL) of a concentrated substrate solution is injected into the cell with the aid of a hypodermic syringe.

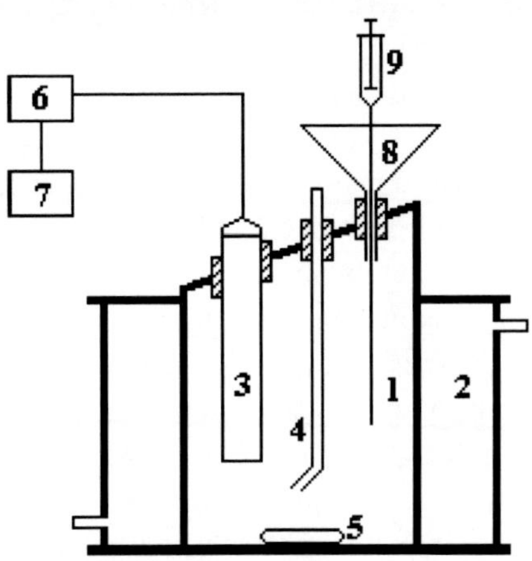

1 – Respirometric cell, 2 – water jacket, 3 – oxygen probe, 4 – aeration frit, 5 – magnetic mixing bar, 6 – oximeter, 7 – computer treatment of data, 8 – expansion funnel, 9 – substrate injection.

Figure 5. Scheme of the respirometer.

The addition of substrate to the respirometric cell causes a temporary increase in the respiration rate. The constant rate represents the total specific respiration rate $r_{X,t}$ [mg g^{-1} d^{-1}] at the substrate concentration S [mg L^{-1}], which corresponds to the substrate dosing into respirometric cell). As the substrate concentration decreases with time, the respiration rate also decreases, getting limited at low concentrations (according to Monod kinetics) and finally returns more or less to the original endogenous rate. Thus, the specific rate of substrate oxidation $r_{X,ox}$ [mg g^{-1} d^{-1}] at substrate concentration S can be computed as follows (Pitter and Chudoba, 1999):

$$r_{X,ox} = r_{X,t} - r_{X,e} \tag{12}$$

where $r_{X,ox}$ [mg g^{-1} d^{-1}] is the specific rate of substrate oxidation at concentration S, $r_{X,t}$ [mg g^{-1} d^{-1}] is the total specific respiration rate and $r_{X,e}$ [mg g^{-1} d^{-1}] is the specific endogenous respiration rate. The specific rate of substrate removal r_X at concentration S can be calculated as follows:

$$r_X = \frac{r_{X,ox}}{OC/S} \tag{13}$$

where r_X [mg g^{-1} d^{-1}] is specific rate of substrate removal at concentration S, $r_{X,ox}$ [mg g^{-1} d^{-1}] is specific rate of substrate oxidation at concentration S, OC [mg L^{-1}] is a net oxygen consumption and S [mg L^{-1}] is substrate concentration.

Assessment of Toxicity

Nitrification and denitrification kinetic tests were also carried out with activated sludge cultivated in a semicontinuous and intermittently aerated completely mixed activated sludge bioreactor.

The toxicity impact of ozonated leachate samples on autotrophic and heterotrophic microorganisms in activated sludge can be assessed from evaluation of these tests. Nitrification bacteria are the most sensitive activated sludge microorganisms and play a principal role for biological removal of nitrogen pollution from leachates. The rate of the first step of the nitrification process can be obtained by evaluation of time courses of ammonium nitrogen concentration. The changes in nitrate nitrogen during the nitrification test serve the calculation of the second step of nitrification rate. The denitrification rate is calculated from decline of nitrate and nitrite concentration values during the test. The results of the test give information about the process performance as well as the availability of ozonated leachates as an organic carbon source for denitrification proess (Derco et al., 2009).

The concentration of biomass in batch systems is measured in order to calculate values of specific nitrification and denitrification rates. These tests were performed in a batch system. After sedimentation the supernatant was decanted and activated sludge was washed using tap water. Ozonated leachate was added into cylinder. Activated sludge was maintained in suspension by aeration during the nitrification test and mechanical mixing during the anoxic test. Oxygen was purged by nitrogen before starting the denitrification test.

Toxicity of some of the samples (LL3, LL4) was also determined prior to and after oxidative treatment, using two standardized tests (Žgajnar-Gotvajn et al., 2009):

1. Toxicity was first determined by *Vibrio fischeri* bioluminescence inhibition test (ISO 11348-3, 2007). Freeze-dried luminescent bacteria *Vibrio fischeri* (DR. LANGE LUMIStox, 2001) are used to measure inhibition of the bioluminescence, which is affected by the toxicity of the sample. The pH of samples was corrected to 7.0 ± 0.2 and samples were diluted by 2% NaCl solution to obtain a series of test concentrations (50 – 0.02 % v/v). Bioluminescence was measured prior and after incubation period of 30 minutes by a LUMIStox luminometer to calculate the % of inhibition in comparison to blank test (2% NaCl).
2. As a second toxicity test, a laboratory method for the determination of oxygen consumption inhibition by activated sludge for carbonaceous and ammonium oxidation (ISO 8192, 2007) was used. Activated sludge from a laboratory municipal wastewater treatment plant (1,500 mg_{MLVSS} L^{-1}) was used. It consumed oxygen due to the presence of easily biodegradable substance – peptone. Addition of a landfill leachate resulted in a decrease of the oxygen consumption rate [mg_{O2} L^{-1} min^{-1}], measured by an oxygen sensor. Tests were conducted with different concentrations of the leachates (1 – 70 % v/v). pH of the test mixtures was in the range of 6.3 to 7.6 and it was not adjusted. Temperature was in the range of 21.3 ± 0.9 ºC. The incubation period was 180 minutes. Adequate consistency of activated sludge was confirmed in a preliminary test with 3.5-dichlorophenol.

Toxicity was determined for both samples (Table 15). The toxic effect in terms of EC (Effective Concentration) values was determined in both tests. $30minEC_{50}$ value means that after 30 minutes of exposure to the sample, 50% of tested population responded to the toxic effect. In the first test this means lower bioluminescence while in the second one slower oxygen consumption. All experiments were conducted twice.

Mathematical Treatment of Experimental Data

The BOD measurements were besides the BOD_5 values determination also used for assessment of parameters of BOD equation. The course of biochemical oxygen demand can be described by a kinetic equation of the first order (Davis and Cornwel, 1991; Weiner and Matthews, 2003):

$$\frac{dL_t}{L_t} = -k\,dt \tag{14}$$

where L_t is the oxygen equivalent of the organics [mg L^{-3}] remaining at time t [d] and k is the reaction rate constant [d^{-1}]. By rearranging and integrating Equation (14) yields:

$$L_t = L_0 \cdot e^{-kt} \tag{15}$$

where L_0 is the oxygen equivalent of organics [mg L^{-3}] at time t = 0. The amount of oxygen used in the consumption of the organics at time t, i.e. BOD_t, is the difference between the initial value of L_0 and L_t:

$$BOD_t = L_0\left(1 - e^{-kt}\right) \tag{16}$$

Equation (16) is called the BOD rate equation and is often written in base 10:

$$BOD_t = L_0\left(1 - 10^{-Kt}\right) \tag{17}$$

The reaction rate constants k and K in Equations (15) and (16) are related:

$$k = 2.303\,K \tag{18}$$

L_0 is often referred to as the ultimate BOD_u, that is, the maximum oxygen consumption when the waste has been completely degraded (Weiner and Matthews, 2003). The parameter values of the BOD equation (Eqn. 16) were determined by evaluation of experimental data obtained by respirometric measurements with OxiTop® Control System (1998).

The parameter values were calculated by the grid search optimization procedure. The residual sum of squares (S_r) between the observed values and the values given by the model, divided by its number of degrees of freedom ν (the number of observations less the number of parameters estimated) was used as the objective function (S_r^2).

Besides the BOD tests, COD analyses also were performed simultaneously with the ozonated leachate samples. Experimental kinetic data were fitted by zero (Eq. (19)), first (Eq. (20)), and second (Eq. (21)) order reaction kinetic models. Besides these single power law models, a combined – "two components" (TCM) kinetic model (Eq. (22)) with the first order reaction kinetics was applied to proportions of easily and slowly oxidizable organics. For a batch reaction system, under the assumption of a constant reaction volume, the following relationships were obtained

$$COD_t = COD_0 - k_0 t \tag{19}$$

$$COD_t = COD_0 \exp(-k_1 t) \tag{20}$$

$$COD_t = \frac{COD_0}{\left(1 + COD_0 k_2 t\right)} \tag{21}$$

$$COD_t = COD_{EO,t} + COD_{SO,t} = \alpha\, COD_0 \exp(-k_{EO} t) + (1-\alpha) COD_0 \exp(-k_{SO} t) \tag{22}$$

where COD_t [g m^{-3}] denotes the value of COD in wastewater in time t, COD_0 [g m^{-3}] the initial value of COD in wastewater, k_0 [g m^{-3} h^{-1}], k_1/h^{-1}, k_2 (g^{-1} m^3 h^{-1}] the rate constants for the zero, first and second order kinetics, respectively, $COD_{EO,t}$ (g m^{-3}] the value of easily oxidizable organics, $COD_{SO,t}$ [g m^{-3}] the value of slowly oxidizable organics, k_{EO} [h^{-1}] the first order reaction rate constant for easily oxidizable organics, k_{SO} [h^{-1}] the first order reaction rate constant for slowly oxidizable organics, and *a* the portion of easily oxidizable organics. Parameter values of the applied kinetic models were calculated by the grid search optimization procedure.

COD reduction by partial oxidation can be evaluated by the Eqs. (23) – (26) (Fiehn et al, 1998):

COD removal in total (αCOD_{oxi}, [-]) could be described as

$$\alpha COD_{oxi} = 1 - \frac{COD_t}{COD_0} \tag{23}$$

COD removal (αCOD_{min}, [-]) due to mineralization could be calculated as

$$\alpha COD_{min} = 1 - \frac{DOC_t}{DOC_0} \tag{24}$$

COD removal ($\alpha COD_{partoxi}$, [-]) due to partial mineralization can be determined as

$$\alpha COD_{partoxi} = \alpha COD_{oxi} - \alpha COD_{min} \tag{25}$$

Degree of effectiveness of partial oxidation ($\mu COD_{partoxi}$, [-]) could be calculated as

$$\mu COD_{partoxi} = \frac{\alpha COD_{partoxi}}{\alpha COD_{oxi}} \tag{26}$$

where αCOD_{oxi} is COD removal in total [-], COD_0 and COD_t are COD values at the beginning and in time t respectively, αCOD_{min} is COD removal due to mineralization, DOC_0 and DOC_t are DOC values at the beginning and in time t respectively, $\alpha COD_{partoxi}$ is COD removal due to partial mineralization [-], can be determined as $O_{3,trans}$ denotes the ozone transferred [g L^{-1}], and $\mu COD_{partoxi}$ is degree of effectiveness of partial oxidation [-]

During ozonation of leachates LL3 and LL4 ozone transferred in the reactor was calculated by macroscopic mass balance across the reactor:

$$O_{3,trans} = Q_g \int_0^t \frac{O_{3,in} - O_{3,out}}{V_R} dt \tag{27}$$

where $O_{3,trans}$ denotes the ozone transferred [g L^{-1}], Q_g the gas flow rate [Nm^{-3} min^{-1}], $O_{3,in}$ the ozone concentration in the gas at the inlet [g Nm^{-3}], $O_{3,out}$ the ozone concentration in the gas at the outlet [g Nm^{-3}] and V_R is the volume of the reactor. A UV detector Life ODU 100 was used to perform analyses of the ozone gas content.

CASE STUDY – RESULTS AND DISCUSSION

The investigated samples were collected at four landfills, two in Slovakia (LL1 and LL2) and two in Slovenia (LL3 and LL4). The samples were significantly different with regard to the content of COD, ammonium nitrogen and degradability. They all represent typical municipal landfill leachates for each particular country.

Landfill Leachates Collected in Slovakia

Several experiments were performed with these two landfill leachates (LL1 and LL2). First experiments were conducted to investigate the potential of biological treatment, i.e. with regard to organics and nitrogen pollutant removal. Reduction of the content of ammonium nitrogen was expected by biological oxidation to nitrite or nitrate and consecutive performance of the denitrification process.

Ozonation of raw leachates as well as ozonation in the presence of Fenton reagents were carried out with leachates LL1 and LL2 for comparative reasons. Ozonation of biologically pre-treated landfill leachate was also carried out. The main goal of additional ozonation was to enhance biodegradability of resistant organics. The other aim was to specify the conditions for partial oxidation of resistant organics. This creates conditions to perform combined ozonation and biological treatment processes enabling mineralization of the major portion of organics contained in the landfill leachate biologically.

Characterization of Landfill Leachates

Basic characteristics of raw landfill leachate LL1 and LL2 collected from municipal waste landfills in Slovakia are summarized in Table 2.

Both landfills are conventional with regard to leachate collection and control of landfill gas emissions. The landfill producing landfill leachate LL1 is supposed to be closed at the end of 2016. Extending the operation of the landfill producing LL2 for the next 20 years is under preparation.

Landfill leachate LL1 is actually treated at an industrial wastewater treatment plant, while leachate LL2 is discharged into the municipal sewer system after pre-treatment to the level of required discharge limits (Table 3). The variation of the landfill leachate LL1 composition within two years is illustrated in Figures 6 to 8. A relatively high content of biologically degradable organics and higher concentrations of ammonium nitrogen content in comparison to LL1 are characteristic for municipal landfill leachate LL2 (young leachate). In addition to these pollution indicators, higher chromium content and high concentration values of dissolved compounds (Table 2) are typical for these leachates, due to its being in an initial acidic phase of stabilization, also indicated by a slightly lower pH. There are tree types of effluent values for discharging landfill leachates LL1 and LL2 summarized in Table 3. The first of them recommend effluent values by the Slovak Government Decree (2004) for discharging into the public sewer system in general after relevant pre-treatment. The second values are effluent standards for discharging landfill leachate according the Slovak Government Regulation (2005).

Table 2. Results of analysis of leachate samples LL1 and LL2 collected from two municipal waste landfills in Slovakia

Parameter Dimension	Values	
Name of the sample	LL1	LL2
BOD_5 [mg L^{-1}]	82 – 516	81 – 2264
COD [mg L^{-1}]	1267 – 8343	2 700 – 4 855
$N-NH_4^+$ [mg L^{-1}]	232 – 576	1 145 – 2 634
$N-NO_3^-$ [mg L^{-1}]	< 0.05	0.05 - 0.42
$P-PO_4^{3-}$ [mg L^{-1}]	3.7	14.5
BOD_5/COD [-]	< 0.10	< 0.30
COD/BOD_5 [-]	> 10	< 3.3
Cl^- [mg L^{-1}]	806 – 916	8 058 – 14 889
SO_4^{2-} [mg L^{-1}]	110 – 124	0 - 597
Dissolved compounds [mg L^{-1}]	6271 – 9180	16 036 – 27 998
Suspended solids [mg L^{-1}]	0 – 50	15 – 106
pH [-]	8.1 – 8.5	7.55 – 7.90
Fe [mg L^{-1}]	1.8 – 3.9	0.74 – 2.45
Cd [mg L^{-1}]	< 0.01	< 0.005
Hg [mg L^{-1}]	0.1 – 0.3	Not available
Cu [mg L^{-1}]	0.01 – 0.4	0.054 – 0.114
Pb [mg L^{-1}]	0.01 – 0.04	0.001 – 0.03
Cr_{tot} [mg·L^{-1}]	0.01 – 0.70	0 – 389
Cr^{6+} [mg L^{-1}]	< 0.001	0 – 18
Q [m^3 d^{-1}]	8.6 – 26.0	43 – 52

It should be mentioned that effluent standards are set in general for landfill leachates without taking into account water flow and quality in receiving water. The three types of effluent values are limits set by a Local Water Management Authority taking into account circumstances for discharging of treated leachates.

Thus, the ′combined approach′ (i.e. emission limit values and water quality objectives mutually reinforce each other) was used to calculate emission limits for discharging LL1 leachate into receiving water in the vicinity of the landfill site. In other words, effluent limits for discharging leachate LL1 were calculated based on mass balance taking into account water quality objectives in receiving water (Regulation, 2005).

Very stringent effluent standard values have been settled for discharge landfill leachate LL1 (Table 3) due to very low water flow in the receiving watercourse. On the other hand, effluent limits for discharging leachate LL2 settled by the Local Water Management Authority, taking into account recommended values by Decree (2004), are more strict due to the fact that it will be discharged only to the municipal sewer system and consequently it will undergo conventional treatment in the municipal wastewater treatment plant. However, there are also relatively very stringent limit settled for ammonium nitrogen.

Table 3. Effluent quality standards and limits for LL1 and LL2

Parameter	Decree 2004 Effluent Recommended	Regulation 2005 Effluent standard	Influent	LL1 Effluent Limit	Influent	LL2 Effluent Limit
T [°C]	40	-	-	-	-	-
pH [-]	6.0 – 9.0	6.0 – 9.0	8.2	-	7.9	6.0 – 9.0
COD [mg L^{-1}]	800	400	4188	85	4855	1200
BOD_5 [mg L^{-1}]	-	25	168	21.5	2264	2200
SS [mg L^{-1}]	500	25		-	106.0	600
N- NH_4^+ [mg L^{-1}]	45	$15/50^{W1}/-^{W2}$	808	5.9	2634	120
N- NO_3^- [mg L^{-1}]	-	-	0.04	24.7	0.4	0
N_{tot} [mg L^{-1}]	70	-	-	-	-	-
P_{tot} [mg L^{-1}]	15	-	3.7	-	6.0	-
DS [mg L^{-1}]	2500	-	627.1	-	27990	3600
CN [mg L^{-1}]	0.2		<0.001	-	-	-
NEL [mg L^{-1}]	10	-	2.700	-	-	-
EL [mg L^{-1}]	80	-	-	-	-	-
AOX [mg L^{-1}]	0.5	1.0	3.100	-	-	-
PAL-A [mg L^{-1}]	10	-	-	-	9.0	10
Hg [mg L^{-1}]	0.05	-	-	-	-	-
Cu [mg L^{-1}]	1.0	0.5	<0.001	-	0.11	-
Ni [mg L^{-1}]	0.2	0.5	0.009	-	-	-
Cr^{6+} [mg L^{-1}]	0.1	-	<0.001	<0.001	18.3	-
As [mg L^{-1}]	0.2	0.15	0.001	0.020	-	-
Zn [mg L^{-1}]	2.0	2.0	0.020	0.010	-	-
Cd [mg L^{-1}]	0.1	0.15	0.104	<0.001	<0.005	-
PAH [mg L^{-1}]	0.05	-	-	-	-	-

Not available

W1 – Winter Temperature 12°C;

W2 – Winter Temperature 9°C.

DS – Dissolved Solids.

SS – Suspended Solids

NEL – Non-polar Extractable Substances.

EL – Extractable Substances.

AOX - Adsorbable Organochlorinated Compounds.

PAL - Surfactants.

CN – Cyanides.

PAH - Polyaromatic Hydrocarbons.

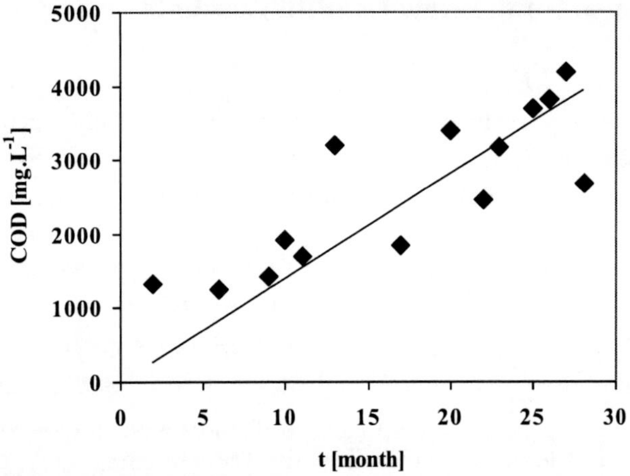

Figure 6. Variation of COD in the landfill leachate LL1.

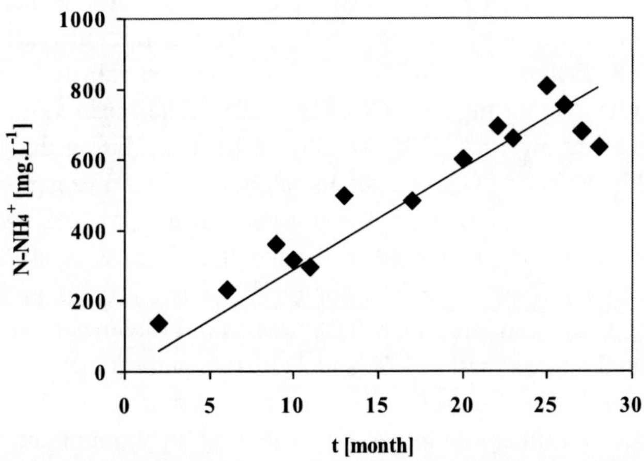

Figure 7. Variation of ammonium nitrogen in the landfill leachate LL1.

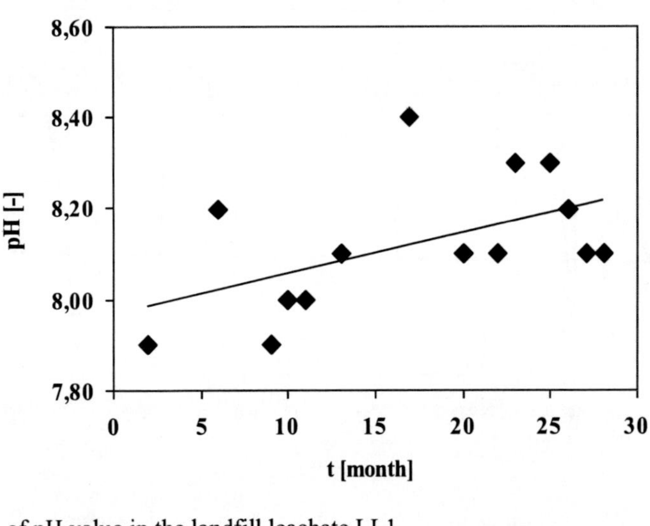

Figure 8. Variation of pH value in the landfill leachate LL1.

Biological Treatment of the Raw Landfill Leachate

Biological Treatment of LL1

Raw landfill leachate (LL1) collected in Slovakia (basic characteristics are shown in Table 2) was treated biologically in the lab-scale intermittently aerated activated sludge lab-scale intermittently aerated completely mixed activated sludge reactor (CMASR) (Figure 4). A high variation of COD and ammonium nitrogen (Figures 6 to 8) was typical for the landfill leachate. The volume of the bioreactor was 4.2 L and the volume of sedimentation tank was 1.0 L. Activated sludge load ≈ 0.1 kg kg^{-1} d^{-1} (BOD_5/MLSS) was maintained in the bioreactor. Porous diffusers were used for oxygen distribution. Oxygen supply was controlled by a timer. An aeration time of 13 min from the total cycle time of 21 min was selected with the aim of maintaining aerobic conditions during this phase of the bioreactor operation. The maintained oxygen profile in the intermittently aerated completely mixed bioreactor is illustrated in Figure 9.

The bioreactor was mechanically stirred continuously in order to maintain activated sludge in suspension also during the anoxic phase of reaction cycle. The operational parameter values and treatment efficiency values in the intermittently aerated CMASR are summarized in Table 4 (Derco et al., 2009). The observed mean value of the specific removal rate of ammonium was 7.1 mg mg^{-1} h^{-1} (N-NH_4/MLSS). The mean value of specific COD removal rate was 76.8 mg mg^{-1} h^{-1} (COD/MLSS). In general, during this treatment phase a reduction of about 35 - 65% of COD content and an ammonium nitrogen removal efficiency of about 85 - 95% were observed depending on composition of the raw landfill leachate (LL1). Similar results were also published in the literature. Renou et al (2008) summarizes COD removal efficiencies from 48 to 69% for different case studies at BOD_5/MLSS load approximately 0.07. They also presented TOC removal efficiencies from 42 to 57% at BOD_5/MLSS value and approximately 0.06 and TOC removal efficiencies in the range from 60 to 80% when maintaining BOD_5/MLSS ratio value about 0.05. However, the effluent concentrations exceeded effluent limit values (Table 3) in ammonium and nitrite/nitrate nitrogen and particularly in terms of COD.

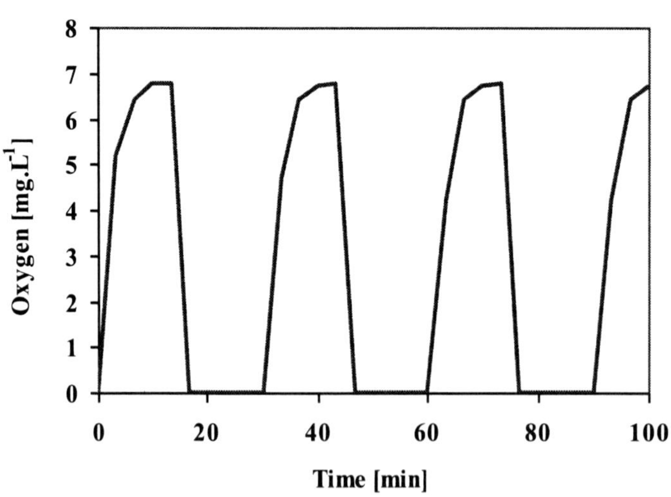

Figure 9. Oxygen profile in intermittently aerated completely mixed bioreactor.

The accumulation of nitrates and particularly nitrites that occurred during the process can be seen in Table 4. This indicates the limitation of the process by deficit of organic carbon. The accumulation of nitrites can be attributed to the occurrence of lower pH values in the bioreactor.

Intermittently aerated CMASR of the same volume (4.2 L) with a biomass carrier was also operated with the aim of improving process performance. Polyurethane biomass carriers (cubes of 2 cm edge) were added into the bioreactor with the aim of increasing biomass concentration and overall treatment efficiency.

Table 4. Parameter values of biological treatment of raw landfill leachate sample LL1

Parameter	Dimension	Influent	Effluent
pH	[-]	8.1	8.6
COD	[mg L^{-1}]	5383	1914
MLSS	[g L^{-1}]	-	0.65
BOD_5	[mg L^{-1}]	38.7	0
$N-NH_4^+$	[mg L^{-1}]	346	25.0
$N-NO_3^-$	[mg L^{-1}]	0	37.3
$N-NO_2^-$	[mg L^{-1}]	0	168.2
E_{COD}	[%]	-	64.4
E_{N-NH4}	[%]	-	92.8
HRT	[h]	-	70
SRT	[d]	-	∞

HRT – Hydraulic retention time.
SRT – Solid retention time.
- Not defined.

Table 5. Parameter values of biological treatment of raw landfill leachate sample LL1

		Sample		
Parameter	Dimension	LL1	LL1 + CH_3OH	Effluent
pH	[-]	8.1	8.6	8.4
COD	[mg L^{-1}]	4188	8928	2140
$N-NH_4^+$	[mg L^{-1}]	680	680	52.3
$N-NO_3^-$	[mg L^{-1}]	0	-	50.4
$E_{COD,LL1}$	[%]	-	-	48.9
E_{N-NH4}	[%]	-	-	92.3
V	[L]	-	-	4.2
Q	[L d^{-1}]	-	-	0.67
HRT	[h]	-	70	92.8
SRT	[d]	-	∞	92.8

HRT – Hydraulic Retention Time.
SRT – Solid Retention Time.
- Not determined.

The volume of the carrier corresponded to 10% from the active volume of the bioreactor. Methanol was added into the influent to the first bioreactor. The amount of methanol corresponded to recommended values of BOD/TKN ratio according Henze at al. (1995).

Characteristics of landfill leachate LL1, influent (LL1 and methanol) and effluent of intermittently aerated CMASR with combined suspended and attached biomass are given in Table 5. The values corresponding to the highest denitrification efficiency observed during this period of study are summarized.

Lower efficiencies of ammonium and nitrate nitrogen removal were observed at the effluent of the bioreactor operated under this mode in comparison to its previous operational mode (Table 4). On the other hand, a significantly higher concentration of ammonium nitrogen is typical for the raw leachate treated in this bioreactor. In addition, an insignificant accumulation of nitrite nitrogen was observed in this operational mode. Thus, a higher removal efficiency of nitrogen compounds was measured at this operational mode in comparison with the previous one. A lower COD removal efficiency of this operational mode, even at lower influent COD concentration values in comparison with previous operational mode (Table 4), can probably be attributed to a lower content of biodegradable organics in the investigated sample of raw leachate.

Similar to the previous operational mode, the effluent concentrations exceeded effluent limit values (Table 3) in ammonium and nitrate nitrogen, particularly in terms of COD. Next, the sequences of two continuously operating single sludge bioreactors of the same volume (4.2 L) were operated in order to improve the performance of the process. Such sequence offers high flexibility with regard to performance of nitrification and denitrification processes (e.g. pre-denitrification, simultaneous denitrification, post-denitrification technology). The pre-denitrification technology was applied during this phase of the research. The first bioreactor was intermittently aerated (Reactor 1, anoxic and oxic phase) and the second one was completely aerated (Reactor 2, oxic phase). The corresponded anoxic volume of the denitrification system was approximately 2.1 L and the aerobic volume was approximately 6.3 L when taking into account the oxygen profiles maintained in the reactors during operation.

No external organic carbon source was added at the beginning of the model operation with the new landfill leachate sample assuming higher content of biodegradable organics in the fresh leachate sample. Nevertheless, the denitrification rate decreased significantly within a few days. Thus, methanol was dosing into influent to the first bioreactor. The amount of methanol corresponded to recommended values of BOD/TKN ratio according to Henze at al. (1995), i.e. 4.0 mL CH_3OH to 1.0 L of LL1.

Serious operational problem occurred approximately 5 days after beginning this operation mode due to fluctuations of pH at 8.5 to 9.0. Thus, methanol was substituted with acetic acid. There was a dual function of acetic acid added into the activated sludge reactor, the first to lower pH and the second one to serve as an organic carbon source for the denitrification process. Characteristics of landfill leachate LL1, influent (LL1 with added acetic acid) and effluents of both consequently connected bioreactors are given in Table 6. The values corresponding to highest denitrification efficiency observed during this period of the study are also summarized. The amount of acetic acid corresponded to recommended values of BOD/TKN ratio according to Henze at al. (1995), i.e. 4.0 mL CH_3COOH to 1.0 L of LL1.

The addition of acetic acid was preferred instead of methanol in order to maintain a convenient pH value in the bioreactor. The contribution of acetic acid to the whole COD value corresponds to 4740 mg L^{-1}. The mean efficiency of complete COD removal after Bioreactor 2 ranged from 27 to 36%. Residual values of acetic acid content in the effluent are considered to be negligible.

Table 6. Parameter values of biological treatment of the raw landfill leachate LL1, influent to the biological treatment plant, effluent after the first phase and effluent after the second phase

Parameter Sample	Dimension	Sample			
		LL1	LL1 + CH_3COOH	Effluent from Bioreactor 1	Effluent from Bioreactor 2
COD	[mg L^{-1}]	4188	8928	2630.9	2750.5
$N-NH_4^+$	[mg L^{-1}]	680	680	42.7	37.9
$N-NO_3^-$	[mg L^{-1}]	-	-	13.5	26.3
$E_{N\text{-}NH4}$	[%]	-	-	-	94.4
$E_{COD,LL1}$	[%]	-	-	-	34.3
pH	[/]	8.1	8.6	8.9	8.7
V	[L]	-	-	2.1	6.3
Q	[L d^{-1}]	-	-	0.33	0.33
HRT	[h]	-	-	152.7	758.2
MLSS	[g $L^{-}]^{1}$	-	-	0.5	0.5
SRT	[d]	-	-	∞	∞

HRT – Hydraulic Retention Time/
SRT – Solid Retention Time/
MLSS – Mixed Liquor Suspended Solids.
- Not determined.

Thus, the total efficiency of COD removal does not include dosed acetic acid. The mean efficiency of ammonium nitrogen removal in both reactors was about 94%. The observed mean value of specific removal rate of ammonium nitrogen removal was 2.8 mg mg^{-1} h^{-1} (N-NH_4/MLSS). The observed mean value of the specific removal rate of nitrate nitrogen removal was 8.1 mg mg^{-1} h^{-1} (N-NO_3/MLSS).

The values of specific removal rates of ammonium as well as nitrate nitrogen removal were calculated taking into account oxic and anoxic hydraulic retention time values. A slightly higher removal efficiency of ammonium nitrogen was measured in comparison with a previous operational mode (Table 5). However, the effluent concentrations exceeded effluent limit values (Table 3) in ammonium and nitrite/nitrate nitrogen and particularly in terms of COD.

Biological Treatment of LL2

The biological treatment of landfill leachate LL2 was examined in laboratory aerobic CMASR. The COD value of raw landfill leachate LL2 samples ranged from 2571 to 2750 mg L^{-1}. The content of ammonium nitrogen was in the range from 3282 to 3425 mg L^{1}. Aerobic conditions were maintained in the bioreactor with the aim of reducing the content of ammonium nitrogen in the landfill leachate. The value of SRT 15 d was maintained in the bioreactor.

The operational parameter values and treatment efficiency values in the aerobic CMASR are summarized in Table 7. The treatment efficiencies of COD removal were in the range 22 to 48%. The decrease of ammonium nitrogen concentration was in the range from 18 to 76%, depending on raw leachate composition. There was a lower COD removal efficiency in comparison to results of leachate LL1 treatment (Table 4) even though significantly lower influent COD values can be attributed to a lower HRT value (1.75 d) during LL2 leachate treatment in comparison to treatment of LL1 leachate (2.9 d). Lower ammonium nitrogen efficiency removal from LL2 leachate in comparison to LL1 leachate treatment (Table 4) can be assigned to significantly higher influent ammonium nitrogen concentration as well as to lower SRT value in comparison to LL1 leachate treatment (Table 4).

In general, the biological treatment of raw LL1 leachate resulted in a significant reduction of ammonium nitrogen content (treatment efficiencies about 95%) compared to the values complying with recommended value for discharge into public sewer system (Decree, 2004). Intermittently aerated CMASR seems to be the most convenient reactor type from those investigated.

Table 7. Parameter values of biological treatment of raw landfill leachate sample LL2

Parameter	Dimension	Sample Influent	Effluent
pH	[-]	8.6	6.8
COD	[mg L^{-1}]	2750	1435
MLSS	[g L^{-1}]	-	1.57
$N-NH_4^+$	[mg L^{-1}]	3425	816
E_{COD}	[%]	-	47.8
$E_{N\text{-}NH4}$	[%]	-	76.2
HRT	[h]	-	42.0
SRT	[d]	-	15

HRT – Hydraulic Retention Time.
SRT – Solid Retention Time.
- Not determined.

The addition of biomass carriers into activated sludge system offers performance of both, nitrification and denitrification processes. However, the effluent concentrations of COD exceeded the effluent limit value (Table 3) recommended for discharge into the sewer system.

Biological pretreatment of LL2 leachate led to complying with effluent limit in COD term (Table 3). On the other hand, post treatment of biologically pre-treated of LL2 samples with regard to decrease of ammonium nitrogen content resulted from the research.

Thus, additional treatment of biologically pretreated raw leachate is required in order to meet effluent limits. An increase of organic carbon amounts for the denitrification process is also required due to the high portion of biologically resistant organic content in the raw landfill leachate. Selected chemical oxidative treatments were investigated as potential options to enhance biodegradability as well as overall treatability of investigated landfill leachates. The results are presented in the following pages.

Chemical Oxidation of the Raw Leachate Samples

The ozonation process and combined ozonation with the Fenton reagents (Fe^{2+} and H_2O_2) were applied in lab-scale to municipal landfill leachates LL1 and LL2. It has been mentioned that leachate LL1 high content of resistant organic pollution is typical (old landfill) as well as a high concentration of ammonium nitrogen (Table 2). On the other hand, very stringent effluent discharge values for these indicators as well as for nitrate nitrogen are settled (5.9 mg L^{-1}) due to very low water flow in the receiving water (Table 3). A relatively high content of biologically degradable organics and higher concentrations of ammonium nitrogen content in comparison to LL1 are characteristic for municipal landfill leachate LL2 (young leachate). In addition to these pollution indicators, higher chromium content and high concentration values of dissolved compounds (28,000 mg L^{-1}) and dissolved inorganic salts (22,400 mg L^{-1}) are typical for these leachates. The LL2 landfill leachate is intended to be discharged into the municipal sewer system after pre-treatment. The effluent discharge limits are specified for all the above-mentioned indicators (Table 3). It should be mentioned that effluent limits for dissolved inorganic salts (3,600 mg L^{-1}) as well as for dissolved compounds (1,500 mg L^{-1}) were not obligatory at the time period when experimental measurements were carried out. On the other hand, a more stringent effluent standard was set for ammonium nitrogen (Table 3). The results of ozonation trials performed with both LL1 and LL2 samples are summarized in Table 8 and shown in Figures 10 and 11. The experiments labeled LL1a and LL2a were carried out without any addition of chemicals. The experiments LL1b and LL2b were performed with addition of Fenton's reagents, i.e. 1.375 g L^{-1} H_2O_2 and 0.45 g L^{-1} Fe (0.075 M H_2O_2 and 0.075 M Fe^{2+}).

The highest COD removal rates were measured within the first 2.5 hours of all experiments. Higher COD removal rates were obvious for LL2 leachate (Figure 10). Insignificant impact of Fenton's reagent addition on COD removal from this leachate was observed.

Table 8. Results of ozonation trials carried out with LL1 and LL2 leachates

t [h]	Parameter [mg L^{-1}]	LL1a	LL1b	Sample LL2a	LL2b
0	COD	2750	2750	2574	2574
	$N-NH_4^+$	424	424	3282	3282
1.5	COD	1973	1375	2391	2272
	$N-NH_4^+$	351	294	3018	2796
2.5	COD	1016	1016	2032	1495
	$N-NH_4^+$	285	280	2898	2773
20.5	COD	965	717	1764	1484
	$N-NH_4^+$	3.2	33.5	2802	2580
40.0	COD	657	628	-	-
	$N-NH_4^+$	3.1	11.8	2796	2250
48.0	COD	239	239	-	-
	$N-NH_4^+$	2.0	10.3	-	-

- Not measured.

On the other hand, a significant enhancement of COD removal rate follows for LL1, particularly within 1.5 and 2.5 hours of ozonation. Significantly higher ammonium removal rates for LL2 (N-$NH_4^+$$_{,2}$) follow in Figure 11 in comparison with LL1 (N-$NH_4^+$$_{,1}$), particularly at a longer ozonation time. An insignificant impact of Fenton´s reagents addition on ammonium removal is obvious from Table 8 and Figure 11. A very close COD effluent value to effluent limit was observed during the ozonation of LL2 leachate in the presence of Fenton reagents.

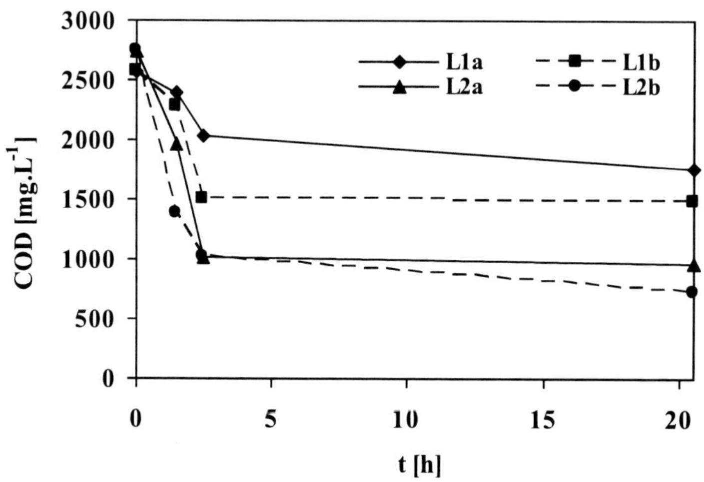

Figure 10. COD decrease with ozonation time.

In conclusion, different reaction rates are characteristic for individual assays. Perspectives of the ozonation process for the treatment of LL1 landfill leachate follows with the results for both, with regards to organic pollutants and ammonium nitrogen, when considering effluent limits for discharging pretreated landfill leachate into the public sewer system.

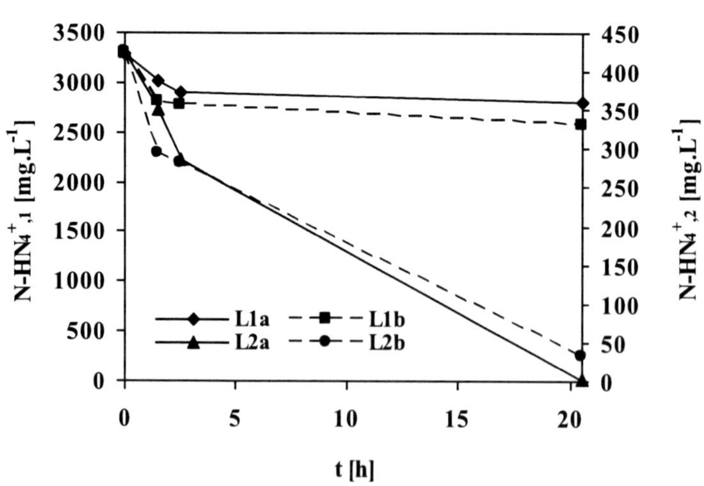

Figure 11. Decrease of ammonium nitrogen with ozonation time.

Complying with the effluent limit in terms of COD for discharging into public sewers was also observed during the LL2 leachate ozonation. However, additional treatment is required to decrease the ammonium nitrogen concentration to the level required for discharging into the public sewer system. Based on the results of the previous investigation, adsorption on activated carbon or zeolite seems to be convenient for additional ammonium nitrogen removal from this landfill leachate.

It should be mentioned that a Fischer Labour ozone generator, with the maximum capacity 625 mg h^{-1}, was applied during this period of the research. In addition to the very low capacity of this ozone generator, the generated ozone was equally distributed into four parallel ozonation columns in which all experiments were carried out simultaneously during this set of experiments. This explains the relatively long ozonation times as given above (Table 8, Figures 10 and 11).

Ozonation of Biologically Pre-Treated Landfill Leachate

As already mentioned, effluent concentrations of ammonium and nitrite/nitrate nitrogen and particularly COD values of biologically treated LL1 leachate does not comply with the limits for discharging biologically treated LL1 leachate neither into receiving water nor into the public sewer system (Table 2, Table 3).

For illustration, the effluent concentration value of COD set by the Local Water Management Authority (Table 3) for the investigated landfill site varies from 85 mg L^{-1} to 185 mg L^{-1} according to the leachate flow rate. Thus, the effluent from the bioreactor, i.e. biologically pre-treated landfill leachate was ozonated with the aim of increasing the biodegradability of the residual COD, which was still too high to be discharged into the watercourse.

The landfill leachate LL1 was subjected to the biological treatment in an intermittently aerated CMASR before performing the ozonation process. The bioreactor was operated as a simultaneous nitrification and denitrification activated sludge process. No withdrawal of excess sludge was performed during the study. The parameters of the raw and biologically pre-treated raw landfill leachate sample (LL1) are summarized in Table 9.

58% COD removal efficiency and 97% removal efficiency of ammonium nitrogen by biological treatment follows from the data in Table 9. These data conform to effluent limits for discharge LL2 leachate into public sewer system (Table 3). When considering ammonium nitrogen it is has a recommended value for discharging industrial and specific wastewater into the public sewer system, which is very close also to the effluent standard value for discharging of such wastewater into rivers (Decree, 2004). However, the effluent does not comply with the effluent standard value in terms of COD. The biodegradability of ozonated leachate was determined via BOD_5 measurement during different time periods. A small amount of an effluent from a lab-scale model of activated sludge process was fed with the landfill leachate and used as bacterial seed. Figure 12 shows the time courses of accumulative amounts of supplied (theoretically produced) ozone and those transferred into fresh water at experimental conditions given above. As can be seen in Figure 12 both accumulative dependencies are linear in time. From this Figure it follows that the rather constant efficiency of the ozone transfer in fresh water is about 71% of the amount theoretically produced. Resulting values from ozonation of biologically pre-treated landfill leachate LL1 are summarized in Table 10.

Table 9. Parameter values of raw and biologically pre-treated raw landfill leachate sample LL1

		Sample	
Parameter	Dimension	Raw	Pre-treated
pH	[-]	7.8	8.5
COD	[mg L]	3870	1615
BOD_5	[mg L]	80	0
$N-NH_4^+$	[mg L]	756	22.0
$N-NO_3^-$	[mg L]	0	24.0

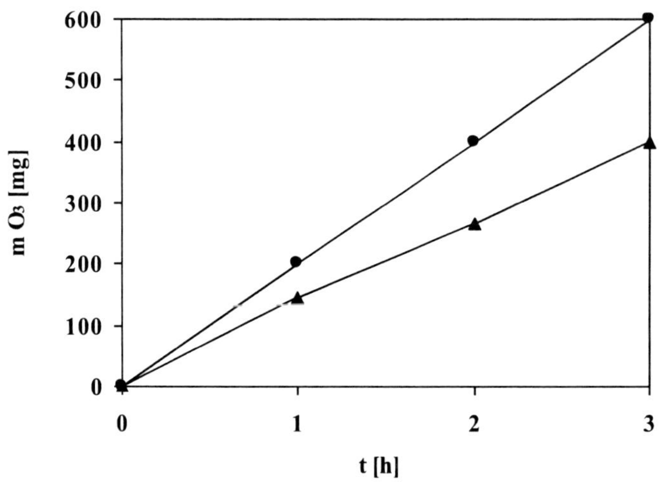

Figure 12. Accumulative amount of generated (▲) and transferred (●) ozone vs. time.

The efficiency of COD removal during ozonation (E(COD), Table 10) trials ranged from 3% to 39% in relation to COD value after biological treatment, i.e. initial COD value for ozonation trials. On the other hand, the efficiency of COD removal increased by 2% to 16% (t = 12.5 hours), i.e. 75% total efficiency of COD removal when considering COD part of the raw LL1 leachate. However, the COD value is still high in comparison to the effluent standard value (Regulation, 2005).

The evolution of COD with the ozonation time is presented in Figure 13. As already mentioned, the landfill leachate was treated by the process of biological nitrification and denitrification with the aim of reducing the high content of ammonium nitrogen and to remove a portion of biodegradable organics prior to ozonation. For each ozonation experiment (a set of ozonation assays at different ozonation times was carried out) the initial COD value was the same. The pH value of biologically pre-treated leachate was about 7.8 and it increased gradually to 8.5 during ozonation.

These relatively small conversions of COD, observed after 10 hours of ozonation, can be generally related to the very high reactivity of organic compounds with ozone. This is followed by the formation of various intermediates of smaller molecular weight such as alcohols, acids, etc. which also contribute to the oxygen demand before their final degradation. On the other hand, they still account for oxygen demand until their final degradation.

Table 10. Resulting values from ozonation of biologically pre-treated landfill leachate LL1

Parameter /t[h]	COD [mg L^{-1}]	BOD_5 [mg L^{-1}]	BOD_5/COD [mg mg^{-1}]	E(COD) [%]
0	**1615.0**	0	0	-
1.0	1560.2	75.4	0.05	3.4
4.0	1420.2	93.7	0.07	12.1
6.0	1375.5	142.5	0.010	14.8
8.0	**1250.0**	136.5	0.011	22.6
9.0	1200.0	119.5	0.099	25.7
10.0	1220.5	242.5	0.199	24.4
10.5	1185.0	244.5	0.206	26.6
11.0	1495.1	227.5	0.152	7.4
11.5	**1250.0**	259.5	0.208	22.6
12.0	**1032.8**	131.0	0.127	36.0
12.5	**985.7**	175.5	0.178	39.0
13.0	**1076.6**	142.0	0.132	33.3
15.0	1235.3	176.0	0.142	23.5
20.0	**1196.0**	127.0	0.106	25.9
25.0	**1056.0**	253.5	0.240	34.6
27.0	**1136.4**	227.0	0.199	29.6

E – Removal Efficiency.

According to the results shown in Figure 13, the highest COD decrease rate was observed during the first 6 to 8 h of ozonation, when the COD removal efficiency achieved about 23%. When the ozonation was prolonged to 12.5 h, the COD removal efficiency ranged between 36% and 39% (the last efficiency value corresponds to 12.5 hours of ozonation). Experimental COD data were fitted by zero, the first, and the second order reaction kinetic models (Eqns. 19-22). Besides these single power law models, a combined ′two components′ (TCM) kinetic model was also used. The statistical and kinetics parameter values are summarized in Table 11. According to the values of correlation coefficient (r_{yx}^2) and the sum of residual squares (S_r^2) the best fit of COD removal data was obtained by the two component model (Eq. 23). The comparison of experimental data and calculated values by TCM is shown in Figure 13.

From the parameter values in Table 11 it follows that the landfill leachate contains about 58% organics in terms of COD, which are oxidable by ozone. In Figure 14 the evolution of BOD_5 values depending on ozonation time is depicted. A significant increase of BOD_5 values for samples ozonated for 10.0 to 11.5 h was observed. The highest BOD_5 value was measured after 11.5 hours of ozonation (Derco et al., 2002).

Figure 15 shows the dependence of BOD_5/COD ratio values vs. ozonation time. The highest increase of biodegradability was observed for the period of ozonation from 10.0 to 11.5 h.

The highest biodegradability of LL1 leachate was achieved after 24 hours of ozonation. However, the additional slow increase BOD_5/COD ratio value observed for prolonged ozonation of the sample to more than 11.5 h is negligible from point operational costs and in general from the point of controlled chemical oxidation of the leachate LL1.

Table 11. Kinetic parameters and statistical values

n	k_n	k_r $[h^{-1}]$	k_s $[h^{-1}]$	a [-]	r_{yx}^2 [-]	S_r^2 [-]
0	27.56	-	-	-	0.236	$2.8\ 10^4$
1	$2.14\ 10^{-2}$	-	-	-	0.406	$2.2\ 10^4$
2	$1.61\ 10^{-5}$	-	-	-	0.512	$1.8\ 10^4$
TCM	-	$4.57.10^{-3}$	$9.61.10^{-2}$	0.695	0.615	$1.4\ 10^4$

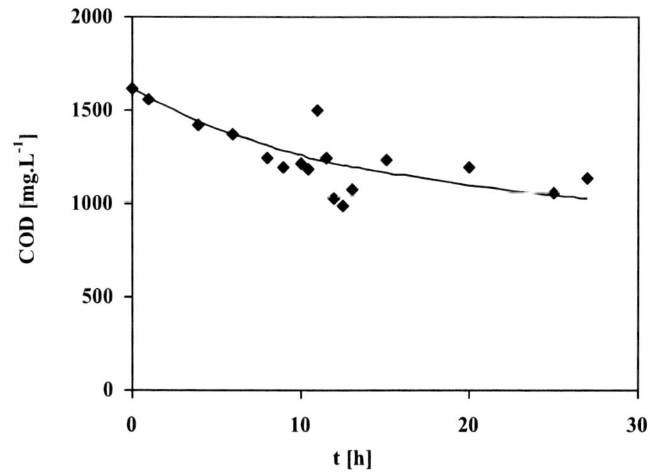

Figure 13. Variations of experimental (♦) and calculated (-) COD values during ozonation.

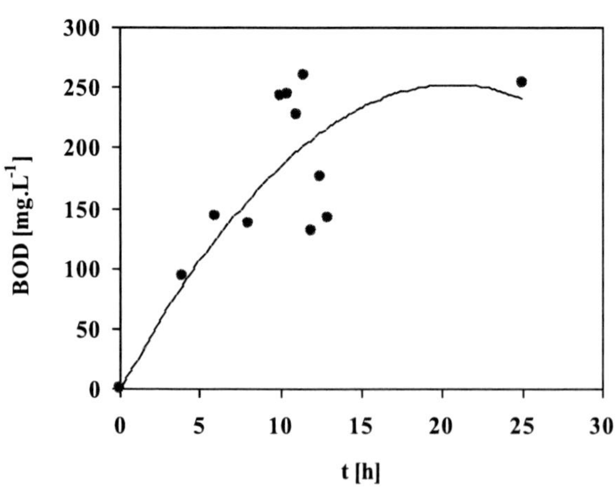

Figure 14. BOD_5 changes with ozonation time.

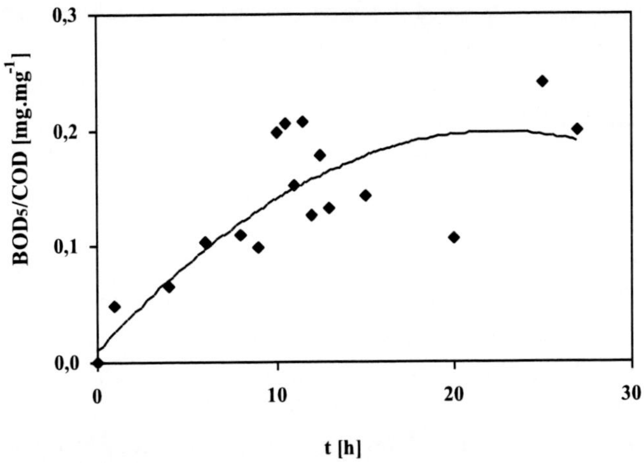

Figure 15. Changes of BOD_5/COD ratio with time of ozonation.

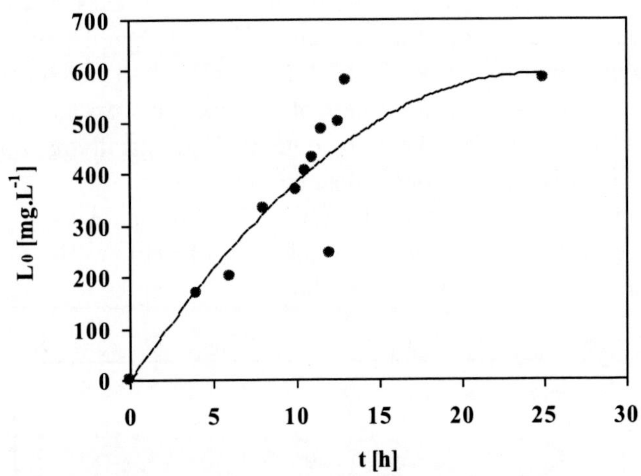

Figure 16. Dependence of ultimate BOD (L_0) on time of ozonation.

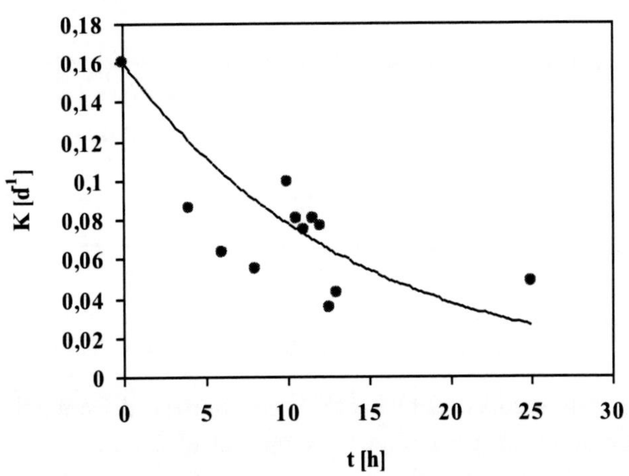

Figure 17. Dependence of rate constant K on time of ozonation.

If the data are presented vice versa it can be shown that the value of the COD/BOD_5 ratio decreased from 132.9 at the beginning of the ozonation trial to 5.0 at the end of the experiment. This clearly shows that the remaining organics are more biodegradable than before the ozonation treatment.

In general, the BOD_5/COD ratio is often not sufficient to appreciate the biodegradability with good accuracy because the activated sludge adaptation could increase the possibility of degradation from recalcitrant organics. From the Equation (17) which follows, that with an increase of ultimate biochemical oxygen demand value L_0 and a decrease of reaction rate K value a residual BOD at the effluent of wastewater treatment plant will increase. Consequently, the oxygen depletion in a water receiver will also increase.

The dependency of L_0 on ozonation time is pictured in Figure 16. The highest values of L_0 and the lowest values of K occur in the region of ozonation time of about 11.5 hours. These results are in very good correlation with the BOD_5 and BOD_5/COD data plotted in Figures 14 and 15 regarding the enhancement of biodegradability by ozonation of biologically pre-treated landfill leachate LL1.

This is evident also from Figure 16 where time dependencies of experimental BOD_5 values and calculated ones using Equation (17) are plotted. It should be mentioned that ozone was produced from pure oxygen by AZCO HTU, 500-DC with the maximum capacity of 250 mg h^{-1} of ozone. Likewise in the former part of the text, this explains relatively the long ozonation times as presented above. The high content of biologically resistant organics in the landfill leachate also contributes to a longer reaction time.

In conclusion, an approximate 60% reduction of COD in biologically pre-treated landfill leachate was observed after about 10 to 11.5 h lasting ozonation. The maximum observed BOD_5 value, measured by the respirometric method was 244 mg L^{-3}.

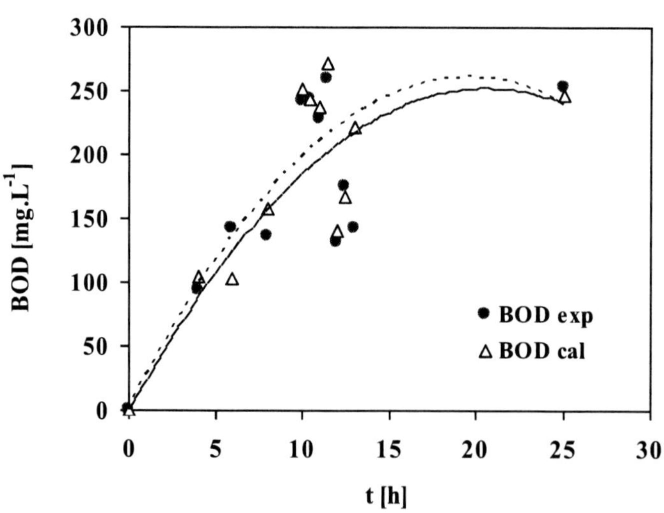

Figure 18. Variations of experimental (—) and calculated (----) BOD_5 values during ozonation.

Biodegradability expressed as a BOD_5/COD ratio increased from zero up to a maximum value of 0.21, corresponding to the ozone consumption of 0.7 mg per mg of COD initially present in the leachate. The highest values of L_0 and the lowest values of K also occur in the region of ozonation time of about 11.5 hours. These results are in very good correlation with the BOD_5 and BOD_5/COD data regarding the enhancement of biodegradability by ozonation of biologically pre-treated landfill leachate LL1. The highest biodegradability of LL1 leachate was achieved after 24 hours of ozonation. However, the additional slow increase BOD_5/COD ratio value observed after prolonged ozonation of the sample to more than 11.5 h is negligible with regard to the increase in operational costs and in general, from the point of controlled chemical oxidation of the leachate LL1.

Assessment of Biotretablity of Ozonated Landfill Leachate

First, a non-adapted activated sludge was used, while in the second experiments a partially adapted sludge was investigated to determine the possible adaptation of the microorganisms to landfill leachate.

Non-Adapted Sludge to Ozonated Landfill Leachate

The effectiveness of ozone treatment for enhancement of the biodegradability of resistant organic pollutants contained in a landfill leachate after biological pre-treatment was also studied. Kinetic tests using activated sludge cultivated in continuous flow or in a semicontinuous activated sludge system were performed in order to evaluate the impact of ozonated landfill leachate LL1 after its biological pre-treatment on activated sludge activity, i.e. possible inhibition impact or biodegradability enhancement. Taking into account the high content of nitrogenous pollutants in reduced form in LL1 leachate, nitrification and denitrification tests were carried out in order to evaluate the possible inhibition/toxic effects of the ozonation process intermediates/products on the processes performance.

Activated sludge cultivated in a continuous flow intermittently aerated completely mixed activated sludge reactor (CMASR) was used in the first series of kinetic tests. Homogeneous distribution of substrate concentration in the whole volume and low concentration values of readily biodegradable as well as those possibly presented toxic compounds is characteristic for microorganisms' cultivation in this type of bioreactor due to the dilution effect. The bioreactor was operated to perform simultaneous nitrification and denitrification processes. The oxygen profile maintained in the bioreactor is illustrated in Figure 19. The aerobic phase took 18 minutes and the anoxic phase occupied 13 minutes from the whole cycle. Landfill leachate LL1 was used as a substrate for cultivation. Parameters of leachate and bioreactor are summarized in Table 4 (previous text).

Effluent from biological pre-treatment (COD = 1650 – 3300 mg L^{-1}; = 20 – 40 mg L^{-1}; = 30 – 180 mg L^{-1}) was collected and stored at laboratory temperature. Aeration was performed during the effluent storage in order to prevent undesirable biochemical reduction processes, e.g. process of desulphation.

Specific endogenous respiration rate was 7.84 mg g^{-1} h^{-1} and the maximum specific exogenous respiration rate (using acetic acid) of 137.5 mg g^{-1} h^{-1} resulted from respirometric measurements performed with the activated sludge.

After biological pretreatment the ozonated leachate was applied for kinetic testing. The parameters of landfill leachate ozonated after biological pretreatment are summarized in Table 12. The experimental data measured during *the first kinetic test* with ozonated landfill leachate after biological pre-treatment are summarized in Table 13. The test was performed after about *one week of cultivation with raw leachate LL1* as a substrate. Time courses of COD, ammonium, nitrite and nitrate nitrogen during the test are shown in Figures 20 and 21.

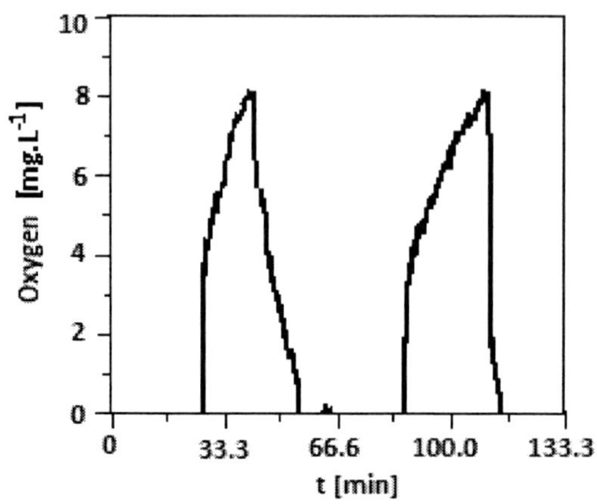

Figure 19. Oxygen profile in intermittently aerated CMASR, used for cultivation of activated sludge for kinetic tests.

Table 12. Parameter value of ozonated landfill leachate sample LL1 after biological pre-treatment

Parameter	Dimension	Value
t	[h]	7.6
pH	[-]	8.1
COD	[mg L^{-1}]	1910.4
BOD_5	[mg L^{-1}]	128.2
$N-NH_4^+$	[mg L^{-1}]	8.9
$N-NO_2^-$	[mg L^{-1}]	13.9
$N-NO_3^-$	[mg L^{-1}]	157.9

Table 13. Experimental values measured during kinetic test – activated sludge cultivated in *intermittently aerated CMASR* (fed with raw landfill leachate)

Parameter / t [h]	$N-NH_4^+$ [mg L^{-1}]	$N-NO_2^-$ [mg L^{-1}]	$N-NO_3^-$ [mg L^{-1}]	COD [mg L^{-1}]
0	8.6	13.5	153.4	1853.7
1	6.9	13.7	159.0	1674.4
2	4.9	14.3	168.2	1435.5
3	2.6	10.0	177.5	1335.0
23	0	7.5	181.1	1195.1

It is obvious from the Table 13 that biological processes could be limited by ammonium nitrogen after 3 hours of duration of the test. However, the evaluation of the process's performance will be done for comparative reasons only for the first two hours.

23% COD and 43% ammonium nitrogen removal efficiencies were measured during the first 2 hours of the test. Total COD removal efficiency during the test performance was 35%.

The value of the specific rate of COD removal during the first two hours of the process was relatively very high, i.e. 394.5 mg g^{-1} h^{-1}.

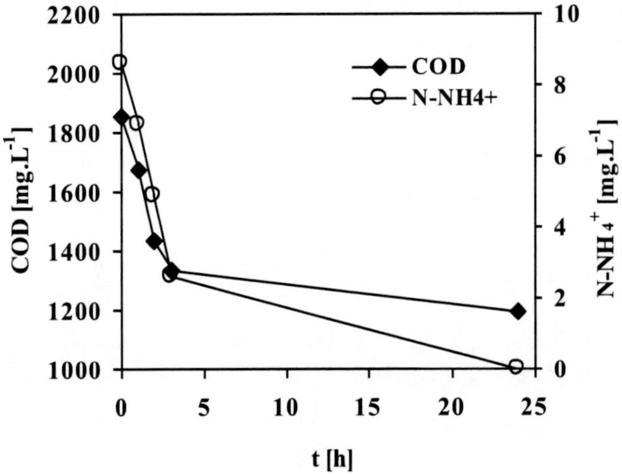

Figure 20. Changes of COD and $N\text{-}NH_4$ contents during kinetic test performed with ozonated landfill leachate after biological pretreatment – activated sludge cultivated in intermittently aerated CMASR.

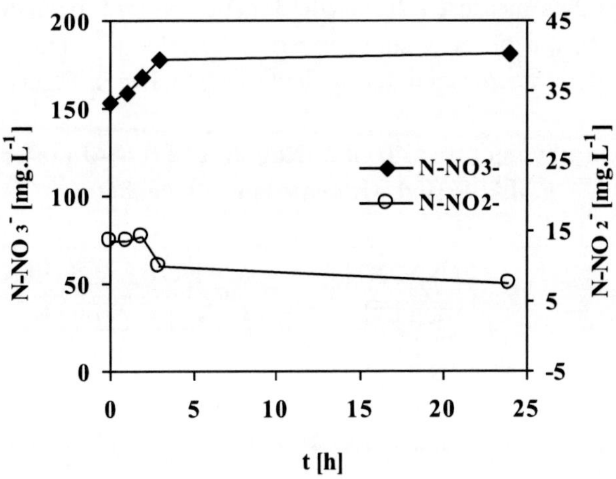

Figure 21. Changes of $N\text{-}NO_3$ and $N\text{-}NO_2$ concentrations during kinetic test performed with ozonated landfill leachate after biological pretreatment – activated sludge cultivated in intermittently aerated CMASR.

Compounds with maximum specific removal rates in the range from 15 to 50 mg g^{-1} h^{-1} can be considered readily degradable by the acclimated mixed culture (Pitter and Chudoba, 1990). However, the authors cite similar values of maximal specific respiration rates, e.g. 317 mg g^{-1} h^{-1} for methanol and cultivation in a completely mixed bioreactor maintaining SRT 0.6 days, 191 – 206 mg g^{-1} h^{-1} for phenol and cultivation in a completely mixed bioreactor maintaining SRT 3.5 days, 205 - 245 mg g^{-1} h^{-1} for sorbitol and cultivation/adaptation in a semicontinuous bioreactor maintaining SRT 9.5 days, or 165 – 314 for glucose and adaptation in a continuous selector system maintaining SRT 4.3 days. The value of the specific rate of ammonium nitrogen removal measured during the first two hours of the process was 1.7 mg g^{-1} h^{-1}. The value of the specific rate of the sum of nitrite and nitrate nitrogen increase measured during the first two hours of the process was 14.7 mg g^{-1} h^{-1}. The parameters of landfill leachate ozonated after biological pre-treatment used for *the second kinetic* test performed with activated sludge cultivated in *intermittently aerated CMASR* are summarized in Table 14. This test was carried out after about two weeks of the CMASR operation.

Table 14. Parameter value of ozonated landfill leachate sample LL1 after biological pre-treatment

Parameter	Dimension	Value
T	[h]	11.5
pH	[-]	8.1
COD	[mg L]	1853.8
BOD_5	[mg L]	150.5
$N-NH_4^+$	[mg L]	22.0
$N-NO_2^-$	[mg L]	0
$N-NO_3^-$	[mg L]	199.8

The experimental data measured during the kinetic test with ozonated landfill leachate, after biological pre-treatment, are summarized in Table 15. Time courses of COD, ammonium, nitrite and nitrate nitrogen during the test are shown in Figures 22 and 23.

Table 15. Experimental values measured during kinetic test – in intermittently aerated CMASR (fed with raw landfill leachate)

Parameter / t [h]	$N-NH_4^+$ [mg L^{-1}]	$N-NO_2^-$ [mg L^{-1}]	$N-NO_3^-$ [mg L^{-1}]	COD [mg L^{-1}]
0	79.9	0	199.8	1853.7
1	68.0	7.2	203.2	1315.8
2	59.0	12.5	208.2	1205.5
3	46.8	14.0	217.3	1120.2
4	33.1	19.8	227.8	1089.5
23	33.0	16.5	229.0	1008.6

35% COD and 26% ammonium nitrogen removal efficiencies were measured during the first 2 hours of the test. The value of the specific rate of COD removal during the first two hours of the process was 356.3 mg g^{-1} h^{-1}. The total COD efficiency removal during the test was 46%. The value of the specific rate of ammonium nitrogen removal measured during the first two hours of the process was 11.5 mg $g^{-1}h^{-1}$.

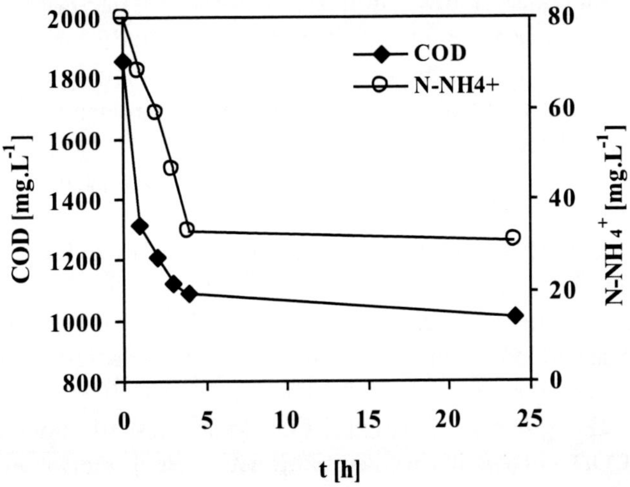

Figure 22. Changes of COD and N-NH_4 contents during kinetic test performed with ozonated landfill leachate after biological pretreatment – activated sludge cultivated in intermittently aerated CMASR.

The value of the specific rate of the sum of nitrite and nitrate nitrogen increase measured during the first two hours of the process was 11.5 mg g^{-1} h^{-1}. The mean value of the specific rate of ammonium removal during the test was 2.2 mg g^{-1} h^{-1}. The value of the specific rate of the sum of nitrite and nitrate nitrogen increase during the whole test was 2.2 mg $g^{-1}h^{-1}$. The total efficiency of ammonium removal was 59%. The total amount of nitrate and nitrite nitrogen corresponds to total ammonium nitrogen removal during the test.

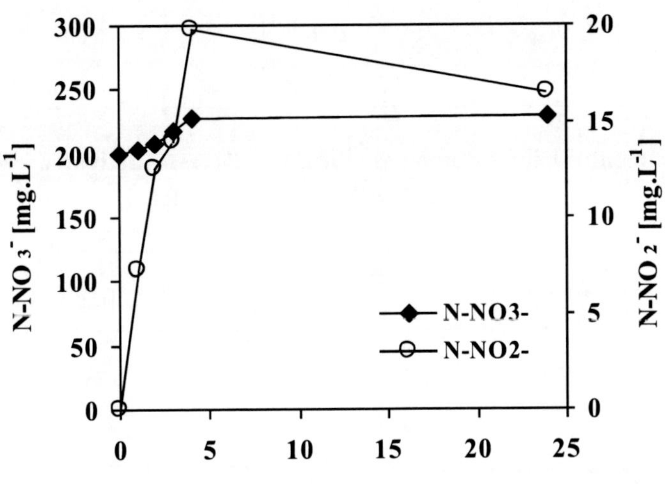

Figure 23. Changes of N-NO_3 and N-NO_2 concentrations during a kinetic test performed with ozonated landfill leachate after biological pretreatment – activated sludge cultivated in intermittently aerated CMASR.

The values of specific rates measured during this test are very close to the values measured with partially acclimated activated sludge (Table 14, Figures 20 and 21). Higher

COD removal efficiency measured within this second test can be attributed to higher BOD_5 value in the ozonated leachate sample. On the other hand, the lower value of ammonium nitrogen removal efficiency during the second test in comparison to results of the first one can be explained by a lower concentration value of ammonium nitrogen more than by one order during the first test performance.

The third aerobic kinetic test was performed after *one month of operation of the CMASR*, i.e. with more or less *adapted activated sludge to raw landfill leachate LL1*. The parameters of landfill leachate ozonated after biological pre-treatment used for this kinetic test are summarized in Table 16.

Fresh ozonated leachate, i.e. without storing was applied for the kinetic test. Ammonium chloride solution was added at the beginning of the kinetic test in order to increase ammonium content (resulting in value of 74.9 mg L^{-1} $N\text{-}NH_4$) and to avoid substrate limitation of nitrification process. The results of this kinetic test are summarized in Table 17. The evolution of COD content during aerobic kinetic test is plotted in Figure 24. High specific removal rates of COD were observed.

Table 16. Parameter value of ozonated landfill leachate sample LL1 after biological pre-treatment

Parameter	Dimension	Value
t	[h]	10.0
pH	[-]	9.3
COD	[mg $L^{-1]}$	1730.0
BOD_5	[mg L^{-1}]	160.0
$N-NH_4^+$	[mg L^{-1}]	15.4
$N-NO_2^-$	[mg L^{-1}]	0
$N-NO_3^-$	[mg L^{-1}]	181.9

Table 17. Experimental values during *the third kinetic test* – activated sludge cultivated in *intermittently aerated CMASR*

Parameter / t [h]	$N-NH_4^+$ [mg L^{-1}]	$N-NO_3^-$ [mg L^{-1}]	$N-NO_2^-$ [mg L^{-1}]	COD [mg L^{-1}]
0	74.9	181.9	0	1730.0
1	73.1	183.5	0.8	1638.7
2	71.2	184.1	1.0	1579.7
3	69.7	186.5	1.1	1469.7
4	67.9	187.5	1.0	1411.1
5	66.1	189.4	1.8	1323.2
7	64,4	191.1	2.2	1237.5
8	64.0	190.0	2.0	1186.7
9	61.8	191.5	2.2	1114.2
10	63.1	192.1	2.0	1061.6
12	60.8	194.5	2.1	1019.3
13	59.1	193.1	2.7	978.4
24	48.8	206.5	3.5	778.8

The value of the specific rate of COD removal during the first two hours of the process was 501mg g^{-1} h^{-1}. Observed COD removal efficiency during 2 hours was 9%. The mean value of specific removal rate of COD during the test was 264.2 mg g^{-1} h^{-1}. About 55% COD removal efficiency was observed during 24 hours. The value of specific rate of ammonium nitrogen removal measured during the first two hours of the process was 12.3 mg g^{-1} h^{-1}. The value of the specific rate of the sum of nitrite and nitrate nitrogen increase measured during the first two hours of the process was 10.7 mg g^{-1} h^{-1}.

The mean value of the specific rate of ammonium removal during the test was 7.3 mg g^{-1} h^{-1} (Figure 25). The value of specific rate of the sum of nitrite and nitrate nitrogen increase during the whole test was 7.8 mg g^{-1} h^{-1}. The total efficiency of ammonium removal was 35%. The total amount of nitrate and nitrite nitrogen corresponds to total ammonium nitrogen removal during the test. The difference 0.5 mg L^{-1} can be attributed to incorporation into a new biomass.

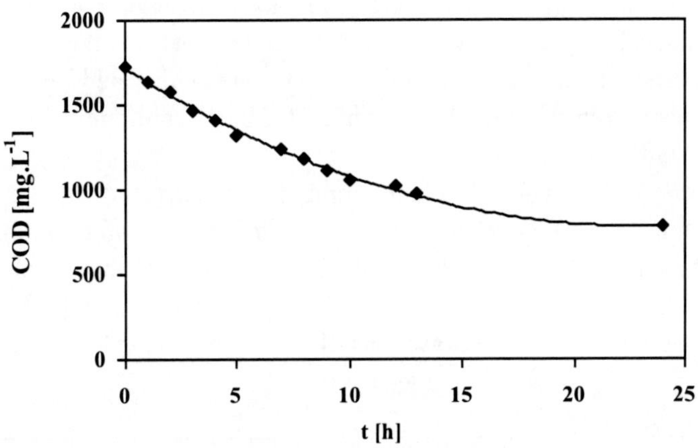

Figure 24. COD changes during the third kinetic test – activated sludge cultivated in intermittently aerated CMASR.

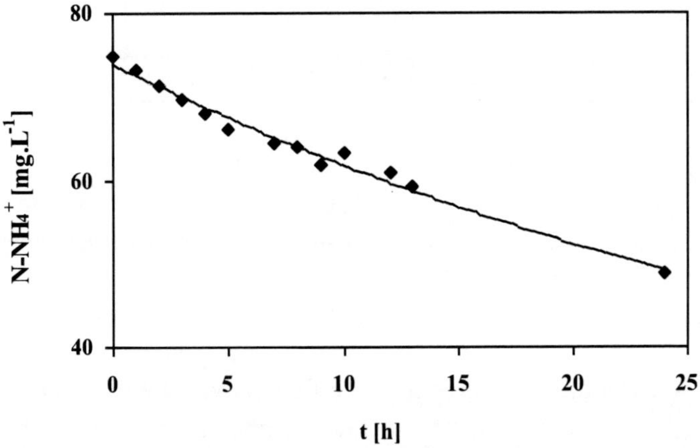

Figure 25. Changes of N-NH_4 during the first kinetic test – activated sludge cultivated in intermittently aerated CMASR.

So, the system can be considered to be in balance with regard to nitrogen compounds. In conclusion, the gradual adaptation of activated sludge to raw LL1 landfill leachate resulted in obtaining activated sludge capable of performing, with high efficiency, organic carbon mineralization and nitrification processes when applying ozonated LL1 leachate after biological pretreatment. The mean value of the specific rate of ammonium removal during the test was 7.3 mg g^{-1} h^{-1}. The value of the specific rate of the sum of nitrite and nitrate nitrogen increase during the whole test was 7.8 mg g^{-1} h^{-1}. COD removal efficiency was 55% and ammonium removal efficiency was 35% during the aerobic kinetic test.

Adapted Sludge to Ozonated Landfill Leachate

The results of kinetic tests carried out with activated sludge adapted to ozonated landfill leachate LL1 were also run. The intermittently aerated CMASR was fed with raw landfill leachate LL1 (COD = 4,000 – 5,000 mg L^{-1}; N-NH_4 = 450 – 700 mg L^{-1}; BOD_5 = 50 – 65 mg L^{-1}). The *effluent* from this reactor *was used for ozonation* and next it was introduced as *influent into a semicontinuous* lab-scale model. Activated sludge cultivated in a semicontinuous lab-scale bioreactor was used for kinetic tests. Unlike intermittently aerated CMASR, the microoganisms of activated sludge are cultivated at variable substrate concentration values within the operation cycle of the semi-continuous bioreactor due to the concentration gradient in the system.

The parameters of ozonated leachate applied for activated sludge acclimation are summarized in Table 18. *Acclimation of activated sludge to ozonated landfill leachate* was carried out during a *one-month period before* the performance of kinetic tests.

Table 18. Parameter value of ozonated landfill leachate sample LL1 after biological pre-treatment

Parameter	Dimension	Value
t	[h]	10.0
pH	[-]	8.4
COD	[mg L^{-1}]	1914
BOD_5	[mg L^{-1}]	150
$N-NH_4^+$	[mg L^{-1}]	17.0
$N-NO_3^-$	[mg L^{-1}]	212

The value of sludge load in the semicontinuous bioreactor with regard to BOD_5 of ozonated leachate was 0.084 kg kg^{-1} d^{-1} (BOD_5/MLSS). The concentration of MLSS decreased from 0.94 to 0.85 g L^{-1} during the test. Experimental values measured during the kinetic test are summarized in Table 19.

An approximate 55% removal efficiency of COD was observed (Figure 26) during 24 hours. The value of the specific rate of COD removal during the first two hours of the process was 264.0 mg g^{-1} h^{-1}. The mean value of the specific removal rate of COD during the test was 576.7 mg g^{-1} h^{-1}. The value of the specific rate of ammonium nitrogen removal measured during the first two hours of the process was 5.5 mg g^{-1} h^{-1}. The value of the specific rate of the sum of nitrite and nitrate nitrogen increase measured during the first two hours of the process was 5.9 mg $g^{-1}h^{-1}$.

Table 19. Experimental values obtained in the second kinetic test – activated sludge cultivated in semicontinuous bioreactor

Parameter / t [h]	$N-NH_4^+$ [mg L^{-1}]	$N-NO_2^-$ [mg L^{-1}]	$N-NO_3^-$ [mg L^{-1}]	COD [mg L^{-1}]
0	50.5	27.1	183.9	1853.8
1.0	44.2	27.0	191.2	1775.4
2.0	40.1	28.3	193.8	1680.8
3.0	35.5	30.9	198.7	1598.9
4.0	31.0	31.5	199.5	1480.9
5.0	26.1	33.8	203.1	1395.7
7.0	23.6	33.1	205.2	1315.5
8.0	19.3	35.5	207.1	1209.5
11.5	11.2	38.2	211.0	959.4
24.0	2.5	43.9	215.4	833.5

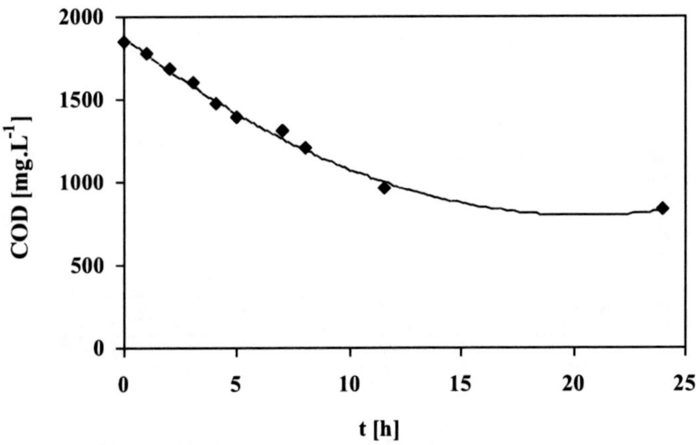

Figure 26. COD removal during the first kinetic test – activated sludge cultivated in semicontinuous bioreactor.

The mean value of the specific rate of ammonium removal during the test was 2.1 mg g^{-1}h^{-1} (Table 19, Figure 27). The value of the specific rate of the sum of nitrite and nitrate nitrogen increase during the whole test was 2.1 mg g^{-1} h^{-1}. The total efficiency of ammonium removal was 95%. The total amount of nitrate and nitrite nitrogen increase during the test is equal to the total ammonium nitrogen removal during the test. So, some organic nitrogen probably participated in covering nitrogen consumption during new microbial cell synthesis.

Very similar COD removal efficiencies follow from the above presented kinetic tests, carried out with adapted and non-adapted activated sludge to ozonated leachate LL1 after biological pre-treatment. The higher value of COD specific removal rate measured with non-adapted sludge can be attributed to higher BOD_5 in the ozonated sample.

On the other hand, a significantly higher ammonium removal rate was observed with activated sludge adapted to ozonated LL1 leachate. However, when considering the significantly lower MLSS value in the previous kinetic test (0.15 g L^{-1}) in comparison to this one (0.94 g L^{-1}) the higher activity of activated sludge microorganisms cultivated in the completely mixed bioreactor and non-acclimated to ozonated leachate can be concluded in comparison to the activity of microorganisms of activated sludge cultivated in the semicontinuous (with concentration gradient) system and underwent adaptation to ozonated leachate.

Some adverse impact of ozonated leachate sample on microorganism activity during adaptation in the semicontinuous/gradient bioreactor also indicate a decrease of the removal rate of ammonium nitrogen and an increase of nitrate/nitrite concentration by 25 to 70% in comparison to specific removal rates measured with non-adapted sludge (Tables 17 and 19) cultivated in homogeneous ambient typical for a completely mixed bioreactor.

In conclusion, very close COD removal efficiencies were observed using both, adapted and to ozonated LL1 leachate non-adapted activated sludge. Significantly higher ammonium nitrogen removal efficiency was measured with sludge adapted to ozonated LL1 leachate after biological pretreatment. However, a serious decrease of specific rates of ammonium nitrogen as well as nitrate/nitrite nitrogen increase were measured with activated sludge adapted to ozonated LL1 leachate in a semicontinuous mixed bioreactor.

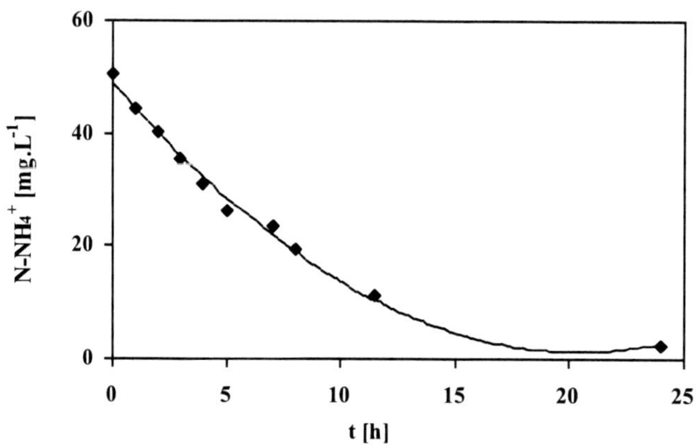

Figure 27. Ammonium nitrogen removal during the first kinetic test – activated sludge cultivated in semicontinuous bioreactor.

A more significant adverse impact of some unspecified ozonation products on oxic processes can be attributed to their higher initial concentration values in a semicontinuous bioreactor in comparison to a completely mixed one.

Table 20. Parameter value of ozonated landfill leachate sample LL1 after biological pre-treatment

Parameter	Dimension	Value
T	[h]	8.0
pH	[-]	8.4
COD	[mg L^{-1}]	2810
BOD_5	[mg L^{-1}]	117
$N - NH_4^+$	[mg L^{-1}]	24.8
$N - NO_3^-$	[mg L^{-1}]	235

Table 21. Experimental values during the kinetic test – activated sludge cultivated in semicontinuous bioreactor (*stored ozonated leachate LL1*)

Parameter / t [h]	$N-NH_4^+$ [mg L^{-1}]	$N-NO_3^-$ [mg L^{-1}]	$N-NO_2^-$ [mg L^{-1}]	COD [mg L^{-1}]
0	51.3	235.4	3.8	2810.3
1.0	46.2	238.2	4.1	2769.5
2.0	41.3	244.1	4.5	2723.9
3.0	35.5	249.0	4.6	2682.4
4.0	30.8	252.4	5.2	2639.2
5.0	26.3	257.1	5.4	2592.1
6.0	21.5	260.0	5.6	2550.0
24.0	0	274.4	9.2	1845.0

The next aerobic test performed was very similar to the previous one. The main difference was in the applied landfill leachate sample. The parameters of ozonated leachate applied for this test are summarized in Table 20. It should be emphasized that the ozonated landfill leachate specified in Table 20 *was stored for nine days* at the laboratory temperature before performing this kinetic test.

An ammonium chloride solution was added at the beginning of the kinetic test (resulting in a value of 51.3 mg L^{-1} N-NH_4) in order to increase the ammonium content and to avoid substrate limitation of nitrification process. The value of organic load with regard to BOD_5 of ozonated leachate was 0.18 kg kg^{-1} d^{-1} (BOD_5/MLSS). The concentration of MLSS decreased from 0.65 to 0.60 g L^{-1} during the test. Experimental values measured during the kinetic test are summarized in Table 21.

About 34% removal efficiency of COD was observed (Figure 28) during 24 hours. The value of the specific rate of COD removal during the first two hours of the process was 66.5 mg g^{-1} h^{-1}. The mean value of the specific removal rate of COD during the test was 61.9 mg g^{-1} h^{-1}. The value of the specific rate of ammonium nitrogen removal measured during the first two hours of the process was 7.7 mg g^{-1} h^{-1}. The value of the specific rate of the sum of nitrite and nitrate nitrogen increase measured during the first two hours of the process was 7.2 mg g^{-1} h^{-1}. The mean value of the specific rate of ammonium removal during the test was 3.3 mg g^{-1} h^{-1} (Table 21, Figure 29). The value of the specific rate of the sum of nitrite and nitrate nitrogen increased during the whole test period was 2.8 mg g^{-1} h^{-1}. The ammonium nitrogen was removed during the test completely. The total amount of nitrate and nitrite nitrogen increase during the test corresponds to the total ammonium nitrogen removal during the test. The difference 0.4 mg g^{-1} h^{-1} can be attributed to ammonium nitrogen consumption for the synthesis of new bacterial cells. In conclusion, a higher ammonium nitrogen removal and nitrate/nitrite increase rates were observed with the sludge adapted to ozonated LL1 leachate after biological pretreatment when applying stored ozonated leachate in comparison to the results of previous kinetic test performed with the fresh (non-stored) ozonated LL1 leachate sample. However, the measured rate values were still lower in comparison to those observed during the kinetic test performed with ozonated LL1 leachate after biological pretreatment with activated sludge cultivated in a completely mixed reactor (CMASR) and adapted to the raw LL1 leachate (Table 16, Figures 24 and 25).

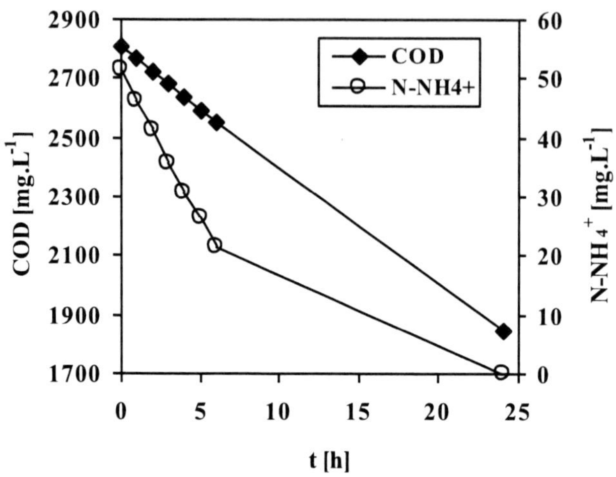

Figure 28. COD removal during the second kinetic test – activated sludge cultivated in semicontinuous bioreactor.

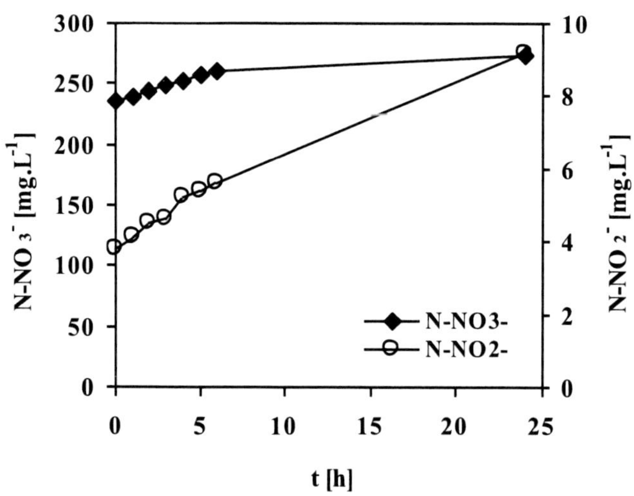

Figure 29. Ammonium nitrogen removal during the second kinetic test – activated sludge cultivated in semicontinuous bioreactor.

Denitrification Tests

Denitrification tests were also performed with the same activated sludge as used in the previous aerobic kinetic test. Activated sludge was cultivated in a semi-continuous reactor and acclimated to ozonated LL1 landfill leachate after biological pre-treatment as described above. The parameters of the ozonated leachate applied for activated sludge acclimation are summarized in Table 22. Ozonated LL1 leachate was used as the source of nitrates and organics during the test, i.e. neither nitrate nor organics were added. Ammonium chloride solution was added at the beginning of the kinetic test (resulting in value of 63.3 mg L^{-1} N-NH_4) in order to increase ammonium content and avoid substrate limitation of the nitrification process. An anoxic environment in a batch test was maintained by purging the reaction mixture with nitrogen. The organic load with regard to BOD_5 of ozonated leachate was 0.18

kg kg^{-1} d^{-1}. Experimental values measured during the denitrification test are summarized in Table 23.

Nitrate/nitrite nitrogen concentration decreased by 42.6 mg L^{-1} within 7 hours, i.e. the specific removal rate of nitrate and nitrite nitrogen was 9.5 mg g^{-1} h^{-1}. Intensive denitrification follows from the balance of nitrite and nitrate nitrogen (N-NO_X) with the mean value of specific removal rate about 10 mg g^{-1} h^{-1}. Measured COD removal efficiency was 17%. The mean experimental value of the specific COD removal rate during the test was 86 mg g^{-1} h^{-1}. The more or less constant COD removal rate during the whole test (Figure 30) testifies to the zero order reaction rate of COD removal. The content of ammonium nitrogen decreased by 8.8 mg L^{-1} during the test, i.e. specific ammonium nitrogen rate 8.4 mg g^{-1} h^{-1} ($N-NH_4^+$ /MLSS) due to synthesis of new microbial cells.

Table 22. Parameter value of ozonated landfill leachate sample LL1 after biological pre-treatment

Parameter	Dimension	Value
t	[h]	7.5
pH	[-]	9.2
COD	[mg L^{-1}]	2324.9
BOD_5	[mg L^{-1}]	117.0
$N-NH_4^+$	[mg L^{-1}]	25.0
$N-NO_3^-$	[mg L^{-1}]	246.6
$N-NO_2^-$	[mg L^{-1}]	23.4

Table 23. Experimental values during denitrification kinetic test – activated sludge cultivated in semicontinuous bioreactor – adapted to ozonated LL1 leachate after biological pretreatment

Parameter / t [h]	$N-NH_4^+$ [mg L^{-1}]	$N-NO_2^-$ [mg L^{-1}]	$N-NO_3^-$ [mg L^{-1}]	COD [mg L^{-1}]
0	63.3	21.7	228.3	2152.7
1.0	61.6	22.8	225.5	2099.0
2.0	60.2	21.0	219.3	2034.5
3.0	59.6	20.7	210.9	1977.5
4.0	58.4	20.1	204.9	1934.0
5.0	58.2	18.4	196.2	1883.9
6.0	56.2	20.1	191.0	1835.6
7.0	54.5	22.6	184.8	1794.0

Time courses of nitrate and nitrite nitrogen concentrations during the denitrification test are plotted in Figure 31. In conclusion, a high removal denitrification rate as well as COD removal rate resulted from the denitrification kinetic test.

The next two denitrification kinetic tests were carried out after another six weeks of activated sludge adaptation.

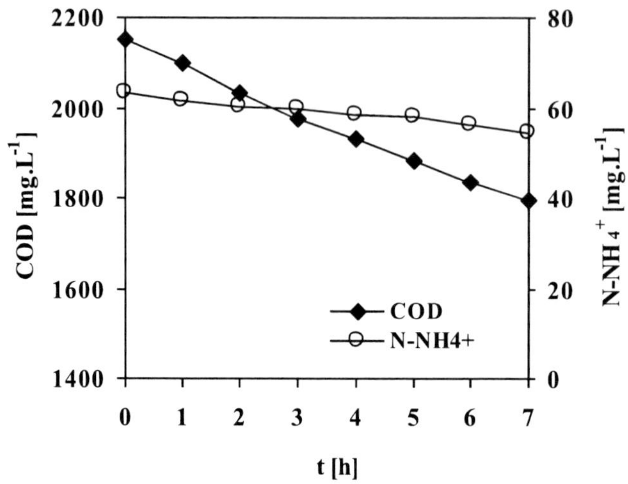

Figure 30. Changes of COD and N-NH_4 during denitrification test – activated sludge cultivated in semicontinuous bioreactor – activated sludge partially adapted on ozonated leachate after biological pretreatment.

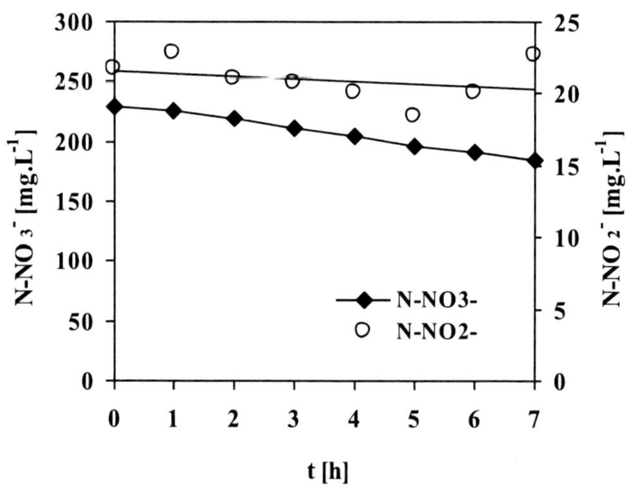

Figure 31. Changes of N-NO_3 and N-NO_2 concentrations during denitrification test test – activated sludge cultivated in semicontinuous bioreactor.

Likewise to previous denitrification test *activated sludge* was cultivated in a *semi-continuous reactor* and *adapted to ozonated LL1 landfill leachate after biological pre-treatment* as described above. The parameters of ozonated leachate applied for activated sludge cultivation and acclimation are summarized in Table 24.

The difference between the two tests is that one test was performed with the addition of organic carbon (4 mL of acetic acid was) at the beginning of the test. The second one was performed without the addition of any organic carbon. Similar to the previous denitrification test, only nitrates contained in ozonated leachate were applied, i.e. no nitrates were added during both denitrification tests. The anoxic environment in the batch test was maintained by purging the reaction mixture with nitrogen.

Table 24. Parameter value of ozonated landfill leachate sample LL1 after biological pre-treatment

Parameter	Dimension	Value
T	[h]	7.0
pH	[-]	8.5
COD	[mg L^{-1}]	2398.6
BOD_5	[mg L^{-1}]	113.0
$N-NH_4^+$	[mg L^{-1}]	28.9
$N-NO_3^-$	[mg L^{-1}]	254.2
$N-NO_2^-$	[mg L^{-1}]	0

No Addition of Organic Carbon

Experimental values measured during the denitrification test with the *addition of organic carbon* are summarized in Table 25.

Table 25. Experimental values during denitrification kinetic test – activated sludge cultivated in semicontinuous bioreactor – adapted to ozonated leachate after biological pre-treatment

Parameter / t [h]	$N-NH_4^+$ [mg L^{-1}]	$N-NO_3^-$ [mg L^{-1}]	$N-NO_2^-$ [mg L^{-1}]	COD [mg L^{-1}]
0	26.8	235.4	0	2220.9
1.0	24.7	223.3	0	2172.4
2.0	22.3	211.9	2.1	2120.3
3.0	20.5	198.8	1.5	2071.8
4.0	18.7	187.8	2.6	2016.9
5.0	16.6	175.4	3.1	1967.1
6.0	14.9	164.5	2.5	1913.7
7.0	13.1	152.4	3.5	1862.3

Nitrate and nitrite nitrogen concentrations decreased by 79.5 mg. $^{-1}$ within 7.0 hours. There was only small increase of nitrite nitrogen observed during the test. Specific removal rate of nitrate/nitrite nitrogen was 20.6 mg g^{-1} h^{-1}. The mean experimental value of specific COD removal rate was 93.1 mg g^{-1} h^{-1} (Figure 32).

The content of ammonium nitrogen decreased by 13.7 mg L^{-1} during the test, i.e. specific ammonium nitrogen rate 3.6 mg g^{-1} h^{-1} ($N-NH_4^+$/MLSS) due to synthesis of new microbial cells. Time courses of nitrate and nitrite nitrogen concentrations during the denitrification test are plotted in Figure 33.

In conclusion, the more than twice higher nitrate/nitrite ammonium nitrogen removal rate was measured with activated sludge adapted to ozonated LL1 leachate after biological pretreatment. The specific COD removal rate was higher by approximately 8% in comparison to non-adapted activated sludge.

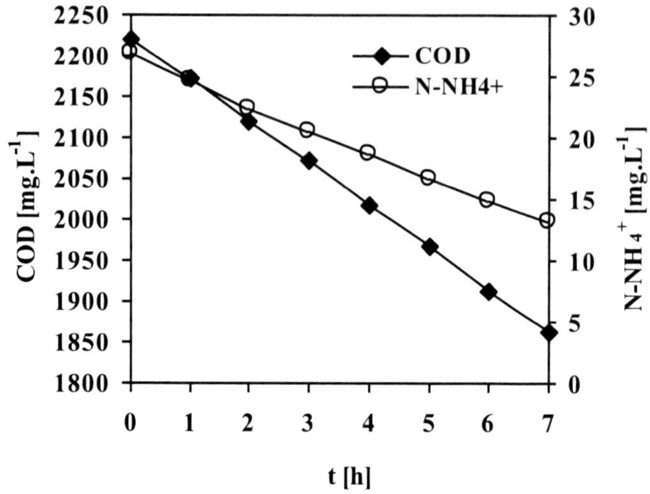

Figure 32. Changes of COD and N-NH_4 during denitrification test – activated sludge cultivated in semicontinuous bioreactor and adapted on ozonated leachate after biological pretreatment.

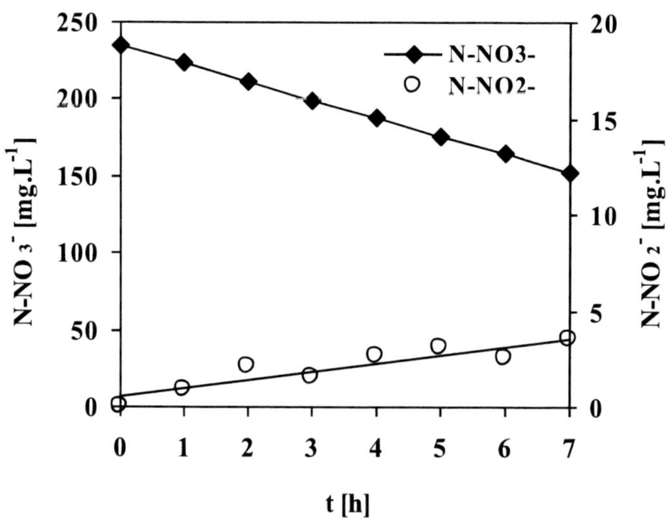

Figure 33. Changes of N-NO_3 and N-NO_2 concentrations during denitrification test – activated sludge cultivated in semicontinuous bioreactor and adapted on ozonated leachate after biological pretreatment.

Addition of Organic Carbon

The next denitrification kinetic test was also performed with the same activated sludge as used in the second denitrification test. Activated sludge was cultivated in a *semi-continuous reactor* and *adapted to ozonated landfill leachate after biological pre-treatment* as described above. The parameters of the ozonated leachate applied for activated sludge adaptation are summarized above in Table 24. Similar to the previous denitrification test, ozonated leachate was used as the source of nitrates and organics during the test, i.e. no nitrates or organics were added. The anoxic environment in the batch test was maintained by purging the reaction mixture with nitrogen. The organic load with regard to BOD_5 of ozonated leachate was 0.21

kg kg^{-1} d^{-1}. Experimental values measured during the denitrificationc test are summarized in Table 26.

Table 26. Experimental values during denitrification kinetic test – activated sludge cultivated in semicontinuous bioreactor – adapted to ozonated leachate after biological pre-treatment

Parameter / t [h]	$N-NH_4^+$ [mg L^{-1}]	$N-NO_3^-$ [mg L^{-1}]	$N-NO_2^-$ [mg L^{-1}]	COD [mg L^{-1}]
0	28.8	235.4	0	6098.2
1.0	27.4	222.0	0	6026.5
2.0	25.6	210.2	0	5957.3
3.0	24.2	196.8	0.11	5887.2
4.0	22.3	184.2	0.30	5828.7
5.0	20.8	171.3	1.05	5768.9
6.0	19.3	159.2	1.80	5714.0
7.0	17.9	146.1	2.1	5653.9

Nitrate and nitrite nitrogen concentrations decreased by 87.2 mg L^{-1} within 7.0 hours. There was only a small increase of nitrite nitrogen observed during the test. The specific removal rate of nitrate/nitrite nitrogen was 22.6 mg g^{-1} h^{-1}. The mean experimental value of the specific COD removal rate was 115.6 mg g^{-1} h^{-1} (Figure 34). The content of ammonium nitrogen decreased by 10.9 mg L^{-1} during the test, i.e. specific ammonium nitrogen rate 2.8 mg g^{-1} h^{-1} ($N-NH_4^+$/MLSS) due to synthesis of new microbial cells. Time courses of nitrate and nitrite nitrogen concentrations during the denitrification test are plotted in Figure 35.

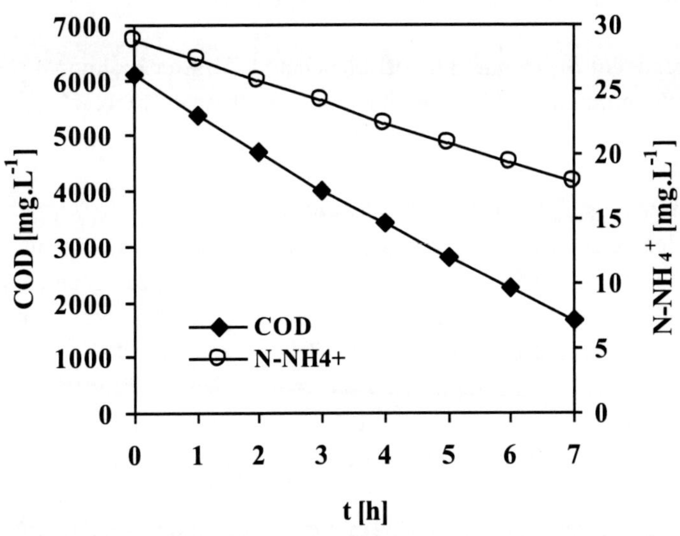

Figure 34. Changes of COD and N-NH_4 during denitrification test – activated sludge cultivated in semicontinuous bioreactor and adapted to ozonated leachate after biological pretreatment – additional organic carbon.

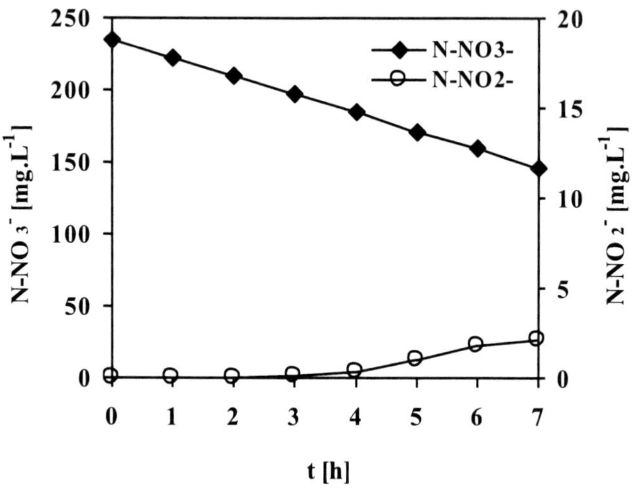

Figure 35. Changes of N-NO_3 and N-NO_2 concentrations during denitrification test – activated sludge cultivated in semicontinuous bioreactor and adapted to ozonated leachate after biological pretreatment – additional organic carbon.

In conclusion, dosing of additional organic carbon enhanced the denitrification specific rate by10%. The COD specific removal rate increased by 19% in comparison to the previous denitrification test (acclimated sludge, no additional organic carbon). No significant limitation of the denitrification process during the previous kinetic test can be concluded, i.e. ozonated LL1 leachate after biological pretreatment contains organics that are available as organic carbon for the denitrification process.

Biological Treatment of Ozonated Landfill Leachate

Biological treatment of ozonated landfill leachate LL1 after biological pre-treatment was also studied. Semicontinuous and completely mixed lab-scale bioreactors were applied for this study.

Biological Treatment in Semicontinuous Systems

The landfill leachate LL1 was subjected to the biological treatment in an intermittently aerated CMASR before performing the ozonation process. The bioreactor was operated as a simultaneous nitrification and denitrification activated sludge process. No withdrawal of excess sludge was performed during the study. The parameters of the raw and biologically pre-treated raw landfill leachate sample (LL1) are summarized in Table 27.

The separation of activated sludge after biological treatment was performed by sedimentation. Thus, micro-flocks of activated sludge were also contained in biologically pre-treated leachate. Biologically pre-treated leachate was stored at laboratory temperature and it was aerated during storage in order to prevent biochemical reduction processes, e.g. desulphatation process. The samples of this leachate were subjected to ozonation. Different ozonation times were applied with the aim of studying the effectiveness of ozone treatment for enhancement of the biodegradability of biorefractory organic pollutants contained in the landfill leachate after biological pre-treatment (Figures. 13 to 18).

Table 27. Parameter values of raw and biologically pre-treated raw landfill leachate sample LL1

		Sample	
Parameter	Dimension	Raw	Pre-treated
pH	[-]	9.1	9.4
COD	[mg L^{-1}]	5382	2511
BOD_5	[mg L^{-1}]	38	0
$N-NH_4^+$	[mg L^{-1}]	345	19.0
$N-NO_3^-$	[mg L^{-1}]	0	173.0

The composite leachate sample after ozonation was used to study biological treatment of the leachate and to perform kinetic tests with activated sludge. The parameters of LL1 leachate ozonated after biological pretreatment are summarized in Table 28.

Two lab-scale semicontinuous activated sludge models were operated in parallel. Model wastewater contained methanol as readily degradable substrate and ammonium (COD = 1810 mg L^{-1}; $N\text{-}NH_4^+$ = 192 mg L^{-1}).

Table 28. Parameter value of ozonated landfill leachate sample LL1 after biological pre-treatment

Parameter	Dimension	Value
t	[h]	1 - 27
pH	[-]	8.2
COD	[mg L^{-1}]	1136
BOD_5	[mg L^{-1}]	66
$N-NH_4^+$	[mg L^{-1}]	15
$N-NO_3^-$	[mg L^{-1}]	185

Activated sludge was cultivated at an organic load of 0.3 kg $kg^{-1}d^{-1}$ (COD/MLSS). A one day operational cycle (23 hours of aeration and 1 hour of sedimentation) was maintained.

Model Substrate

The reference experimental data measured during one cycle in the model fed *with model substrate* are summarized in Table 29. The measurements were performed after 5 days of operating the lab-scale model. Time courses of ammonium and nitrate nitrogen during the test are shown in Figure 36.

The value of the specific rate of ammonium nitrogen removal measured during the first two hours of the process was 10.1 mg g^{-1} h^{-1}. The mean value during the whole reaction period was 1.2 mg g^{-1} h^{-1}. The value of the specific rate of nitrate nitrogen increased measured during the first two hours of the process was 4.1 mg g^{-1} h^{-1}. The mean value during the whole reaction period was 1.0 mg g^{-1} h^{-1}. The efficiency of ammonium nitrogen removal was 19% and the efficiency of COD removal was 80%. The mean value of the specific COD removal rate during the whole reaction period was 70.3 mg g^{-1} h^{-1}.

Table 29. Experimental values during operational cycle test – activated sludge cultivated in semicontinuous bioreactor (*model substrate*)

Parameter / t [h]	$N-NH_4^+$ [mg L^{-1}]	$N-NO_3^-$ [mg L^{-1}]	COD [mg L^{-1}]
0	196.7	10.2	2637
0.5	182.5	12.8	-
1	178.5	16.5	-
2	170.5	20.8	-
23	158.3	40.2	534.1

?- Not measured.

The experimental data measured during one cycle in the lab-scale semicontinuous model fed *with raw LL1 landfill leachate* are summarized in Table 30. The measurements were carried out after 5 days of operation of the lab-scale model and feeding with the raw LL1 leachate.

Time courses of ammonium and nitrate nitrogen during the test are shown in Figure 37. The value of the specific rate of ammonium nitrogen removal measured during the first two hours of the process was 17.8 mg g^{-1} h^{-1}. The mean value during the whole reaction period was 4.6 mg g^{-1} h^{-1}. The value of the specific rate of nitrate nitrogen increase measured during the first two hours of the process was 16.7 mg g^{-1} h^{-1}.

The mean value during the whole reaction period was 3.5 mg g^{-1} h^{-1}. The efficiency of ammonium nitrogen removal was 19%.

In conclusion, higher values of nitrification reaction rate by 11% were measured in comparison to those observed in the semicontinuous model fed with model substrate. This can be attributed to a lower organic load in the bioreactor fed with raw landfill leachate.

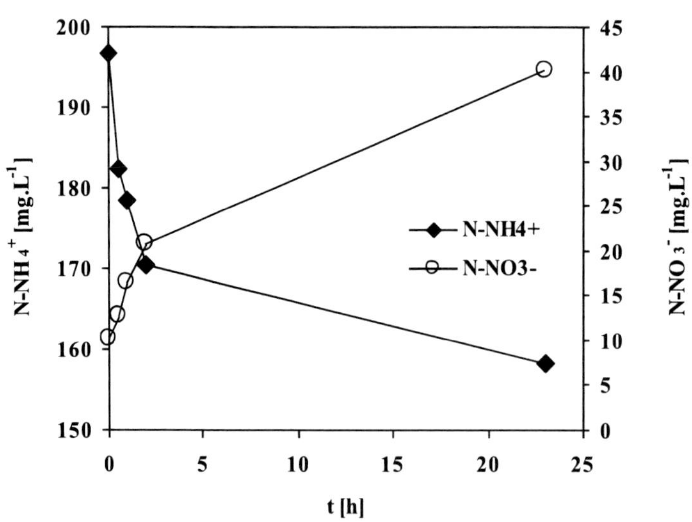

Figure 36. Changes of $N-NO_3$ and $N-NH_4$ concentrations during kinetic test performed with model substrate – activated sludge cultivated in semicontinuous bioreactor.

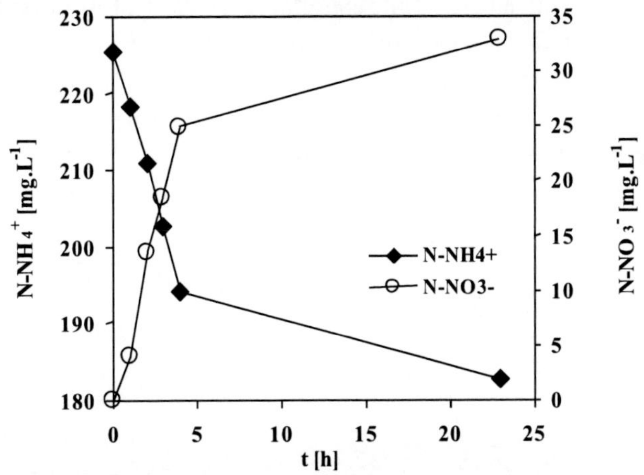

Figure 37. Changes of $N\text{-}NO_3$ and $N\text{-}NH_4$ concentrations during kinetic test performed with model substrate – activated sludge cultivated in semicontinuous bioreactor.

Table 30. Experimental values during operational cycle – activated sludge cultivated in semicontinuous bioreactor fed with *raw LL1 leachate*

Parameter / t [h]	$N-NH_4^+$ [mg L^{-1}]	$N-NO_3^-$ [mg L^{-1}]	COD [mg L^{-1}]
0	225.4	0	1654
1	218.4	4.0	-
2	211.0	13.5	-
3	202.8	18.4	-
4	194.2	24.9	1555
23	182.7	32.9	-

- Not measured.

Non-Adapted Activated Sludge

The next kinetic test was carried out *with ozonated landfill leachate*. The parameters of landfill leachate ozonated after biological pretreatment are summarized in Table 31.

Table 31. Parameter value of ozonated landfill leachate sample LL1 after biological pre-treatment

Parameter	Dimension	Value
t	[h]	6
pH	[-]	8.1
COD	[mg L^{-1}]	2511
BOD_5	[mg L^{-1}]	130
$N-NH_4^+$	[mg L^{-1}]	15
$N-NO_3^-$	[mg L^{-1}]	173

Ammonium nitrogen was added at the beginning of the test in order to avoid substrate limitation. The experimental data measured during one cycle in the model fed with raw landfill leachate are summarized in Table 32.

Table 32. Experimental values during one operational cycle – activated sludge cultivated in semicontinuous bioreactor (ozonated leachate after biological treatment)

Parameter / t [h]	$N-NH_4^+$ [mg L^{-1}]	$N-NO_2^-$ [mg L^{-1}]	$N-NO_3^-$ [mg L^{-1}]	COD [mg L^{-1}]
0	25.6	35.1	166.8	777.0
1	20.6	32.4	175.8	689.5
2	20.8	33.5	184.2	546.8
3	18.9	37.2	193.6	479.5
4	20.9	32.4	211.0	426.5
23	23.1	42.9	229.4	398.4

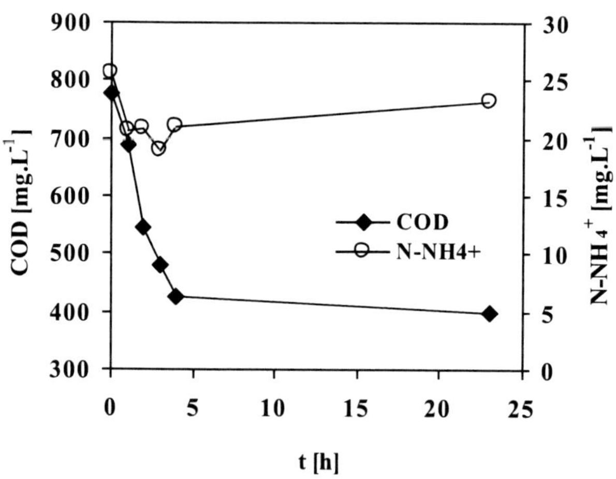

Figure 38. Changes of COD and N-NH_4 contents during kinetic test performed with ozonated landfill leachate after biological pretreatment – activated sludge cultivated in semicontinuous bioreactor.

It should be stressed that the experimental model was fed with model substrate, i.e. there was not an adaptation of activated sludge to LL1 leachate before performing the kinetic measurements. Time courses of ammonium COD, nitrite and nitrate nitrogen during the measurements are shown in Figures 38 and 39.

The value of the specific rate of ammonium nitrogen removal measured during the first two hours of the process was 3.4 mg g^{-1} h^{-1}. The mean value during the whole reaction period was 0.2 mg $g^{-1}h^{-1}$. The value of the specific rate of nitrate/nitrite nitrogen increase measured during the first two hours of the process was 11.3 mg g^{-1} h^{-1}. The mean value during the whole reaction period was 4.4 mg $g^{-1}h^{-1}$. There seems to be an imbalance regarding ammonium, nitrite and nitrate nitrogen transformations in the process.

A significantly higher increase of nitrite and nitrate nitrogen concentrations when very low rate of ammonium nitrogen concentration decrease was observed can be explained by ammonification of organic nitrogen contained in ozonated landfill leachate including microflocs of activated sludge remaining after gravitational separation of activated sludge from the effluent of biological treatment of raw landfill leachate.

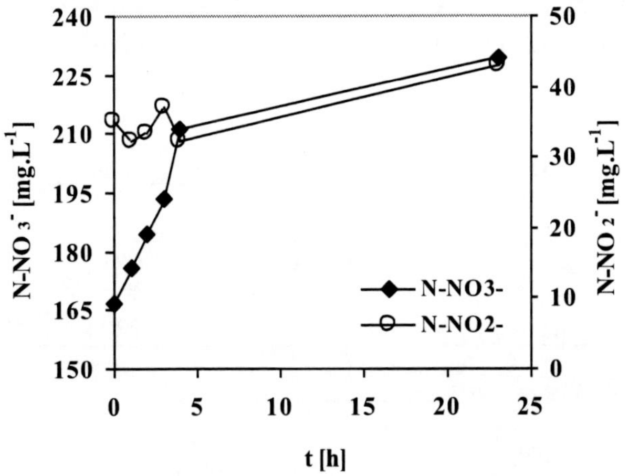

Figure 39. Changes of N-NO_3 and N-NO_2 concentrations during kinetic test performed with ozonated landfill leachate after biological pretreatment – activated sludge cultivated in semicontinuous bioreactor.

The value of the specific rate of COD removal during the first two hours of the process was 164.9 mg g^{-1} h^{-1}. The mean value during the whole reaction period was 23.6 mg g^{-1} h^{-1}.

In conclusion, a lower ammonium nitrogen decrease was observed in comparison to both above measurements. On the other hand, similar to previous measurements, higher values of the nitrate/nitrite concentration increase were observed in comparison to those measured with the model substrate. This can be attributed to a lower organic load in the bioreactor fed with raw landfill leachate.

Activated Sludge after Two Weeks of Adaptation

The next kinetic measurements were performed in a semicontinuous bioreactor with *partially adapted activated sludge to ozonated landfill leachate.* Activated sludge was adapted to ozonated LL1 leachate during two weeks before performing measurements during one of the operational cycles. The parameters of landfill leachate ozonated after biological pretreatment are summarized in Table 33. Time courses of COD and ammonium nitrogen are plotted in Figure 40.

Time courses of nitrite and nitrate nitrogen during the test are shown in Figure 41. 35% COD removal efficiency was observed during the reaction cycle. The value of the specific rate of COD removal during the first two hours of the process was 157.5 mg g^{-1} h^{-1}. The mean of the specific rate of COD removal during the whole reaction period was 26.6 mg g^{-1} h^{-1}. The value of the specific rate of ammonium nitrogen removal measured during the first two hours of the process was 2.5 mg g^{-1} h^{-1}. The mean value during the whole reaction period was 0.5 mg g^{-1} h^{-1}. The value of the specific rate of nitrate/nitrate nitrogen increase measured during the first two hours of the process was 2.0 mg g^{-1} h^{-1}. The mean value during the whole reaction period was 1.6 mg g^{-1} h^{-1}. The increase of nitrate nitrogen concentration corresponds more or less to the decrease of nitrite and ammonium concentration values during the reaction cycle.

Table 33. Experimental values during one cycle – activated sludge cultivated in semicontinuous bioreactor (*ozonated leachate after biological treatment*)

Parameter / t [h]	$N-NH_4^+$ [mg L^{-1}]	$N-NO_2^-$ [mg L^{-1}]	$N-NO_3^-$ [mg L^{-1}]	COD [mg L^{-1}]
0	13.1	40.3	170,2	1614.6
1	10.0	40.4	173,5	1435.3
2	8.5	32.8	181.5	1375.5
3	6.4	25.1	190.2	1210.9
23	3.1	0	220.0	1045.8

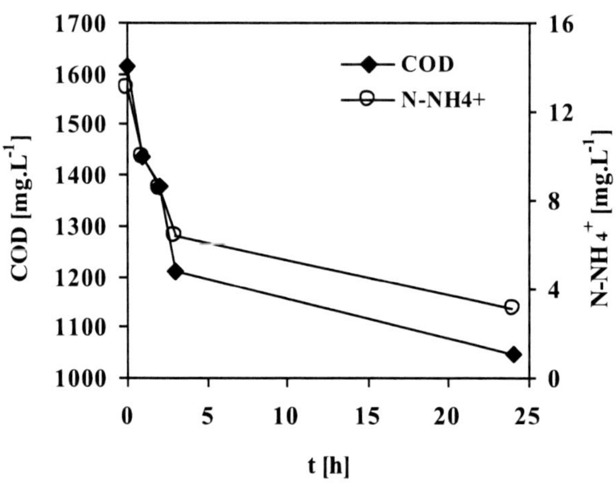

Figure 40. Changes of COD and N-NH_4 contents during kinetic test performed with ozonated landfill leachate after biological pretreatment – activated sludge cultivated in semicontinuous bioreactor.

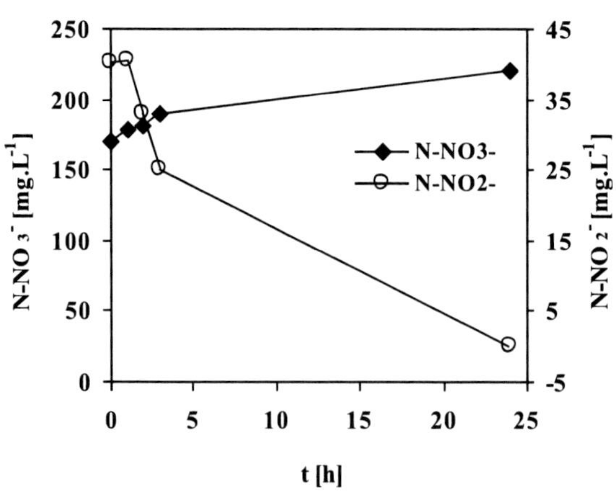

Figure 41. Changes of N-NO_3 and N-NO_2 concentrations during kinetic test performed with ozonated landfill leachate after biological pretreatment – activated sludge cultivated in semicontinuous bioreactor.

In conclusion, the higher COD specific removal rate of 12% was measured in comparison to the previous operational phase with model substrate. The significantly higher specific removal rate of ammonium nitrogen and the lower nitrate/nitrite nitrogen specific removal rate were measured during this phase of operation of the semicontinuous model.

Activated Sludge after Four Weeks of Adaptation

Similar measurements were carried out after the next two weeks of continuing adaption of activated sludge to ozonated LL1 landfill leachate. The parameters of landfill leachate ozonated after biological pretreatment are summarized in Table 34. Ammonium nitrogen was added in order to increase the initial value to 55 mg L^{-1}. Time courses of COD and ammonium nitrogen are plotted in Figure 42. Time courses of nitrite and nitrate nitrogen during the test are shown in Figure 43.

28% COD and 64% ammonium nitrogen removal efficiencies were observed during one reaction cycle. The value of the specific rate of COD removal during the first two hours of the process was 108.7 mg g^{-1} h^{-1}. The mean of the specific rate of COD removal during the whole reaction period was 42.5 mg g^{-1} h^{-1}. The value of the specific rate of ammonium nitrogen removal measured during the first two hours of the process was 21.3 mg g^{-1} h^{-1}.

Table 34. Experimental values during one cycle – activated sludge cultivated in semicontinuous bioreactor (*ozonated leachate after biological treatment*)

Parameter / t [h]	$N-NH_4^+$ [mg L^{-1}]	$N-NO_2^-$ [mg L^{-1}]	$N-NO_3^-$ [mg L^{-1}]	COD [mg L^{-1}]
0	55.0	30.9	152.4	1913.5
1	30.5	30.7	136.9	1853.0
2	31.6	33.9	137.3	1793.9
3	27.6	35.3	140.3	1674.4
4	24. 6	34.7	137.8	1614.6
5	19.7	36.8	149.6	1554.8
23	18.3	50.0	159.3	1375.5

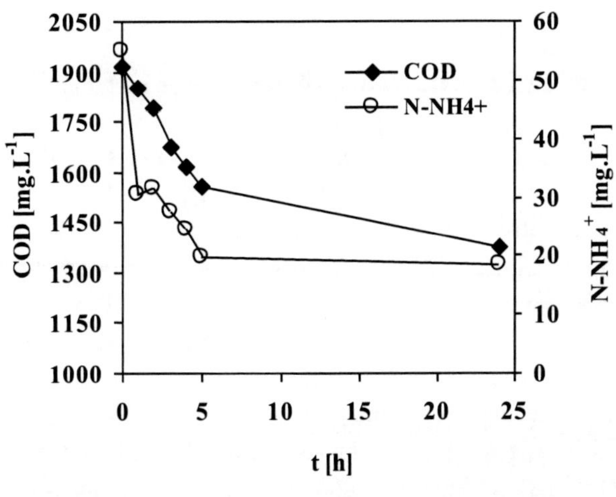

Figure 42. Changes of COD and $N-NH_4$ contents during kinetic test performed with ozonated landfill leachate after biological pretreatment – activated sludge cultivated in semicontinuous bioreactor.

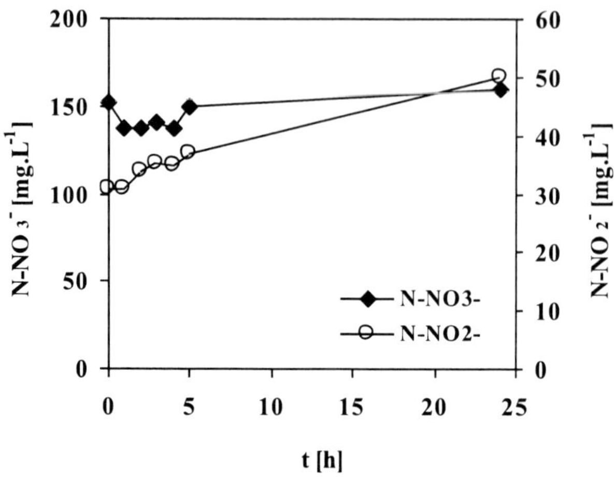

Figure 43. Changes of N-NO_3 and N-NO_2 concentrations during kinetic test performed with ozonated landfill leachate after biological pretreatment – activated sludge cultivated in semicontinuous bioreactor.

The mean value during the whole reaction period was 2.9 mg g^{-1} h^{-1}. The mean value of the specific rate of nitrate/nitrite nitrogen increase measured during the whole reaction period was 2.5 mg g^{-1} h^{-1}. The increase of nitrate nitrogen concentration corresponds more or less to the decrease of nitrite and ammonium concentration values during the reaction cycle.

In conclusion, the specific removal rate of COD increased by 63% and the specific removal rate of nitrate/nitrite nitrogen removal increased by 64% in comparison to the previous operational phase with activated sludge partially adapted to ozonated LL1 leachate after biological pretreatment. In general, treatment of ozonated LL1 leachate (after biological pretreatment) in a semicontinuous bioreactor enables attainment of an effluent value in ammonium nitrogen that also complies with the effluent discharge limit calculation taken into account with every week's flow and composition of water in the receiving river (Regulation, 2005). However, effluent COD values are even higher than recommended values for discharges into the public sewer system (Decree, 2005).

Biological Treatment in Intermittently Aerated CMASR

The sequence of two continuous flow single sludge bioreactors of the same volume (4.2 L) was operated in order to investigate the influence of partially ozonated landfill leachate after previous biological pretreatment. The pre-denitrification technology was applied during this phase of the research. The first bioreactor was intermittently aerated (Reactor 1, anoxic and oxic phase) and the second one was completely aerated (Reactor 2, oxic phase). The corresponding anoxic volume of the denitrification system was approximately 2.1 L and aerobic volume was approximately 6.3 L when taking into account the oxygen profiles maintained in the reactors during operation.

Reactor 1 was fed by the mixture of the raw leachate LL1 and ozonated leachate after biological pretreatment ($LL1_{ozon}$) in (1:1) volumetric ratio. Methanol was also added (2 mL of CH_3OH per liter of the raw leachate) as an external organic carbon source into the storage reservoir of the mixture of raw and ozonated LL1. Characteristics of landfill leachate LL1, ozonated leachate $LL1_{ozon}$, influent (LL1+ $LL1_{ozon}$ + CH_3OH) and effluent of the series of the bioreactors are given in Table 35. The values are also summarized corresponding to the highest denitrification efficiency observed during this period of study (bioreactor 2).

The contribution of acetic acid to the whole COD value corresponds to 2370 mg L^{-1}. The mean efficiency of COD removal after Bioreactor 2 was 33%. Residual values of acetic acid content in the effluent are considered to be negligible. So, the total efficiency of COD removal does not include dosed acetic acid. The efficiency of ammonium nitrogen removal in the plant was about 46%. The portion of $LL1_{ozon}$ was increased during the next phase of the study due to high effluent ammonium nitrogen concentration. Thus, the Bioreactor 1 was fed by the mixture of the raw leachate LL1 and ozonated leachate after biological pretreatment $LL1_{ozon}$ in a 1:2 volumetric ratio. Methanol was also added (4 mL of CH_3OH per liter of the raw leachate) as an external organic carbon source into the storage reservoir of the mixture of raw and ozonated LL1. Characteristics of the landfill leachate LL1, ozonated LL1 ($LL1_{ozon}$), influent (LL1+ $LL1_{ozon}$ + CH_3OH) and effluent of the series of the bioreactors are given in Table 36. There are also summarized values corresponding to the highest denitrification efficiency observed during this period of study. The contribution of acetic acid to the whole COD value corresponds to 4740 mg L^{-1}. The mean efficiency of complete COD removal after Bioreactor 2 was 39%. Residual values of acetic acid content in the effluent are considered to be negligible. Thus, the total efficiency of COD removal does not include dosed acetic acid. The efficiency of ammonium nitrogen removal in the plant was about 75%. Different ozonated landfill leachate $LL1_{ozon}$ with significantly higher nitrogen pollution (about 45% in comparison to previous phases) was applied during the third phase of operation of the series of CMASRs.

Table 35. Parameter values of biological treatment of the raw landfill leachate L1, ozonated LL1, influent to the biological treatment plant and effluent of the plant

Parameter Sample	Dimension	Sample			
		LL1	$LL1_{ozon}$	Influent to Bioreactor 1 LL1 + $LL1_{ozon}$ (1:1) + CH_3OH	Effluent from Bioreactor 2
COD	[mg L^{-1}]	3827	1555	5061	1794
$N-NH_4^+$	[mg L^{-1}]	756	52,2	404	219
$E_{N\text{-}NH4}$	[%]	-	-	-	45.8
$N-NO_3^-$	[mg L^{-1}]	-	-	-	0.7
pH	[/]	8.2	8.5	7.8	7.4
V	[L]	-	-	2.1	6.3
Q	[L d^{-1}]	-	-	1.7	1.7
HRT	[h]	-	-	14.8	44.5
MLSS	[g $L]^1$	-	-	0.9	0.9
SRT	[d]	-	-	6.5	19.5

Legend: HRT – hydraulic retention time; SRT – solid retention time.
MLSS – mixed liquor suspended solids; - Not determined.

Table 36. Parameter values of biological treatment of the raw landfill leachate L1, ozonated LL1, influent to the biological treatment plant and effluent of the plant

Parameter	Dimension	Sample			
Sample		LL1	$LL1_{ozon}$	Influent to Bioreactor 1 LL1 + $LL1_{ozon}$ (1:2)+ CH_3OH	Effluent from Bioreactor 2
COD	[mg L^{-1}]	3827	1235	6839	1973
$N-NH_4^+$	[mg L^{-1}]	756	49.0	284.7	70,1
$E_{N\text{-}NH4}$	[%]	-	-	-	75.4
$N-NO_3^-$	[mg L^{-1}]	-	-	-	1.6
pH	[/]	8.2	8.5	8.4	8.3
V	[L]	-	-	2.1	6.3
Q	[L d^{-1}]	-	-	3.4	3.4
HRT	[h]	-	-	14.8	44.5
MLSS	[g $L^{-}]^{1}$	-	-	0.9	0.9
SRT	[d]	-	-	6.5	19.5

Legend: HRT – hydraulic retention time; SRT – solid retention time.
MLSS – mixed liquor suspended solids; - Not determined.

Table 37. Parameter values of biological treatment of the raw landfill leachate L1, ozonated LL1, influent to the biological treatment plant and effluent of the plant

Parameter	Dimension	Sample			
Sample		LL1	$LL1_{ozon}$	Influent to Bioreactor 1 LL1 + $LL1_{ozon}$ (1:2) + CH_3OH	Effluent from Bioreactor 2
COD	[mg L^{-1}]	3827	1615	7092.3	1734.2
$N-NH_4^+$	[mg L^{-1}]	756	90.7	312.5	22.1
$E_{N\text{-}NH4}$	[%]	-	-	-	95.2
$N-NO_3^-$	[mg L^{-1}]	-	485	323.3	24.0
pH	[/]	8.2	8.5	8.4	8.3
V	[L]	-	-	2.1	6.3
Q	[L d^{-1}]	-	-	3.9	3.9
HRT	[h]	-	-	12.9	38.8
MLSS	[g $L^{-}]^{1}$	-	-	0.9	0.9
SRT	[d]	-	-	6.5	19.5

Legend: HRT – hydraulic retention time; SRT – solid retention time.
MLSS – mixed liquor suspended solids; - Not determined.

The portion of $LL1_{ozon}$ was the same as in previous phase. Thus, Reactor 1 was fed by the mixture of the raw leachate LL1 and ozonated leachate after biological pretreatment $LL1_{ozon}$ in a 1:2 volumetric ratio. Methanol was also added (4 mL of CH_3OH per liter of the raw leachate) as an external organic carbon source into the storage reservoir of the mixture of raw and ozonated LL1. Characteristics of landfill leachate LL1, ozonated LL1 ($LL1_{ozon}$), influent ($LL1+ LL1_{ozon} + CH_3OH$) and effluent of the series of the bioreactors are given in Table 37. There are also summarized values corresponding to highest denitrification efficiency observed during this period of study. The contribution of acetic acid to the whole COD value corresponds to 4740 mg L^{-1}. The mean efficiency of complete COD removal after Bioreactor 2 was 50%. Residual values of acetic acid content in the effluent are considered to be negligible. Thus, the total efficiency of COD removal does not include dosed acetic acid. The efficiency of ammonium nitrogen removal in the plant was about 95%.

In conclusion, during the biological treatment of raw LL1 leachate in combination with ozonated LL1 leachate (after biological pretreatment) in an intermittently aerated CMASR, COD removal efficiency was in the region of 33% to 50% and observed ammonium nitrogen removal efficiency was in the region of 46% to 95% depending on the leachate composition and operational parameter values. Thus, effluent from this bioreactor complies with ammonium nitrogen with effluent limits recommended by Decree (2004) for industrial effluents discharging into a public sewer system.

In addition, the effluent ammonium nitrogen is very close to the effluent limit value (Table 3) calculated base of the emission-imission principle, taking into account the receiving water quantity and quality. On the other hand, the combination of biological and ozonation oxidative processes have not offered the results conveying with effluent discharge limits in terms of COD. For comparison, Rennou et al. (2008) presented COD removal efficiencies from 46% to 97% and 87% efficiency of ammonium nitrogen removal in the activated sludge reactor.

Landfill Leachate Collected in Slovenia

The basic characteristics of raw landfill leachates sampled in Slovenia are summarized in Table 38. The investigated landfill leachates were collected at a large municipal landfill serving 350,000 inhabitants. It is composed of two parts, old and new. At the old part deposition of the wastes started in 1964 and it was closed in 1987. It covers 47.5 ha. Afterwards, the new part opened and up to 2003, 3.000.000 m^3 of wastes were deposited there. There is still space for 2.100.000 m^3 of waste (Urh, 2007). Due to intensive implementation of waste separation it is expected that the landfill will be operational until 2024. The landfill has a collection system for biogas as well as for landfill leachate, which has been treated in a complex wastewater treatment plant prior to its release into the sewage system leading to a common municipal wastewater treatment plant of conventional type.

A LL4 sample was collected at a retention basin for leachates formed from the part of the landfill, closed in 2003, and it was expected to have characteristics of the old – mature leachate while the sample LL3 was collected at the retention basin of leachates from the active part of the landfill, so it is expected to have characteristics of fresh – young leachate. The amount of LL4 leachate varies from 70,666 to 142,144 m^3 $year^{-1}$ (2003 – 2006), while for the LL3 leachate only 11,611 to 21,795 m^3 $year^{-1}$ was released in the same time period. The Fresh LL3 sample was much more polluted than the mature LL4 indicating successful partial stabilization of the waste within the body of the landfill. The COD of the LL3 was 2790 mg L^{-1}, BOD_5 was approximately 230 mg L^{-1} and as a result, poor biodegradability could be expected. It also contained a lot of organic and ammonium nitrogen as well as chloride (Table 38).

The mature LL4 landfill leachate still contains significant amounts of ammonium nitrogen and chloride, but the organic pollution was significantly decreased. Some authors made predictions on the basis of several years of monitoring and have shown that the concentration of chloride in the leachate decreased slowly from 1000 – 1200 mg L^{-1} to below 200 mg L^{-1} in 20 years, while the decrease in the ammonium N concentration is much faster; from 1000 mg L^{-1} to almost 0 mg L^{-1} in 10 years (Sizirici and Tansel, 2010).

The same trend was also observed in our experiments if it would be assumed that mature landfill leachate LL4 would actually be comparable to LL3 exposed to waste stabilization processes in the landfill for four years. In aerobic conditions within of the body of the landfill there is sometimes no mechanism present for nitrogen transformation and so high concentrations could also be expected in aged wastes. Land-filled wastes will continue to produce high ammonium strength leachate for many years after closure (Tyrel et al., 2002).

Table 38. Average values of physico-chemical analysis of the raw landfill leachates (LL3, LL4) collected in Slovenia (N = 2)

Parameter	Value	
Name of the sample	LL3	LL4
pH [-]	8.2 – 9.4	8.1 – 9.1
COD [mg L^{-1}]	2720 - 2790	490 - 590
DOC [mg L^{-1}]	600 - 700	90 - 93
IC [mg L^{-1}]	909 - 1050	431 – 457
BOD_5 [mg L^{-1}]	204 - 230	25 - 28
BOD_5/COD	0.07 - 0.08	0.04 - 0.06
N-NH_4^+ [mg L^{-1}]	463 -1249	316 - 456
N-NO_3 [mg L^{-1}]	0.65 - 2.3	0.43 - 30.9
Organic N [mg L^{-1}]	166 - 408	36 - 70
Cl^- [mg L^{-1}]	1020 - 1790	417 – 667

Mature LL4 was also poorly biodegradable (BOD_5/COD = 0.04-0.06), but this was also true for fresh leachate LL3. This poor biodegradability of LL3 could be a consequence of organic waste separation prior to landfilling, where actually all easily biodegradable substances (remains of food, yard trimmings, etc.) were removed. As a result such fresh leachate expressed some characteristics of the mature one, with a lower amount of pollution and without easily biodegradable substances. Generally, fresh leachate from traditional landfill could contain up to 10,000 mg L^{-1} of BOD_5 (Tyrel et al., 2002).

Toxicity of the leachates to luminescent bacteria *Vibrio fischeri* and microorganisms of activated sludge is presented in Table 39. In the test with activated sludge the common response of heterotrophic and nitrifying microorganisms was measured (Žgajnar Gotvajn et al, 2009b).

Fresh, more polluted landfill leachate LL3 was more toxic to all tested organisms than the mature one (LL4), because of the lower 30minEC_{50} values. High toxicity of the LL3 sample could be a consequence of organics (COD, DOC) present in the sample, as well as a high concentration of ammonium N. Puijgagut et al. (2005) have measured a significant short-term impact of NH_4^+-N at concentrations above 50 mg L^{-1} to a mixed community of activated sludge, but the sludge was able to recover shortly. Our test was accomplished in 30 minutes, so acute toxicity of LL3 containing up to 1020 mg L^{-1} of ammonia could be the result of a toxic ammonium shock. Concentration of the chloride was not high enough (bellow 6,500 mg L^{-1}) to cause significant toxic impact to activated sludge and it also did not affect salt tolerant *Vibrio fischeri* (Chen et al., 2003).

Table 39. Toxicity of the raw landfill leachates LL3 and LL4 to microorganisms of activated sludge and luminescent bacteria *Vibrio fischeri*

Sample	EC_{50} values	Activated sludge Microorganisms	*Vibrio fischeri*
LL3	30minEC_{50} (% v/v)	13 ± 4	12 ± 2
LL4	30minEC_{50} (% v/v)	70 ± 3	46 ± 1

Lower toxicity of LL4 in both toxicity assessment tests could also be explained by the lower presence of ammonium N and organics. But on the other hand, the difference in toxicity could also be a consequence of different organic substances, present in LL3 and LL4. However, it was not possible to identify the group of substances, responsible for the inhibitory effects.

Fenton Oxidation of LL3 and LL4

Fenton´s oxidation experiments with fresh municipal landfill leachate LL3 were performed using 1/13 molar ratio of reagents (0.3 M Fe^{2+}/4.0 M H_2O_2). The same ratio was also used for mature LL4 sample. COD concentrations (mg L^{-1}) versus time of Fenton´s oxidation for both samples are presented in Figure 44.

In the case of fresh LL3 leachate 37% of COD removal was achieved. The efficiency of COD removal of mature LL4 leachate reached 86%. The rate constant values and values of corresponded to the descriptions of experimental data by applied kinetic models are given in Table 40. Zero order kinetics of organic components in terms of COD removal by Fenton oxidation was the best described by the above-mentioned kinetic models. It can be seen from Table 40 that a good fit of experimental data for LL3 was also achieved by the first order kinetics and the combined kinetic model. Other authors (Zhang et al., 2006) also observed the first order COD and TOC removal reaction kinetics. Their results indicated that rate constants would decrease with the increase of initial COD or TOC strength. When fresh and mature landfill leachate treatment efficiency (Table 40) was compared, higher removal rates were measured with oxidation of fresh LL3 landfill leachate. The same was also observed by other authors (Hermosilla et al., 2009) and it is clearly a consequence of the presence of readily biodegradable substances in the fresh leachate.

In Table 41 the toxicity of landfill leachates after Fenton oxidation is presented. Toxicity of fresh LL3 to *Vibrio fischeri* decreased a bit during treatment. 30minEC_{50} of the raw leachate was 12% v/v, and after treatment 20% v/v. However, its toxicity to activated sludge decreased significantly, it was no longer toxic after oxidative treatment.

This could be explained as a consequence of oxidative removal of toxic organics as well as effective removal of ammonium nitrogen during the Fenton procedure, where pH at the end of experiment was being increased up to 12 to remove excess hydrogen peroxide (Žgajnar Gotvajn et al, 2009c). Under such conditions NH_4^+-N is converted into $NH_{3(g)}$ and stripped out of the system (Puigagut, 2003).

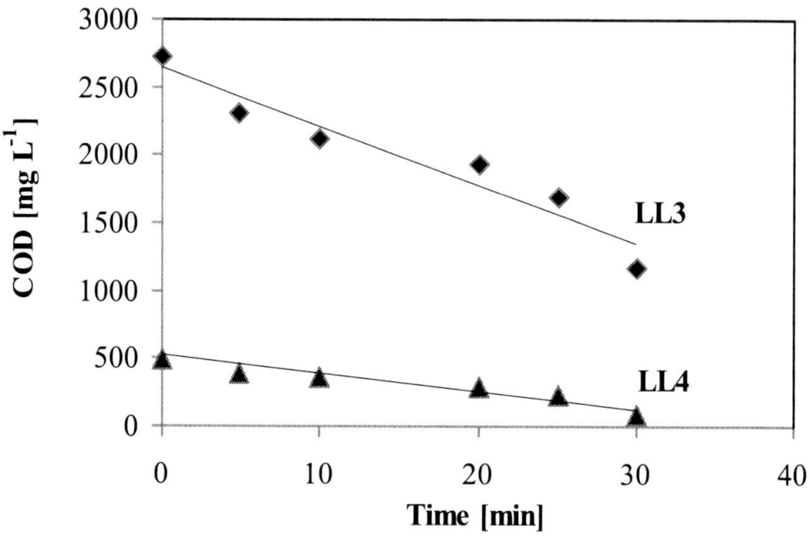

Figure 44. COD concentrations during Fenton's oxidation of fresh LL3 and mature LL4 municipal landfill leachate (LL3: ♦ experimental, -- calculated; LL4: ▲ experimental, -- calculated).

Table 40. Kinetic parameters and statistical characteristics values in the Fenton´s oxidation of LL3 and LL4

		Parameter/Value	
Sample	N	k_n^*	S_r^2 [-]
LL3	0	30.07 [g m^{-3} h^{-1}]	0.6656 x10^5
	1	1.25 x 10^{-2} [h^{-1}]	0.7576 x10^5
	2	5.08 x10^{-6} [g^{-1} m^3 h^{-1}]	0.8479 x 10^5
	1/TCM	/	0.7576 x 10^5
	A	0.305	/
	$k_{1,eo}$	1.25 x 10^{-2} [h^{-1}]	/
	$k_{1,so}$	1.25 x 10^{-2} [h^{-1}]	/
LL4	0	13.26 g m^{-3} [h^{-1}]	0.2531 x 10^3

/ Not calculated.

Final ammonium concentration was 45 ± 5 mg L^{-1}. The same was also noticed with mature LL4 landfill leachate, where final ammonium concentration was even lower (15 ± 2 mg L^{-1}). Biodegradability of the sample LL3 in terms of BOD_5/COD did not increase, because BOD_5 determined in the final was below 5 mg L^{-1}. In the case of LL4 BOD_5 remained in the range of 25 ± 2 mg L^{-1}. But due to 86% of COD removal its biodegradability significantly increased (0.39), because remaining COD was only 64 mg L^{-1}.

It could be concluded, that Fenton oxidation of fresh leachate (LL3) was less effective than for the mature one (LL4), because treatment efficiency expressed as a % of removal was lower, reduction of toxicity was also less noticeable, and it did not increase biodegradability of the fresh leachate.

Table 41. Toxicity of the landfill leachates LL3 and LL4 to microorganisms of activated sludge and luminescent bacteria *Vibrio fischeri* after Fenton oxidation

Sample	EC_{50} values	Activated sludge microorganisms	*Vibrio Fischeri*
LL3	30minEC_{50} (% v/v)	> 100	20 ± 3
LL4	30minEC_{50} (% v/v)	> 100	> 100

Ozonation of LL3 and LL4

COD content in fresh (LL3) and mature (LL4) municipal landfill leachate versus time of ozonation are presented in Figure 45. The combined kinetic model obtained the best fit of COD experimental data. The rate constant values and values of corresponding to the descriptions of COD measurements by the applied model are given in Table 42.

As can be seen from Figure 45, the highest COD removal in the case of LL3 was observed during the first 30 to 40 minutes of ozonation. The COD conversion of LL3 achieved about 40% after 30 minutes of treatment. Further increase in reaction time led only to a low increase in treatment efficiency.

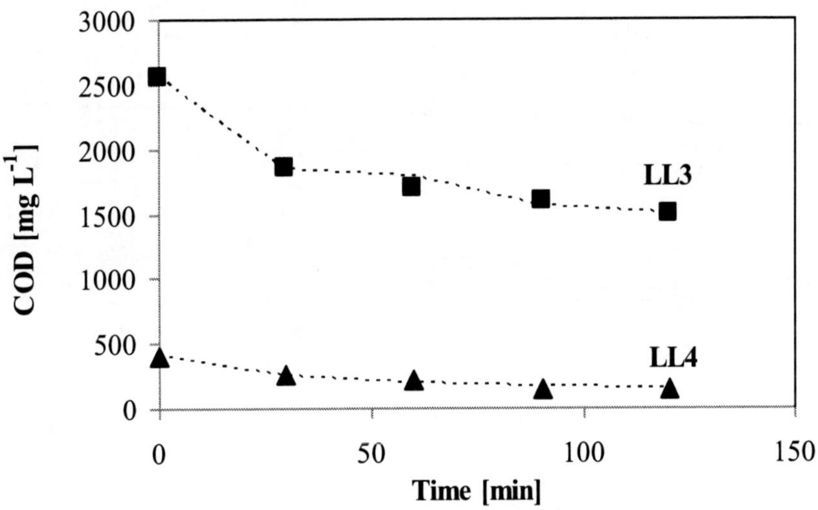

Figure 45. COD concentrations during ozonation of fresh LL3 and mature LL4 municipal landfill leachate (LL3: ♦ experimental, -- calculated; LL4: ▲ experimental, -- calculated).

The concentration of organic N was reduced 28%, its final concentration was 293 mg L^{-1}. The concentration of ammonium was also reduced for 23%, its concentration after treatment was still 356 mg L^{-1}. The LL4 treatment was the most effective in the first few minutes of ozonation. The conversion of COD achieved about 65% and it did not increase with time. Other authors also reported comparable treatment efficiency (48%) for mature landfill leachate (da Silva, 2008) when a high ozone dose was (3 g O_3 L^{-3}) employed. Wang et al., (2004) obtained 47% COD removal efficiencies for aged landfill leachate at an ozone dose of

2.6 g L^{-3}. The doses of ozone used in the experiment correspond approximately to 3.0 respectively 2.6 hours of ozonation when applying our experimental condition values. The average ozone absorption corresponds to 1.0 mg per each mg of reduced COD. We have also removed 43% of organic N and 32% of ammonium nitrogen from LL4. The dependencies of COD removal (Eqn. 23) and COD partial mineralization (Eqn. 24) portions of organics in fresh LL3 leachate on ozonation time are shown in Figure 46. It is obvious from the Figure that all reduced COD due to ozone oxidation was more or less mineralized.

The dependencies COD removal (Eq. 23) and COD partial mineralization (Eq. 24) proportions of organics contained in mature LL4 leachate on ozonation time are pictured in Figure 46. Higher reducing rates and final proportions of both reduced COD and partially mineralized COD can be seen in this Figure in comparison to fresh LL4 leachate (Figure 47).

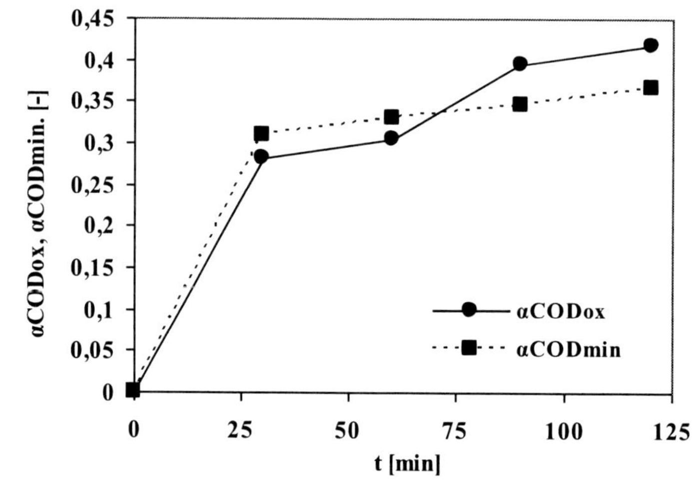

Figure 46. Influence of ozonation time on COD removal (• αCOD_{oxi}) and mineralization (■ αCOD_{min}) – fresh LL3 leachate.

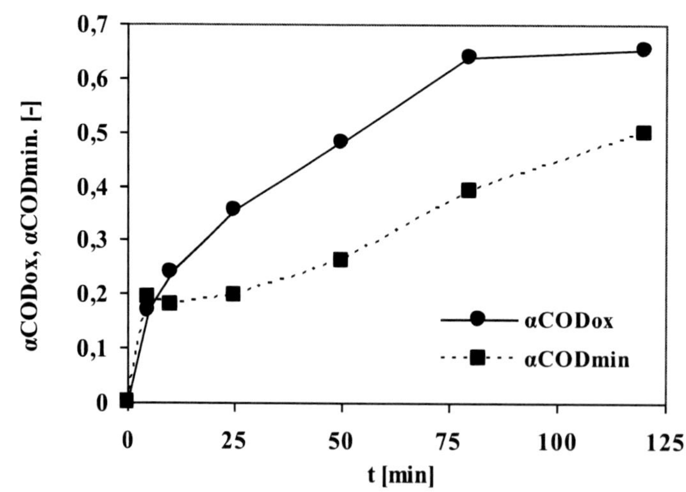

Figure 47. Influence of ozonation time on COD removal (• αCOD_{oxi}) and mineralization (■ αCOD_{min}) – mature LL4 leachate.

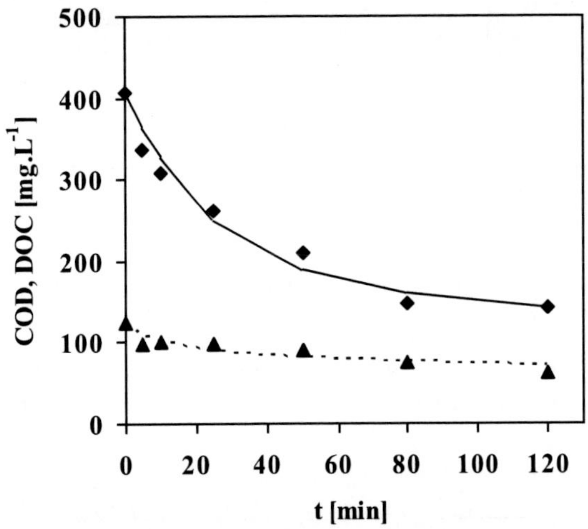

Figure 48. COD and DOC removal in mature municipal landfill leachate (LL4) with ozone after two weeks of sample storage: COD (♦ experimental, — calculated) and DOC (▲experimental, -- calculated).

In addition to this, there are higher differences between COD removal and COD mineralization, i.e. there is a higher ozone treatment potential for an enhancement of biological degradability of organics contained in mature LL4 leachate in comparison to fresh LL3 leachate. Figure 48 shows the time courses of COD and DOC removal during ozonation of mature LL3 landfill leachate that was carried out after two weeks of its storage in a refrigerator at 8°C. Longer reaction times were applied in comparison to the ozonation experiment with non-stored LL4 leachate. The highest removal rate of organics (both in COD and DOC terms) was observed during the first 60 min of ozonation. The conversion of COD was about 45% after one hour of ozonation. The COD removal efficiency reached about 69% after three hours. Similar to fresh landfill leachate, very similar time dependencies of COD and DOC were obtained.

Table 42. Kinetic parameters and statistical values for combined TCM model (Eq. (22)) – ozonation of mature LL4 leachate after two weeks of storage of the sample

Parameter	Calculated on the basis of removal	
	DOC	COD
	Eq. (22)	Eq. (22)
a [-]	0.5379	0.4508
k_{eo} [$h^{-1]}$	1.8×10^{-2}	1.6×10^{-2}
k_{so}/ [$h^{-1]}$	1.3×10^{-3}	1.7×10^{-3}
S_r^2 [-]	0.9×10^{1}	0.3×10^{3}

On the other hand, significant differences in the ozonation potential of stored and non-stored mature landfill leachate (Figures 45 and 46) were observed. Stored LL3 degraded up to 69% while the non-stored one degraded only by 55%. Values of kinetic parameters and statistical characteristics corresponding to the best fit of the experimental data are summarized in Table 42. Lower values of the portion of readily degradable organics as well as of the kinetic rate constants follow from Tables 40 and 42 for the stored leachate sample. It can be concluded that different reactions took place during the ageing of mature landfill leachate and these could affect the treatment efficiency.

Figure 49 shows the courses of COD removal (Eq. 12) and COD partial mineralization (Eq. 13) portions of organics contained in mature LL4 leachate dependent on ozonation time. Higher COD reducing rates are characteristic for non-stored mature LL4 leachate (Fig. 49). On the other hand, higher partially mineralized COD evolution rate follows for the stored mature LL4 leachate. The final portions of both COD removal (Eq. 23) and COD partial mineralization (Eq. 24) proportions of organics contained in stored mature LL4 leachate are higher (Figure 49) in comparison to non-stored leachate (Figure 47) due to the significantly longer reaction times applied in this case.

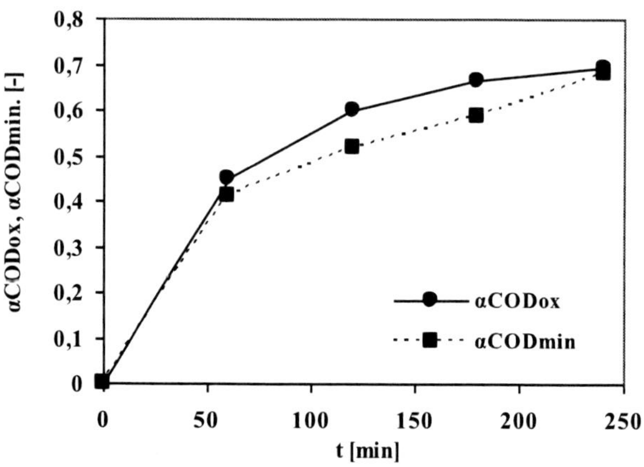

Figure 49. Influence of ozonation time on COD removal (● αCOD_{oxi}) and mineralization (■ αCOD_{min}) – mature LL4 leachate after two weeks of sample storage.

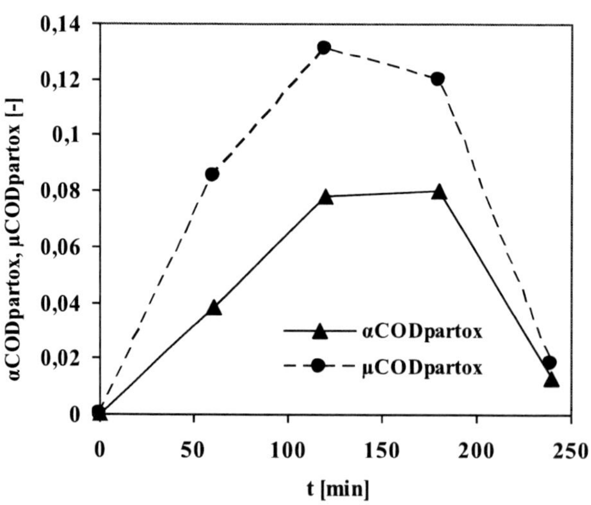

Figure 50. Influence of ozonation time on COD partial oxidation (▲) $\alpha COD_{parcoxi}$ and degree of effectiveness of partial oxidation(●) $\mu COD_{parcoxi}$ – mature LL4 leachate after two weeks of sample storage.

The dependency of COD reduction due to partial oxidation (Eq. 25) and the degree of effectiveness of partial oxidation (Eq. 26) on ozonation time for mature LL4 leachate after two weeks of sample storage are pictured in Figure 50. The highest effectiveness of partial mineralization corresponds to ozonation time in the region from 120 to 180 minutes. On the other hand, very low partial oxidation was observed at lower and longer ozonation times (Figure 50). Toxicity of the leachates was also determined after 60 minutes of ozonation (Table 44).

Table 43. Kinetic parameters and statistical characteristic values for combined (TCM) model of ozonation of fresh (LL3) and mature (LL4) leachate

	Parameter/Value			
Sample	a	$k_{1,eo}$ [h^{-1}]	$k_{1,so}$ [h^{-1}]	S_r^2 [-]
LL3	0.2560	$7.74.10^{-2}$	$1.94.10^{-3}$	$0.3224.10^{4}$
LL4	0.5729	$4.10.10^{-2}$	$1.73.10^{-3}$	$0.3305.10^{3}$

Toxicity of the LL3 sample to *Vibrio fischeri* decreased significantly during ozonation. After 60 minutes 30minEC_{50} was 45% v/v, which is significantly lower in comparison to the raw fresh leachate LL3 (30minEC_{50} = 12% v/v, Table 43).

Regarding toxicity to *Vibrio fischeri*, ozonation treatment was more effective than Fenton´s oxidation in treatment of fresh municipal leachate. The same also happened with mature LL4 leachate. Its toxicity to *Vibrio fischeri* decreased during ozonation; after 60 minutes of ozonation 30minEC_{50} was above 100% v/v. None of the sample was toxic to microorganisms of activated sludge after ozonation.

Table 44. Toxicity of the landfill leachates LL3 and LL4 to microorganisms of activated sludge and luminescent bacteria *Vibrio fischeri* after ozonation

Sample	EC_{50} values	Activated sludge microorganisms	*Vibrio fischeri*
LL3	30minEC_{50} (% v/v)	> 100	45 ± 2
LL4	30minEC_{50} (% v/v)	>100	> 100

During ozonation, BOD_5 of the LL3 was reduced 50% to 115 mg L^{-1}, final COD was approximately 1,500 mg L^{-1} and biodegradability as BOD_5/COD remained practically identical to the raw leachate (0.07/0.08). BOD_5 of LL4 after ozonation was 19 mg L^{-1}, and because COD was reduced to 171 mg L^{-1}, the BOD_5/COD ratio increased (0.11) and in comparison to the raw sample (0.06) the increase of biodegradability could be noticed. It could be concluded that both applied oxidation processes increased biodegradability of the mature landfill leachate LL4. However, ozonation of fresh leachate (LL3) was less effective than the mature one, because treatment efficiency expressed as a % of removal was lower, reduction of toxicity was also less noticeable, and it did not increase biodegradability of the fresh leachate.

The same improved efficiency of ozonation and Fenton oxidation of mature leachates in comparison to fresh ones was also noticed by other authors (Cortez et al., 2011). They performed experiments with mature leachate and in an optimized Fenton procedure they removed 46% of COD (86% in our case) and they increased the biodegradability of BOD_5/COD from 0.01 to 0.11, while in our case the final ratio was 0.39. They also oxidize mature leachate with ozone and 72% of COD removal was obtained followed by an increase in biodegradability up to 0.24. The leachate they investigated contained higher COD (743 mg L^{-1}) and also more ammonium nitrogen (714 mg L^{-1}). In our case ozonation was less effective with 65% of COD removal, followed by a minor increase in biodegradability.

CONCLUSION

The aim of our work was to integrate biological and chemical methods for landfill leachate treatment. The efficiencies of different chemical and biological oxidation processes for the treatment of municipal landfill leachates were evaluated. The samples from four different landfills were investigated, two collected in Slovakia and two in Slovenia. Fenton's process and ozonation were selected as chemical oxidation processes and their feasibility for landfill leachate treatment was studied. Kinetics and treatment efficiency were determined. The study of influence of chemical oxidation processes on toxicity and biodegradability enhancement of leachates was also involved. As a comparison, biological treatment processes were applied for treatment of raw, as well as chemically pretreated landfill leachates.

The mature raw landfill leachate LL1 collected in the Slovakia landfill was biologically pretreated in the lab-scale intermittently aerated activated sludge process. COD content was reduced about 35 - 60% and ammonium nitrogen removal efficiency was about 85 - 95%, depending on the composition of the raw landfill leachate. After biological treatment, landfill leachate was chemically treated by ozonation. As a result, a significant increase of biodegradability of organics contained in the landfill leachate was observed. Approximately 39% reduction of COD in biologically pre-treated landfill leachate corresponded to the highest biodegradability observed during ozonation. The efficiency of COD removal increased by 2 - 16% after ozonation in comparison to the initial COD value.

The total efficiency of COD removal by ozonation after biological pre-treatment varied from 60 to 75%. Insignificant toxicity of ozonation products to activated sludge microorganisms was confirmed by nitrification and denitrification kinetic tests as well as from results of operation of continuously flow intermittently aerated completely mixed activated sludge system. The potential of biological processes for treated ozonated landfill leachate LL1 followed the results of nitrification and denitrification tests performed with ozonated landfill leachate. The mean value of specific nitrification rate was 5 $mg.g^{-1}.h^{-1}$ and the mean value of specific denitrification rate of 10 $mg.g^{-1}.h^{-1}$ was observed.

The efficiency of Fenton's process and ozonation for the treatment of fresh (LL3) and mature (LL4) municipal landfill leachate from municipal waste disposal in Slovenia was also evaluated. Both samples had a high organic load, high toxicity and low biodegradability. Fresh, more polluted landfill leachate was more toxic than mature ones.

Experimental results indicated that the Fenton's oxidative pretreatment of heavily polluted landfill leachate was more effective for mature landfill leachates. For this sample up to 88% of organics could be removed, while in the case of fresh leachate only 46% of COD could be removed. Zero or first order kinetics provided the best description of organics (in terms of COD and DOC) removal by Fenton's oxidation for both landfill leachates. Higher reaction rates of Fenton's oxidation were achieved with fresh leachate samples.

The efficiency of initial organics removal during ozonation was about 70%, both in COD and DOC terms. In case of fresh LL3 leachate only 42% of COD were removed. The best fits of COD and DOC experimental data from oxidation of fresh and mature LL4 leachates with ozone were obtained applying the combined kinetic model. Toxicity was reduced more effectively than in the Fenton's oxidation. Similar to the Fenton reaction, ozonation was also assessed as appropriate for the removal of organics, while it did not improve the biotreatable potential of the leachates.

Acknowledgments

The authors wish to thank the Slovak and the Slovenian Research Agencies for the financial support of the bilateral Slovak-Slovene project SK-SI-0019-10/BI-SK-11-12-012. Part of this work was done with the support from VEGA Grant 1/0734/12. We would also like to thank Mrs. Ingrid Kučerová, Mrs. Lucia Císarová, Mrs. Mária Valičková and Mrs. Živana Urh for their excellent technical assistance.

References

Alvares, A.B.C., Diaper, C., Parsons, S.A. (2001). Partial oxidation by ozone to remove recalcitrance from wastewaters – a review. *Environmental Technology*, *22*, 409-427.

Anderozzi R. et al. (1999) Advanced oxidation processes (AOP) for water purification and recovery. *Catalysis Today*, 53, 51 - 59.

APHA, AWWA, WEF (2005). In A. E. Greenberg, L. S. Clesceri & A. Eaton (Eds.), *Standard methods for the examination of water and wastewater* (22nd ed.). Washington, DC, USA: APHA.

Araki, H., Koga, K., Inomae, K., Kusuta, T., Away, Y. (1990). Intermittent aeration for nitrogen removal in small oxidation ditches. *Wat. Sci. Tech. 34*, 131-138.

Arslan-Alaton, I. (2003). The effect of pre-ozonation on the biocompatibility of reactive dye hydrolysates. *Chemosphere*, *51*(9), 825-833.

Bell, B. A., Hardcastle G. J. (1984). Treatment of high-strength industrial waste in a continuously fed, intermittently operated activated sludge system. *Journal WPCF*, *56,* 1160-1164.

Bigot, V., Luck, F., Pailard, H., Wagner, A. (1995). Industrial application of advanced oxidation for landfill lechate treatment. In *proceedings of the 12th world congress of the international ozone association*, 15–8 May, Lille, France, 439-450.

Bila, D. M., Montalvão, A.F., Silva, A.C., Dezotti, M. (2005). Ozonation of a landfill leachate: evaluation of toxicity removal and biodegradability improvement. *Journal of Hazardous Materials*, *117*, 235–242. DOI: 10.1016/j.jhazmat.2004.09.022.

Beltran-Heredia, J., Torregrosa, J., Domingues, R., Garcia, J (2000). Treatment of black-olive wastewaters by ozonation and aerobic biological degradation. J. Wat. Res., 34, 3515-3522.

Cortez, S., Teixeira, P., Oliveira, R., Mota, M. (2011). Evaluation of Fenton and ozone-based advanced oxidation processes as mature landfill leachate pre-treatment. *Journal of Environmental Management, 92*, 749-755.

Čech, J.S., Chudoba, J., and Grau, P. (1985). Determination of kinetic constants of activated sludge microorganisms. *Wat. Sci. Technol., 17* (2/3), 259-272.

Čech, J.S., Farkač J., Chudoba, J. (1984b). Simple respirometric method for estimation of kinetic constant of activated sludge microorganisms. *Vodní hosp., B 34*, 215-221. In Czech.

Davis, M.L., Cornwel, D.A. (1991). *Introduction to Environmental Engineering.* McGraw-Hill, Inc. New York.

Decree (2004). Ministry of Environment of the Slovak Republic Decree No 55/2004 wherewith is established appurtenances operational regulations of public water supply and sewer systems. In Slovak.

Derco, J., Králik, M., Hutnan, M., Bodík, I., Cernák, R. (1994). Modelling of the carrousel plant. *Wat. Sci. Tech., 30*, 345-354.

Derco, J., Kovács, A., Králik, M., Shansab, S. (1999). Modelling of an intermittently aerated completely mixing activated sludge reactor. *Mathematical and computer modelling of dynamical systems 5*(4) 313-336.

Derco, J., Gulyásová, A., Horňák, M. (2002). Influence of ozonation on biodegradability of refractory organics in a landfill leachate. *Chemical Papers, 56*(1) 41- 44.

Derco, J., Mencáková, A., Almásiová, B. (2009). Utilisation of ozone for treatment of landfill leachate. *Chem. Listy, 103*, 581-588. In Slovak.

Derco, J., Žgajnar Gotvajn, A., Zagorc-Končan, J., Almásiová, B., Kassai, A. (2010). Pretreatment of landfill leachate by chemical oxidation processes. *Chemical Papers, 64* (2) 237–245 DOI: 10.2478/s11696-009-0116-5.

Deng, Y., Englehardt, J.D. (2006). Treatment of landfill leachate by the Fenton process. *Water Research, 40*, 3683–3694. DOI: 10.1016/j.watres.2006.08.009.

Deng, Y. (2007). Physical and oxidative removal of organics during Fenton. *Journal of Hazardous Materials*, 146, 334-340.

Eckenfelder, W.W., and Musterman, J.L., (1995) Activated sludge treatment of industrial wastewater. Technomic Publishing Co., Inc. Basel, Switzerland.

Esplugas, S., Marco, A., Saum, G. (1996). Oxidation technologies for water and wastewater treatment, Goslar, Germany.

Goi, A., Trapido, M. (2002). Hydrogen peroxide photolysis, Fenton reagent and photo-Fenton for the degradation of nitrophenols: a comparative study. *Chemosphere, 46*, 913–922. DOI: 10.1016/S0045-6535(01)00203-X.

Gordon, S. A., Chughtaia, A. R., Smith, D. M. (2000). Revisiting iodometric estimation as a reference method for ozone in laboratory gas streams. *American Laboratory News, 32* (9), 12-13.

Gulyas, H., Bockelmann, D., Hemerling, L., Bahnemann, D., Sekoulov, I. (1994). Treatment of recalcitrant organic compounds in oil reclaiming wastewater by ozone/hydrogen and UV/titanium dioxide. *Wat. Sci. Tech. 29*, 129-132.

Gulyas, H., Bismarck, R. von, and Hemerling, L. (1995). Treatment of industrial wastewaters with ozone/hydrogen peroxide. *Wat. Sci. Tech.* 32(7), 127-134.

Gulyas, H. (1997). Processes for the removal of recalcitrant organics from industrial wastewaters. *Wat. Sci. Tech. 36* (2-3) 9-16.

Hart, E. J. (1983) Molar absorptivities of ultraviolet and visible bands of ozone in aqueous solution. *An. Chem.*, *120*, 109 – 109.

Heduit, A., Duchene Ph, Sintes L. (1990). Optimization of nitrogen removal in small activated sludge plants. *Wat. Sci. Tech. 22*, 123-130.

Henze, M., Harremoës, P., Jansen, J., Arvin, E. (1995). *Wastewater Treatment. Biological and Chemical Processes*. Springer-Verlag Berlin Heidelberg.

Hermosilla, D., Cortijo, M., Huang, C.P. (2009). Optimizing the treatment of landfill leachate by conventional Fenton and photo-Fenton processes. *Science of the total environment, 407*, 3473–3481. DOI: 10.1016/j.scitotenv.2009.02.009.

Hoigné, J., Bader, H. (1983). Rate constants of reactions of ozone with organic and inorganic compounds in water. I. Non-dissociating organic compounds. *Water Research, 17,* 173-183.

Chaturapruek, A., Visvanathan, C., Ahn, K.H. (2005). Ozonation of membrane bioreactor effluent for landfill leachate treatment. *Environmental Technology*, *26*, 65–73. DOI: 10.1080/09593332608618583.

Chen, G.-H., Wong, M.-T., Okabe, S., Watanabe, Y. (2003). Dynamic response of nitrifying activated sludge batch culture to increased chloride concentration, *Wat. Res.*, *37*(13), 3125-3135.

ISO 5813 (1983). *Water quality – Determination of dissolved oxygen – Iodometric method.* Geneve, Switzerland: International Organisation for Standardisation.

ISO 5663 (1984). Water quality – Determination of Kjeldahl nitrogen – Method after mineralization with selenium. Geneve, Switzerland: International Organisation for Standardisation.

ISO 7150-1 (1984). *Water quality – Determination of ammonium – Part 1: Manual spectrometric method.* Geneve, Switzerland: International Organisation for Standardisation.

ISO 6060 (1989). *Water quality – Determination of chemical oxygen demand.* Geneve, Switzerland: International Organisation for Standardisation.

ISO 10304-1 (1992). Water quality – Determination of dissolved fluoride, chloride, nitrite, orthophosphate, bromide, nitrate and sulphate ions using liquid chromatography of ions – *Part 1: Method for water with low contamination.* Geneve, Switzerland: International Organisation for Standardisation.

ISO 11348-2 (1998). Water quality – Determination of the inhibitory effect of water samples on the light emission of *Vibrio fischeri* (Luminescent bacteria test) – *Part 2: Method using liquid-dried bacteria*. Geneve, Switzerland and: International Organisation for Standardisation.

ISO 8245 (1999). Water quality – Guidelines for the determination of total organic carbon (TOC) and dissolved organic carbon (DOC). Geneve, Switzerland: International Organisation for Standardisation.

ISO 5815-1 (2003). Water quality - Determination of biochemical oxygen demand after n days (BOD_n) – *Part 1: Dilution and seeding method with allylthiourea addition*. Geneve, Switzerland: International Organisation for Standardisation.

ISO 11348-3 (2007). Water quality - Determination of the inhibitory effect of water samples on the light emission of *Vibrio fischeri* (Luminiscent bacteria test). Geneve, Switzerland: International Organisation for Standardisation.

ISO 8192 (2007) Water quality - Test for inhibition of oxygen consumption by activated sludge for carbonaceous and ammonium oxidation. Geneve, Switzerland: International Organisation for Standardisation.

Kim, J.-S., Kim, H.-Y., Won, C.-H., Kim, J.-G. (2001). Treatment of leachate produced in stabilized landfills by coagulation and Fenton oxidation process. *Journal of the Chinese Institute of Chemical Engineers*, *32*, 425–429.

Klimiuk, E., Kulikowska, D. (2006). Organics removal from landfill leachate and activated sludge production on SBR reactors, *Waste Management*, 26, 1140-1147.

Kurniawan, T. A., Lo, W.-H., Chan, G.Y.S. (2006). Physico-chemical treatments for removal of recalcitrant contaminants from landfill leachates. *Journal of Hazardous Materials*, *129*, 80–100. DOI: 10.1016/j.jhazmat.2005.08.010.

Kurniawan, T.A., Lo, W.H., Chan, G.Y.S. (2006). Physico-chemical treatments for removal of recalcitrant contaminants from landfill leachates, *Journal of Hazardous Materials*, *B129*, 80-100.

Lamb, J.C., Westgardth, W.C., Rogers, J.L., and Vernimmen, A.P. (1964). A technique for evaluating the biological treatability of industrial wastes. *J. Wat. Pollut. Contr. Fed. 36*, 1263-1273.

Ledakowicz, S. (1999a). Integrated treatment of industrial wastewater by chemical and biological oxidation process. *Proceedings of the 3rd European Meating on 'Chemical Industry and Environment'*. Krakow, Poland, 1. - 3. September, 285-292.

Ledakowicz, S., (1999b). Optimisation of oxidants dose for combined chemical and biological treatment of textile wastewater. *Water Research 33*, 2511-2516.

Lopez, A., Ricco, G., Mascolo, G., Tiravanti, G., Di Pinto, A., C. and Passino, R., (1998). Biodegradability enhancement of refractory pollutants by ozonation: a laboratory investigation on an azo-dyes intermediate. *Wat. Sci. Tech.* 38 (4-5), 239-245.

Lopez, A., Pagano, M., Volpe, A., Di Pinto, A.C., (2004). Fentons pretreatment of mature landfill leachate. *Chemosphere 54*, 1005-1010.

Nakajima, J., Kaneko, M. (1991). Practical performance of nitrogen removal in small-scale sewerage treatment plants operated in intermittent aeration mode. *Wat. Sci. Tech. 23*, 709-718.

Neczaj, E., Okoniewska, E., Kacprzak, M. (2005). Treatment of landfill leachate by sequencing batch reactor, *Desalination*, 185, 357-362.

Neyens, E., Baeyens, J. (2003). A review of classic Fenton's peroxidation as an advanced oxidation technique. *Journal of Hazardous Materials, 98,* 33–50. DOI: 10.1016/S0304-3894(02)00282-0.

Nowell, L. H., Hoigné, J. (1987). Interaction of iron (II) and other transition metals with aqueous ozone. 8th Ozone World Congress, Zurich, September 1987.

Painter, H.A. (1995). Detailed review paper on biodegradability testing. OECD Guidelines for the testing of chemicals, OECD, Paris.

Pignatello, J., Oliveros, E. and MaCkay, A. (2006). Advanced oxidation processes for organic contaminant destruction based on the Fenton reaction and related chemistry, *Crit. Rew. Environ. Sci. Technol,* 36(1), 1-84.

Pitter, P., Chudoba, J. (1990). *Biodegradabillity of organic substances in the aquatic environment.* CRC Press, Inc., New York, USA.

Prousek, J., (1996). Advanced oxidation processes for water treatment. Chemical processes. *Chem. Listy, 90*, 229-237.

Puigagut, J., Salvado, H., Garcıa, J. (2005). Short-term harmful effects of ammonia nitrogen on activated sludge microfauna, *Water Research*, *39*, 4397–4404.

Ratanatamskul, C., Auesuntrachun, P. (2009). Removal of COD and colour from old-landfill leachate by advanced oxidation processes. *International journal of environment and waste management*, *4*, 470–480. DOI: 10.1504/IJEWM.2009.027410.

Regulation (2005). Regulation of the Government of the Slovak Republic N° 296/2005 on Indicators of admissible pollution of wastewaters disposed into recipients.

Renou, S., Givaudan, J.G., Poulain, S., Dirassouyan, F., Moulin, P. (2008). Landfill leachate treatment: Review and opportunity. *Journal of Hazardous Materials*, *150*, 468–493. DOI: 10.1016/
j.jhazmat.2007.09.077.

Rice, R. G., (1997). Applications of ozone for industrial wastewater treatment – A review. *Ozone Science and Engineering*. 18, 477-515.

Robinson, H.D., Mris, P.J., 1983, The treatment of leachates from domestic wastes in landfills. I. Aerobic biological treatment of a medium strength leachate, *Water Research*, 11, 1537-1548.

Scott, Jon P., Ollis, David F. (1995). Integration of chemical and biological oxidation for water treatment: review and recommendations. *Environmental Progress*, 14 (2), 88-103.

Silva, A. C., Dezotti, M., Sant'Anna, G.L., Jr. (2004). Treatment and detoxification of a sanitary landfill leachate. *Chemosphere*, *55*, 207–214. DOI: 10.1016/j.chemosphere. 2003.10.013.

Sizirici, B., Tansel, B. (2010). Projection of landfill stabilization period by time series analysis of leachate quality and transformation trends of VOCs. *Waste Management, 30*, 82-91.

Spanjer, H., Vanrolleghem, P. A., Olson, G., Dold, P. L. (1998). *Respirometry in control of the activated sludge process: Principles. Scientific and technical report No. 7.* IWA. J. W. Arrowsmith Ltd, Bristol, England.

Staehelln, J., Hoigné, J. (1982). Decomposition of ozone in water: Rate of initiation by hydroxide ions and hydrogen peroxide. *Environmental Science and Technology,* 16 (10), pp. 676-681.

System OxiTop® Control system (1998). *Operating manual.* WTW D 82362 Weilheim.

Tizaoui, C., Bouselmi, L., Mansouri, L, Ghrabi, A. (2007). Landfill leachate treatment with ozone and ozone/hydrogen peroxide systems. *Journal of hazardous materials, 140*, 316–324. DOI: 10.1016/j.jhazmat.2006.09.023.

Tyrel, S.F., Leeds-Harrison, P.B., Harrison, K.S. (2002). Removal of ammoniacal nitrogen from landfill leachate by irrigation onto vegetated treatment planes. *Water Research, 36*(1), 291-299.

Urh, Ž. (2007). Ozonation of municipal landfill leachate, Diploma thesis, University of Ljubljana, *Faculty of Chemistry and Chemical Technology*, Ljubljana (In Slovene).

Volk Ch. et al., (1997). Comparison of the effect of ozone, ozone - hydrogen peroxide system and catalytic ozone on the biodegradable organic matter of afulvic acid solution. *Water. Research*, *31*(3) 650-656.

Wang, F., El-Din, M.G., Smith, D.W. (2004). Ozonation of biologically treated landfill leachate: Treatment efficiency and molecular size distribution analysis. *Environmental Technology, 25*(11), 1277–1283. DOI: 10.1080/09593332508618371.

Wang, F., Smith, D. W., El-Din, M.G. (2006). Aged raw landfill leachate: Membrane fractionation, O_3 only and O_3/H_2O_2 oxidation, and molecular size distribution analysis. *Water Research, 40*, 463–474. DOI: 10.1016/j.watres. 2005.11.038.

Weiner, R.F., Matthews, R.A. (2006). *Environmental Engineering.* Elsevier Science. Burlinghton. USA.

Žgajnar Gotvajn, A., Zagorc-Koncan, J., Derco, J., Almásiová, B., Kassai, A. (2009). Oxidative pretreatment of fresh and mature landfill leachate. *Journal of advanced oxidation technologies, 12* (1), 71-80, ISSN 1203-8407.

Žgajnar Gotvajn, A., Derco, J., Tišler, T., Cotman, M., Zagorc-Koncan, J. (2009). Removal of organics from different types of landfill leachate by ozonation. *Wat. Sci. Tech., 60*(3), 597-602.

Zgajnar Gotvajn, A., Zagorc-Koncan, J., Cotman, M. (2009). Comparison of different methods for removal of metals from landfill leachate. In *Proceedings of the 12th International waste management and landfill symposium*, 5–9 October 2009. S. Margherita di Pula, Italy, 1-10.

Zhang, H., Choi, H.J., Huang, C.-P. (2006). Treatment of landfill leachate by Fenton's reagent in a continuous stirred tank reactor. *Journal of Hazardous Materials, 136*, 618 623. DOI: 10.1016/j.jhazmat.2005.12.040.

In: Landfills
Editors: G. B. C. Cabral and B. A. E. Botelho

ISBN: 978-1-62257-121-5

Chapter 2

LANDFILL LEACHATE TREATMENT: A PORTUGUESE CASE STUDY

Susana Cortez[1], Pilar Teixeira[2], Rosário Oliveira[3] and Manuel Mota[4]
IBB – Institute for Biotechnology and Bioengineering, Centre of Biological Engineering, University of Minho, Campus de Gualtar, Portugal

ABSTRACT

Although several countries have been introducing more stringent laws to reduce the amount of waste to be landfilled, in an attempt to maximize recycling and materials recovery, landfilling is still the most generalized practice for municipal solid waste (MSW) treatment. The generation of a high-polluted wastewater leaching from a sanitary landfill - the leachate - is inevitable and its treatment constitutes an important burden in landfill management, even after landfill closure.

The selection and design of the management and treatment of a landfill leachate has to take into consideration factors such as leachate characteristics, effluent discharge alternatives, technological possibilities, final discharge requirements and economical aspects.

Careful site management can increase the quality and reduce pollution potential of the formed leachate, however it does not allow total pollutant removal.

The leachate complexity and variability prevents the formulation of general treatment recommendations of universal validity. Techniques successfully developed for one site might not necessarily being applicable elsewhere. In addition, treatment methods that work well at young landfills are likely to become progressively less effective with the tip's age. Therefore, the leachate treatment system must be flexible enough to

[1] susana_cortez@deb.uminho.pt
[2] pilar@deb.uminho.pt
[3] roliveira@deb.uminho.pt
[4] corresponding author: mmota@deb.uminho.pt, phone: (+351) 253 604 405, fax: (+351) 253 678 986, postal address: IBB – Institute for Biotechnology and Bioengineering, Centre of Biological Engineering, University of Minho, Campus de Gualtar, 4710 – 057 Braga, Portugal.

produce the same effluent quality despite the sudden and large variations in the leachate strength.

Currently, the treatment systems used for leachate treatment include biological, physico-chemical and a combination of these processes. The enforcement of environmental discharge rules and the ageing of landfill sites with more stabilized leachate in different countries have put pressure on operators to implement better processes for the reduction of pollutant. New treatment alternatives and conventional technology improvements, mainly specialized on tertiary treatment, have been developed in recent years.

This work presents an overview of the most common processes used in landfill leachate treatment, their drawbacks and advantages. As the removal of nitrogenous compounds (mainly ammonium and nitrate) to values below the discharge limits has been one of the major problems in leachate treatment plants operating in Portugal, particular attention is focused on the Portuguese leachate treatment or pre-treatment plants. Furthermore, the results obtained by our group with a lab-scale anoxic rotating biological contactor to remove the nitrate excess, from a real landfill leachate, collected in a Portuguese municipal landfill before being discharged to the municipal sewer, are presented. The results achieved so far point out this process as a promising alternative polishing treatment.

Keywords: Biological, physico-chemical and combined treatment; landfill leachate; municipal solid waste (MSW); sanitary landfill

1. Landfill Leachate Definition, Generation and Composition

Sanitary landfilling is still the most widely applied method for the disposal of municipal solid waste (MSW) all around the world due to its economic advantages. In fact, landfilling must be accepted as inevitable because every treatment method leaves waste fractions (for example ashes and slag), which ultimately must be landfilled.

Despite the evolution of landfill technology, from open uncontrolled dumps to highly engineered facilities designed to eliminate or minimize the potential adverse impact of the waste on the surrounding environment, the generation of leachate remains predictable [1]. Moreover, landfills constructed and filled in the past still produce leachate and will probably do so for several decades [2–4], which still requires monitoring and treatment.

Therefore, the treatment of landfill leachate is and will be in the foreseeable future a major environmental concern. This is evident from the increasing number of publications on the generation, collection, storage, composition and treatment of landfill leachate during the period between 1980 and 2010. From Figure 1 it can be seen that in the 1980s, there were less than 25 related articles published annually. Recently, the number in the ISI Web of Knowledge database overcame 300 articles.

Landfill leachate can be defined as the liquid resulting from percolation of water and liquid waste through deposited solid waste, containing a cocktail of organic and inorganic compounds, biodegradable or non-biodegradable, in dissolved or suspended forms. Around 100 hazardous compounds have been identified in the landfill leachate. Some of them may be found at extremely high concentrations, presenting a cumulative, threatening and detrimental effect to the surrounding environment. In short, those compounds can be divided into four main groups: dissolved organic matter; inorganic macro-components, such as ammonium (NH_4^+), calcium (Ca^{2+}), sodium (Na^+), iron (Fe^{2+}), sulphate (SO_4^{2-}), chloride (Cl^-); heavy metals, such as cadmium (Cd^{2+}), chromium (Cr^{3+}), copper (Cu^{2+}), lead (Pb^{2+}), nickel (Ni^{2+}); and xenobiotic organic compounds (XOCs). Dissolved organic matter covers a wide range of organic species, from methane (CH_4), volatile fatty acids (VFA) to more recalcitrant humic substances found in landfill leachate and is quantified as chemical oxygen demand (COD) and/or total organic carbon (TOC). Xenobiotic organic compounds resulting from household or industrial chemicals are present in very low concentrations (usually lower than one milligram per liter level). These compounds include, among others, a variety of aromatic hydrocarbons, phenols and chlorinated aliphatics [5].

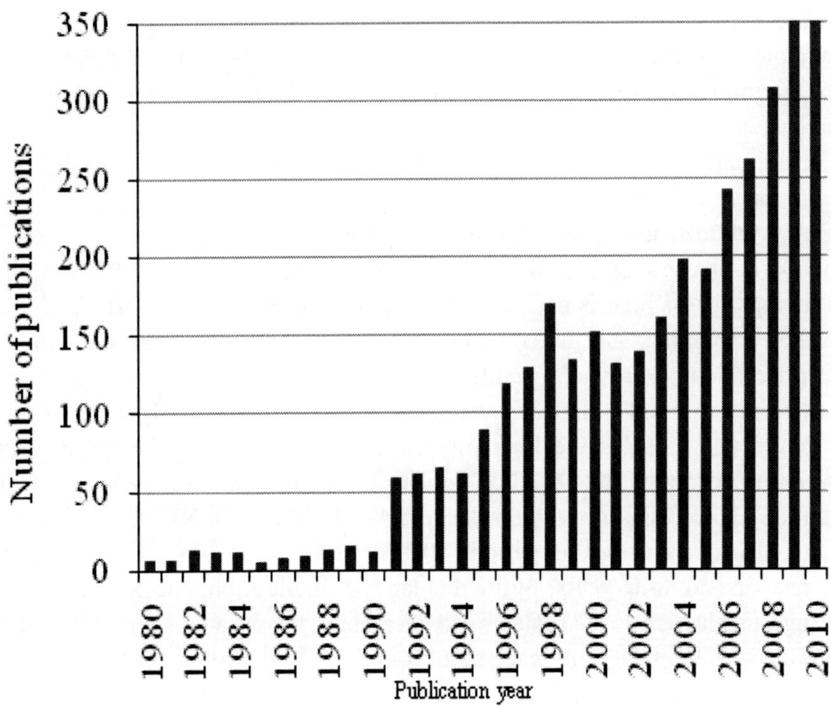

Figure 1. Number of publications concerning landfill leachate during the period 1980-2010 (source: ISI Web of Knowledge).

Landfill leachate flow rate varies from site to site, seasonally within each site [6], and is influenced by many factors: climatic and hydro-geological conditions, site operations and management, refuse characteristics and internal landfill processes. These factors can be divided into those that contribute directly to leachate production (rainfall, snowmelt, surface runoff, evaporation, evapotranspiration, groundwater intrusion, initial waste water content, irrigation, recirculation, liquid waste co-disposal and refuse decomposition) and those that affect leachate distribution within the landfill (refuse pre-treatment, compaction, permeability, particle size, density, settlement, vegetation, cover, sidewall, liner material, gas and heat generation and transport) [7]. The volume of landfill leachate produced is small compared to other wastewaters [8].

The composition of landfill leachate is closely related to the quantity produced, and can exhibit considerable spatial and temporal variations depending upon site operations and management practices (such as refuse pre-treatment, irrigation, and recirculation), refuse characteristics (composition and age) and internal landfill processes (such as hydrolysis, adsorption and biodegradation) [6,7]. The physicochemical and biological processes that occur within the waste and landfill age are usually the major determinants of leachate composition [9].

In general, the composition of the landfill leachate is characterized by the basic parameters: COD, TOC, five-day biochemical oxygen demand (BOD_5), BOD_5/COD ratio (biodegradability), total Kjeldahl nitrogen (TKN), nitrogen-ammonium ($N\text{-}NH_4^+$), pH, total dissolved solids (TDS), turbidity, alkalinity, color, conductivity, salts, xenobiotic organic substances and/or heavy metals content. Leachate toxic composition may also be evaluated by different toxicological tests, which provide indirect information on the content of pollutants that may be harmful to severe test organisms (such as micro-algae, duckweed, rotifers or luminescent bacteria).

Table 1 summarizes the major characteristics of leachate according to landfill site and landfill age.

After being landfilled, the solid waste decomposes through different phases. The first one, which only lasts a few days or weeks, corresponds to the aerobic degradation of organic matter. As soon as the oxygen is depleted, the degradation proceeds anaerobically [4,10]. In young landfills containing substantial amounts of biodegradable organic matter, a rapid anaerobic fermentation takes place, resulting in low molecular weight compounds such as volatile fatty acids (VFAs) [11]. This early phase of a landfill's lifetime is called the acidogenic phase and, besides the high organic fraction of VFAs, the leachate produced presents weak to strongly acidic pH, high COD and TOC concentrations, and high biodegradability (BOD_5/COD ratio in the range 0.4-0.7) [4,10,12]. $N\text{-}NH_4^+$ and heavy metals concentrations also rise during this period [10]. It has been established that the acidogenic phase may last one to four years. With the landfill maturation, the methanogenic phase begins. During this phase, VFAs are degraded decreasing the organic strength in the leachate, biogas is produced (CH_4, CO_2) and leachate becomes neutral or alkaline. With the increasing of age (>10 years), the matured leachate evolves towards a high strength of ammonium nitrogen, a moderate strength in COD, low concentrations of heavy metals and a low BOD_5/COD ratio (most often close to 0.1) [4,13]. Typically, most of the organic materials present in the mature leachate have a high molecular weight and are recalcitrant compounds such as humic- and fulvic-like acids [11]. It is important to note that the different degradation phases are not definitive, since dynamic variations in the environmental conditions result in changes in their distribution and rates within the landfill. For example, under certain conditions, the acid-formers may outgrow methane-formers, leading to acidification and interruption of the anaerobic digestion process [14].

Recently, it has been observed that even leachate from young landfills contains low organics (COD and TOC) concentration [12,18]. This can be due to leachate recirculation that started to be practiced in many landfills.

The data presented in Table 1 prove that landfill leachate constitutes a serious pollution hazard to the surrounding soil, and ground or surface water. The understanding, monitoring, management and treatment of quantity and quality of landfill leachate during operation and after landfill closure are of great importance.

Table 1. Mean composition of leachate at different landfill sites and ages

Landfill site	Spain [15]		China [16]		Greece [17]	
Parameter \ Landfill age	Young	Mature	Young	Mature	Young	Mature
pH	8.3	8.5	6.7	8.8	6.2	7.9
Conductivity (mS·cm-1)	32.7	8.0	31.5	29.1	29.5	18.3
TDS (g·L-1)	16	4	-	-	51	9.6
BOD5 (mg·L-1)	1000	43	39900	149	26800	1050
COD (mg·L-1)	6119	837	53200	1863	70900	5350
BOD5/COD	0.2	0.05	0.75	0.08	0.4	0.2
TOC (mg·L-1)	1481	223	-	-	-	-
TKN (mg·L-1)	-	-	-	-	3400	1100
N-NH4+ (mg·L-1)	1965	200	2760	796	3100	940
Alkalinity (mg CaCO3·L-1)	14443	2204	-	-	12880	4950
Color (units PtCo)	-	-	-	-	10550	7600
Turbidity (NTU)	-	-	-	-	1700	340
PO43- (mg·L-1)	-	-	1.63	1.86	n.d.*	3.0
SO42- (mg·L-1)	-	-	163	18	1600	210
Cl- (mg·L-1)	4430	1335	6150	8250	3260	4120

*n.d.- not detected.

2. LANDFILL LEACHATE MANAGEMENT AND TREATMENT TECHNOLOGIES

The selection and design of the management and treatment of a landfill leachate has to take into consideration factors such as leachate characteristics, effluent discharge alternatives, technological possibilities, final discharge requirements and economical aspects [19,20].

The most common options of leachate management include recirculation and recycling of the leachate back into the landfill, combined treatment with municipal wastewater, use of systems such as constructed wetlands, leachate evaporation or evapotranspiration (through vegetation) and on-site treatment followed by discharge. Leachate recirculation is being increasingly used since it accelerates the stabilization of the landfilled municipal solid waste and increases methane production. However, recycling leachate invariably results in accumulation of ammonia, which is nearly inert under anaerobic conditions. This high-strength wastewater requires intensive treatment before its safe disposal [21,22]. The treatment of landfill leachate together with municipal wastewater is not recommended since it may contribute to increase total pollutant load in sewage (such as XOCs and nitrogen-ammonium), which can be adsorbed into sludge particles and transferred to the sludge processing systems [23]. Furthermore, landfills are generally located far from metropolitan areas and sewage treatment plants, making this option increasingly expensive. Where land is available at low-cost, systems such as constructed wetlands, are attractive alternatives for landfill leachate management. Constructed wetlands are not likely to provide pollutants removal below detection for the major constituents of a specific leachate, but may very well do so for trace constituents [24]. Leachate evaporation or evapotranspiration (through vegetation) during warm periods may produce a high quality condensate, being residuals a small fraction of the original leachate volume [25]. Careful site management can increase the quality and reduce the pollution potential of the formed leachate, however it does not allow total pollutant removal. The application of a general strategy for on-site landfill leachate treatment is hampered by its great diversity. Techniques successfully developed for one site might not necessarily being applicable elsewhere. In addition, treatment methods that work well at young landfills are likely to become progressively less effective with the tip's age [6]. Biological treatments of landfill leachate have been shown to be very effective in removing organic matter in the early stages when the BOD_5/COD ratio of the leachate is high, but this ratio decreases with the age of the landfill and the process is less effective with time, when the presence of recalcitrant organic matter is higher. Accordingly, a variety of physico-chemical processes have been developed and used to treat leachate with this type of matter [26]. Therefore, the leachate treatment system must be flexible enough to produce the same effluent quality despite the sudden and large variations in the leachate strength [27,28]. Currently, the treatment systems used for leachate treatment include biological, physico-chemical and a combination of these processes. Alvarez-Vazquez et al. [29] estimated that the treatment systems used for leachate treatment were 72% biological, 11% flocculation/coagulation, 5% membrane filtration, 4% air stripping, 4% chemical oxidation and 4% adsorption on activated carbon. The enforcement of environmental discharge rules and the aging of landfill sites with more and more stabilized leachate in different countries have put pressure on operators to implement better processes for the reduction of pollutants [30]. New treatment alternatives and conventional technology improvements, mainly specialized on tertiary treatment, have been developed in recent years. This section provides an up to date overview of landfill leachate treatment.

2.1. Biological Treatment Systems

Biological treatment technologies have been shown to be feasible, simple, effective and economical methods to degrade biodegradable organic and nitrogenous matter of landfill leachate [4,31,32]. In aerobic processes, microorganisms degrade leachate organic compounds to carbon dioxide, water and microbial biomass and reduce nitrogen-ammonium to nitrogen-nitrate by nitrification. Nitrogen-nitrate is then oxidized to nitrogen gas under an anoxic environment and with external carbon addition during denitrification. Biological anaerobic processes are also used for removal of organic pollutants, mainly from high strength leachate, presenting some advantages such as biogas (mainly methane and carbon dioxide) generation and low biomass production, and for the conversion of ammonium to nitrogen gas in a process called ANAMMOX (anaerobic ammonium oxidation), which allows energy and cost savings [33].

It is worth mentioning that parameters such as food-microorganisms ratio (F/M), hydraulic retention time (HRT) and environmental factors such as pH or mixing are usually controlled to insure the optimum growth of the complex mixed microbiological populations and, consequently, to optimize biological treatment [1]. For instance, phosphorus deficiency in leachate hampers the production of microorganisms and consequently the treatment performance [34]. It is thus common to overcome this deficiency by adding phosphorus. Likewise, since most denitrifiers are heterotrophs, external carbon is usually added following nitrification, mainly when treating mature leachate, whose organic load is resistant to biodegradation. Biological treatment of landfill leachate includes suspended-growth biomass systems such as lagooning, activated sludge (AS) and sequencing batch reactors (SBRs), as well as attached-growth biomass systems such as trickling filters (TFs), rotating biological contactors (RBCs) and suspended carrier biofilm reactors (SCBRs). The latter mentioned systems, using biofilm, present higher volumetric conversion rates, lower hydraulic retention time, higher resistance to toxic agents, lower sensitivity to environmental conditions (such as temperature, pH and metabolic products) and less sludge production [1,35,36]. As a consequence, these systems are of particular interest for landfill leachate nitrification due to a high resistance to shock of organics and nitrogen loadings.

2.1.1. Suspended-Growth Biomass Systems

2.1.1.1. Activated Sludge

Conventional activated sludge (AS) processes use an aerobic and/or anoxic biological tank followed by a settlement chamber for sludge separation. Part of this sludge is then recycled to the tank to provide biomass to treat the new leachate. In the past, these traditional treatment schemes, proved to exhibit satisfactory and consistent leachate treatment performance in terms of biodegradable organic carbon and nutrients removal, providing enough hydraulic residence time [37]. However, over time and with landfill aging, many operational problems such as control of biomass settling, loss of slow growing organisms (for example, nitrifiers) able to remove the biodegradable pollutants at low kinetics, high-energy consumption, excess of sludge production and microbial inhibition by toxic compounds led to the abandonment of this type of reactor for landfill leachate treatment [1,19,38]. There are very few recent works available concerning landfill leachate treatment by conventional activated sludge methods (Table 2). Increasing attention has been given on the inclusion of plastic carriers [11], bentonite [39] or granular activated carbon [40] into the AS reactor. Such modifications can improve settling sludge properties, allowing retaining the microorganisms in the system and, in the case of activated carbon, enhancing adsorption of the substrate, resulting in more effective nitrogen-ammonium, metal and COD removal. Nevertheless, the operational cost of leachate treatment by this modified activated sludge system is notably increased [1].

2.1.1.2. Sequencing Batch Reactors

The sequencing batch reactor (SBR) is a variation of the conventional activated sludge system in which biological oxidation and sludge separation are carried out in the same tank and environmental conditions are controlled using fill and draw operations at distinct time intervals [31,37]. The periodic and controlled change of process conditions, such as concentration of oxygen, makes the SBR technology attractive for aerobic, anoxic and anaerobic processes and has resulted in a wide application for landfill leachate treatment. Typical performances of SBRs are reported in Table 3. To enhance the degradation of organic constituents, as well as to reduce the nitrogen content of a landfill leachate, small improvements to the SBR system have been tried: Zoubolis et al. [45] analyzed the addition of enzymes while Uygur and Kargi [46] studied the introduction of powdered activated carbon (PAC) in a SBR. Both solutions achieved interesting results but, inevitably, represent additional costs.

Table 2. Landfill leachate treatment using AS reactors

Landfill site and age, reference	Leachate feeding characteristics				Reactor operational parameters					Performance
	pH	COD ($mg \cdot L^{-1}$)	BOD_5/COD	$N\text{-}NH_4^+$ ($mg \cdot L^{-1}$)	T (°C)	Type of operation and scale	HRT (d)	SRT (d)	OLR (g COD·$L^{-1} \cdot d^{-1}$)	
Moyer (Philadelphia, USA), mature leachate, [41]	7	733	0.02	354	-	Aerobic, lab-scale (6L aeration tank+1.4L settling tank)	2	30	-	100 and 40% for $N\text{-}NH_4^+$ and COD removal, respectively
Burns Bog (British Columbia, USA), mature leachate, [42]	7.5	-	-	200-1200	20	Anoxic/aerobic, pilot-scale (5L anoxic tank+10L aeration tank+4L settling tank)	0.14	13	-	-
Kimpo (Korea), mature leachate, [43]	-	3130	0.44	-	rt	Aerobic, lab-scale	3	-	-	69 and 98% for COD and BOD_5 removal, respectively
Djebel Chekir (Tunisia), young leachate, [44]	7.8	3200	0.51	1750	rt	Aerobic, pilot-scale (14L aeration tank+settling tank)	5.5	-	4.5	92 and 75% for $N\text{-}NH_4^+$ and COD removal, respectively

SRT: sludge retention time; OLR: organic loading rate; rt: room temperature.

Table 3. Landfill leachate treatment using SBRs

Landfill site and age, reference	Leachate feeding characteristics				Reactor operational parameters					Performance
	pH	COD ($mg \cdot L^{-1}$)	BOD_5/COD	$N\text{-}NH_4^+$ ($mg \cdot L^{-1}$)	T (°C)	Type of operation and scale	HRT (d)	SRT (d)	Cycle	
Jiangmen (China), mature leachate, [47]	8.8	3000	0.22	1200	et	Aerobic/anoxic, full-scale (1200 m^3)	-	20	24 h (1 h feed, 4 h mixing, 5 h aeration, 2 h mixing, 3 h aeration, 8 h settle and 1 h decant)	99 and 76% for $N\text{-}NH_4^+$ and COD removal, respectively

Landfill site and age, reference	Leachate feeding characteristics				Reactor operational parameters					Performance
	pH	COD ($mg \cdot L^{-1}$)	BOD_5/COD	$N\text{-}NH_4^+$ ($mg \cdot L^{-1}$)	T (ºC)	Type of operation and scale	HRT (d)	SRT (d)	Cycle	
Wysieka (Poland), mature leachate, [48]	-	680	0.16	312	rt	Aerobic/anoxic, lab-scale (6 dm^3)	3	-	24 h (0.25 h feed, 3 h mixing, 17.5 h aeration, 3 h settle and 0.25 h decant)	100, 49 and 93% for $N\text{-}NH_4^+$, COD and BOD_5 removal, respectively
	-	680	0.16	312	rt	Aerobic/anoxic, lab-scale (6 dm^3)	2	-	24 h (0.25 h feed, 3 h mixing, 17.5 h aeration, 3 h settle and 0.25 h decant)	100, 48 and 93% for $N\text{-}NH_4^+$, COD and BOD_5 removal, respectively; occasionally a higher amount of $N\text{-}NH_4^+$ was observed
Turkey, mature leachate, previously coagulated and air stripped, [46]	7-7.5	5750	-	185	25	Anaerobic/aerobic/anoxic, lab-scale (5 dm^3)	-	10	21 h (1 h anaerobic, 1 h anoxic, 2 h aerobic, 1 h anoxic and 1 h aerobic × 3 cycles)	31 and 62% for COD and BOD_5 removal, respectively
Thessaloniki (Greece), mature leachate, [45]	7.5	15000	0.37	1800	40-50	Aerobic/anoxic, lab-scale (8 dm^3)	20	-	16 h (2 h feed and mixing, 6 h aeration, 1 h mixing, 6 h aeration, 1 h settle and decant)	60-70, 40 and 70% for $N\text{-}NH_4^+$, COD and BOD_5 removal, respectively
Heimifeng (China), mature leachate, [49]	7.7	3876	0.14	1451	30	Anaerobic/aerobic/anoxic, lab-scale (3 dm^3)	-	-	12 h	90, <6.7 and >95% for $N\text{-}NH_4^+$, COD and BOD_5 removal, respectively
Wuhan (China), mature leachate, [50]	8.1	2400	0.15	2470	rt	Aerobic/anoxic, lab-scale (10 dm^3)	-	20	23.25 h (feed, 8 h aeration, 3 h mixing, 6 h aeration, 5 h mixing, 1 h settle and 0.25 h decant)	79, 77 and 86% for $N\text{-}NH_4^+$, COD and BOD_5 removal, respectively

et: environment temperature.

2.1.1.3. Lagooning

Aerated and non-aerated lagoons are interesting options for on-site treatment of landfill leachate since they offer low operational and maintenance costs, can effectively remove pathogens, organic and inorganic compounds and support significant fluctuations of influent concentration and strength, typical of landfill leachate [51]. Aerated lagoon systems are similar to activated sludge systems except that no recycle is used. In many cases lagoons are operated as a leachate pre-treatment step prior to disposal into municipal sewers, other on-site leachate treatments or recycling into landfills.

Although a considerable amount of lagoons has been employed on biological leachate treatment, few studies relative to the utilization of these systems for this purpose are available in the current literature (Table 4).

Table 4. Landfill leachate treatment using aerated and non-aerated lagoons

Landfill site and age, reference	Leachate feeding characteristics				Lagoon operational parameters			Performance
	pH	COD ($mg \cdot L^{-1}$)	BOD_5/ COD	$N\text{-}NH_4^+$ ($mg \cdot L^{-1}$)	T (°C)	Type of operation and scale	HRT (d)	
Penzberg (Germany), mature leachate, [52]	-	136-1980	-	-	et	(1) Aerobic lagoon, full-scale (2000 m^3)	-	≈50% for $N\text{-}NH_4^+$
Tre Monti (Italy), mature leachate, [51]	8.4	5050	0.25	1330	et	(2) Anaerobic lagoons (4.4 and 5.2m depths) (3) facultative aerobic lagoons (1.5 and 2.1 depths), full-scale	32 (anaerobic lagoons) 148 (facultative aerobic lagoons)	77, 40 and 64% for $N\text{-}NH_4^+$, COD and BOD_5 removal, respectively
Colchester (United Kingdom), mature leachate, [53]	7.2	1740	-	965	et	(4) Facultative aerobic lagoons, full-scale (60-80 m^3)	11-254	99 and 75% for $N\text{-}NH_4^+$ and COD removal, respectively

2.1.1.4. Anaerobic Digesters

The anaerobic digestion (AD) treatment of leachate enables the process initiated in the landfill to be concluded [6]. An interesting feature of anaerobic treatment is the possibility of using the biogas produced to warm the digester that usually works at 35 °C.

The performance of some conventional anaerobic suspended-growth reactors is presented in Table 5. The efficiencies reported demonstrate that the anaerobic digester is an economic and reliable process, particularly suitable for dealing with high organic matter concentration from young landfills. Obviously, anaerobic processes do not remove ammonium and therefore always need downstream stripping or, more frequently, nitrification [29].

Table 5. Landfill leachate treatment using conventional ADs

Landfill site and age, reference	Leachate feeding characteristics				Reactor operational parameters				Performance
	pH	COD ($mg \cdot L^{-1}$)	BOD_5/COD	$N\text{-}NH_4^+$ ($mg \cdot L^{-1}$)	T (°C)	Scale	HRT (d)	OLR (g $COD \cdot L^{-1} \cdot d^{-1}$)	
Coruña (Spain), young leachate, [54]	7.3	4100	-	800	37	Lab-scale (0.9 L)	4	1.1	47% COD removal
	7.3	4100	-	800	20	Lab-scale (0.9 L)	4	1.1	29% COD removal
Kaushung (Taiwan), young leachate, [55]	6.3-6.4	22750	0.7	171	35	Lab-scale (2 L)	15	1.51	93 and 97% for COD and BOD_5 removal, respectively; 770 mg $N\text{-}NH_4^+ \cdot L^{-1}$ in the effluent
Shiraz (Iran), young leachate, [56]	6.3	50534	0.7	1460	32	Pilot-scale (150 L)	15	3.37	79 and 75% for COD and BOD_5 removal, respectively

2.1.1.5. Upflow Anaerobic Sludge Blanket Reactors

In the upflow anaerobic sludge blanket (UASB) reactor, a high concentration of active anaerobic biomass is held in suspension in the reactor by hydraulic design and occurs the formation of granular biomass. The upward motion of methane and carbon dioxide gas bubbles imparts mixing in the sludge bed. Over the years, several UASB reactors have been shown to be efficient in the treatment of the MSW leachate [57]. From Table 6 it can be seen that the average performance of COD removal efficiency was always higher than 69% for temperatures in the range 19-42 °C.

Kettunen and Rintala [58] observed similar COD removal efficiency at low temperatures (13-23 °C) in the treatment of landfill leachate in a UASB reactor. This surprising result suggests that high treatment efficiency can be achieved without heating the reactor, providing an interesting cost-effective option. Recently, Liu et al. [57] applied a modified reactor of the traditional UASB, the expanded granular sludge bed (EGSB), which operates at much higher superficial velocities and height/diameter ratio for the treatment of a fresh leachate. Under the proposed optimal conditions (HRT=2.83 d and OLR=22.5 g $COD \cdot L^{-1} \cdot d^{-1}$) and at 35 °C, the lab-scale EGSB reactor removed 94-96% of the COD from the raw leachate (COD=66000-68000 $mg \cdot L^{-1}$).

2.1.1.6. Membrane Bioreactors

There is a growing interest in using the membrane bioreactor (MBR), which comprises a bioreactor generally containing suspended microorganisms combined with a membrane module [29,64,65]. The MBR has a low footprint, can be operated at very long sludge age (SRT) and with high biomass concentration, and can achieve an excellent clarified effluent with low sludge waste production [29,66]. Contrary to conventional systems, organisms such as nitrifiers or organisms that are able to degrade slowly biodegradable substances are not washed out of the system and no loss of process activity occurs. Such advantages turn the MBR an attractive alternative for organic matter removal and nitrification of young and even mature landfill leachate [29,65,66]. Naturally, the denitrification process will also occur in the MBR if an intermittent aeration it is adapted to the system. MBRs use ultrafiltration (UF) and/or microfiltration (MF) membranes for the complete retention of sludge [64]. The membrane can be fitted either outside or within (submerged) the bioreactor, and both configurations have been employed for landfill leachate treatment [29]. The major disadvantages of membrane bioreactors are the high capital and operating costs. Operational data and performance of different kinds of MBRs are detailed in Table 7.

Pirbazari et al. [67] tested the combination of a lab-scale MBR with powdered activated carbon (PAC), blending adsorption, biodegradation and membrane filtration processes for the treatment of landfill leachate. The process efficiencies were in the range of 95-98% in terms of TOC reduction and exceeded 97% for specific organic pollutants.

2.1.2. Attached-Growth Biomass Systems

2.1.2.1. Fixed-Bed Reactors

Trickling filters (TFs) and submerged biological filters (BFs) represent the main fixed-bed reactors that have been investigated for organic and nitrogen removal from MSW leachate. In these reactors, biological reactions take place in the biofilm developed on the immobilized support material. Trickling filters are commonly aerobic systems whereas biofilters operate under aerated or non-aerated conditions (Table 8). Biofilters are being increasingly used instead of TFs.

As part of biofilm technology, fixed-bed reactors provide good removal efficiencies, even with effluents with low BOD_5/COD ratio, are resistant to toxic substances and tolerate many of the inhibitors usually contained in leachate [68]. Under certain conditions, a biofilm, with a great diversity of microorganisms performing nitrification and denitrification, occurs in the same reactor. Furthermore, biological treatment and solids separation occur in the same reactor eliminating the requirement for separate secondary clarification and minimizing footprint. An important drawback of fixed-bed reactors is the added cost of the support.

As it can be seen in Table 8, Gálvez et al. [74] observed that the aerated biofilter was more effective than the non-aerated reactor in the removal of COD and BOD_5. Anaerobic biological treatment is generally more susceptible to poisoning by toxic substances, especially heavy metals, phenols, ammonium and chlorinated organic compounds. Aerobic microorganisms have the capacity to acclimatize to the presence of certain toxic organic substances, which can oxidize in some cases, and even to the presence of heavy metal ions, although in the latter case the toxic substances are not biologically oxidized, but rather absorbed by the biological flocks. Di Iaconi et al. [38] used a biofilter with aerobic granular biomass working in the fill and draw mode - a sequencing batch biofilter granular reactor- for the treatment of a mature landfill leachate. This system revealed to be an effective technology, removing about 80% at least up to 1.1 g $COD{\cdot}L^{-1}{\cdot}d^{-1}$.

The hybrid bed filter (HBF) is another relevant modification of biofilters. It consists of an upflow sludge blanket at the bottom and an anaerobic filter on top, combining the advantages of UASB and BF reactors, while minimizing their limitations.

Table 6. Landfill leachate treatment using UASB reactors

Landfill site and age, reference	Leachate feeding characteristics				Reactor operational parameters				Performance
	pH	COD ($mg \cdot L^{-1}$)	BOD_5/COD	$N\text{-}NH_4^+$ ($mg \cdot L^{-1}$)	T (ºC)	Scale	HRT (d)	OLR (g $COD \cdot L^{-1} \cdot d^{-1}$)	
Nepean (Canada), young leachate, [59]	6.9-9.0	4800-9840	0.86	-	35	Lab-scale (6.2 L)	1	4.8-9.8	78% COD removal
Turkey, young leachate, [60]	-	20000	-	679	37-42	Lab-scale (2.5 L)	1.25	16	27 and 79% for $N\text{-}NH_4^+$ and COD removal, respectively
Bucaramanga (Colombia), young leachate, [61]	6.5-7.5	3273	>0.5	-	-	Lab-scale (5 L)	2.3	0.33	62% COD removal
Asturias (Spain), young leachate, [62]	-	11939	-	1679	35	Lab-scale (2 L)	1.6	7.5	80% COD removal; 1665 mg $N\text{-}NH_4^+ \cdot L^{-1}$ in the effluent
Quito (Ecuador), young leachate, [58]	7.1	26000	0.5	-	19	Lab-scale (2.5 L)	1.2	21	69% COD removal
	7.1	26000	0.5	-	30	Lab-scale (2.5 L)	1.2	21	72% COD removal
Istambul (Turkey), medium-age leachate, [63]	7.8-8.4	5500-12070	0.23-0.48	1725-2536	35	Lab-scale (9 L)	3	1.84-4.03	8 and 52% for $N\text{-}NH_4^+$ and COD removal, respectively

Table 7. Landfill leachate treatment using MBRs

Landfill site and age, reference	Leachate feeding characteristics				Reactor characteristics		Membrane properties		Operational parameters				Performance
	pH	COD ($mg \cdot L^{-1}$)	BOD_5/COD	$N\text{-}NH_4^+$ ($mg \cdot L^{-1}$)	Type of reactor	Type of operation and scale	Type of membrane, module and configuration	Operated pressure (bar)	T (ºC)	HRT (d)	SRT (d)	OLR	
Thessaloniki (Greece), mature leachate, [37]	8.4	2456	-	238	Sequencing batch reactor (cycle 12 h)	Aerobic/anoxic, lab-scale (5L)	Polymeric ultrafiltration, hollow fiber submerged	0.1-0.5	-	10	No waste sludge removal took place	-	≈100% for $N\text{-}NH_4^+$ and $N\text{-}NO_3^-$, 40-60% for COD removal
Italy, mature leachate, [66]	7.8-8.2	4224	0.33	1547	Activated sludge	Aerobic, pilot-scale (500 L)	Ceramic ultrafiltration, tubular external	1.5	30-37	-	>45	-	45% COD removal

Table 7. (Continued)

Landfill site and age, reference	Leachate feeding characteristics				Reactor characteristics		Membrane properties		Operational parameters				Performance
	pH	COD ($mg \cdot L^{-1}$)	BOD_5/ COD	$N\text{-}NH_4^+$ ($mg \cdot L^{-1}$)	Type of reactor	Type of operation and scale	Type of membrane, module and configuration	Operated pressure (bar)	T (°C)	HRT (d)	SRT (d)	OLR	
Jebel Chakir (Tunisia), young leachate, [69]	7	41000	-	-	Activated sludge	Anaerobic, pilot-scale (50 L)	Polymeric ultrafiltration, tubular external	0.2-3	37	7	-	6.27 g $COD \cdot L^{-1} \cdot d^{-1}$	91% COD removal
Thailand, young leachate, [70]	6.8-7.0	12000	0.65	1000	Activated sludge	Aerobic, lab-scale (6 L)	Ceramic ultrafiltration, hollow fiber submerged	0.18-0.6	45	1	-	7.2-8.5 g $BOD \cdot L^{-1} \cdot d^{-1}$	79 and 76% for $N\text{-}NH_4^+$ and COD removal, respectively

Table 8. Landfill leachate treatment using fixed-bed reactors

Landfill site and age, reference	Leachate feeding characteristics				Reactor characteristics		Support material properties		Operational parameters			Performance
	pH	COD ($mg \cdot L^{-1}$)	BOD_5/ COD	$N\text{-}NH_4^+$ ($mg \cdot L^{-1}$)	Type of reactor, mode of operation	Type of operation and scale	Type of material and media	Media dimensions	T (°C)	HRT (d)	OLR (g $COD \cdot L^{-1} \cdot d^{-1}$)	
Essex (United Kingdom), mature leachate, [71]	≥7.5	-	-	233	TF, leachate continuously recirculated	Aerobic, full-scale (16500 dm^3)	Corrugated plastic cylinders	Diameter-3 cm; specific surface area-230 $m^2 \cdot m^{-3}$	19.7	0.2-0.6	-	≈100, 87 and 86% for $N\text{-}NH_4^+$, BOD_5 and SS removal, respectively
France, mature leachate, [72]	8.0-8.5	850-1350	0.1-0.2	295	TF, leachate continuously recirculated	Aerobic, full-scale (16500 dm^3)	Plastic fragments	Particles size-10-10 mm	20	4.5	-	52% COD removal
Coruña (Spain), young leachate, [54]	8.2	4200	-	700	BF, continuously recirculated	Anaerobic, lab-scale (2 dm^3)	Corrugated plastic rings	Specific surface area-500 $m^2 \cdot m^{-3}$	37	1.8	1.92	65% COD removal

Landfill site and age, reference	Leachate feeding characteristics				Reactor characteristics		Support material properties		Operational parameters			Performance
	pH	COD (mg·L^{-1})	BOD_5/ COD	$N\text{-}NH_4^+$ (mg·L^{-1})	Type of reactor, mode of operation	Type of operation and scale	Type of material and media	Media dimensions	T (ºC)	HRT (d)	OLR (g COD·L^{-1}·d^{-1})	
Jyväskylä (Finland), mature leachate, [34]	6.5-7.0	230-510	-	60-170	Upflow BF, leachate semi-continuously fed	Aerobic, lab-scale (9.4 dm^3)	Ceramic fragments	Particles size-16-32 mm	25	3.8	-	60-90 and 26-62% for $N\text{-}NH_4^+$ and COD removal, respectively
United Kingdom, mature leachate, [73]	7.8	787	0.1	581	BF, leachate continuously recirculated	Aerobic, pilot-scale (770 dm^3)	Crystal quartz sand	Diameter-1.5-2.2 mm	15-19	-	0.1	99, 36, 93 and 73% for $N\text{-}NH_4^+$, COD, BOD_5 and SS removal, respectively
Granada (Spain), mature leachate, [74]	7.8	787	0.1	581	BF, continuously recirculated	Aerated, leachate lab-scale (5.7 dm^3)	Ceramic fragments	Diameter-6-8 mm	-	0.7	25.1	67, 87 and 27% for COD, BOD_5 and VSS removal, respectively
	7.8	787	0.1	581	BF, continuously recirculated	Non-aerated, leachate lab-scale (5.7 dm^3)	Ceramic fragments	Diameter-6-8 mm	-	0.7	25.1	18, 16 and 38% for COD, BOD_5 and VSS removal, respectively

Table 9. Landfill leachate treatment using RBCs

Landfill site and age, reference	Leachate feeding characteristics				Reactor characteristics		Support material properties		Operational parameters			Performance
	pH	COD ($mg \cdot L^{-1}$)	BOD_5/COD	$N\text{-}NH_4^+$ ($mg \cdot L^{-1}$)	Type of operation and scale	No. stages, rotational speed (rpm)	Type of material, media and quantity/stage	Media dimensions	T (°C)	HRT (d)	SOL (g $COD \cdot m^{-2} \cdot d^{-1}$)	
Moyer (Philadelphia, USA), mature leachate, [81]	7.5-8	358	-	154	Aerobic, lab-scale (3.42 dm^3)	3; 2.3	Flat plastic discs, 2 discs/stage	Diameter-23.8 cm; area of discs/stage-0.187 m^2	22	1.05	6.1 g $COD \cdot m^{-2} \cdot d^{-1}$	≈97, 36 and 81% for $N\text{-}NH_4^+$, COD and BOD_5 removal, respectively; 140 mg $N\text{-}NO_3^- \cdot L^{-1}$ in the effluent
Bucaramanga (Colombia), young leachate, [61]	7.8-8.8	2500-9000	-	-	Aerobic, pilot-scale (20 dm^3)	4; 6	Perforated acetate discs, 10 discs/stage	Diameter-30 cm	-	1	24.7 g $COD \cdot m^{-2} \cdot d^{-1}$	52% COD removal
Bartoszyce (Poland), mature leachate, [36]	-	-	-	243	Aerobic, pilot-scale	2; 7	-	Area of discs/stage-0.67 m^2	20	1	-	≈100% $N\text{-}NH_4^+$ removal; 222 mg $N\text{-}NO_3^- \cdot L^{-1}$ in the effluent
Portugal, mature leachate, [84]	6.8	1116.8 (with additional C source)	-	98	Anoxic, lab-scale (2.5 dm^3)	1; 4	Flat plastic discs, 8 discs/stage	Diameter-13 cm	28	0.42	-	≈80 and 84% for $N\text{-}NO_3^-$ and COD removal, respectively; $N\text{-}NH_4^+$ decreased in the effluent

(SOL: superficial organic load).

Table 10. Landfill leachate treatment using SCBRs

Landfill site and age, reference	Leachate feeding characteristics				Reactor characteristics		Support material properties		Operational parameters			Performance
	pH	COD ($mg \cdot L^{-1}$)	BOD_5/COD	$N-NH_4^+$ ($mg \cdot L^{-1}$)	Type of operation and scale	Mode of operation	Type of material, media and quantity	Media dimensions	T (ºC)	HRT (d)	OLR ($g COD \cdot L^{-1} \cdot d^{-1}$)	
Hyllstofta (Sweden), mature leachate, [85]	7.5	800-1300	-	460-600	Aerobic, lab-scale (0.6 dm^3)	Continuous mode	Plastic cylinders, 60% of reactor volume	Diameter-9.1 mm; length-7.2 mm; specific surface area-500 $m^2 \cdot m^{-3}$	20	5	-	99% $N-NH_4^+$ removal
Japan, mature leachate, [86]	7.5-8.0	-	-	150	Aerobic, lab-scale (1.2 dm^3)	Continuous mode, air-lift	Plastic gel cubes, 15% of reactor volume	Size-3 mm	10	0.2	-	83% $N-NH_4^+$ removal; 90 mg $N-NO_2^- \cdot L^{-1}$, 15 mg $N-NO_3^- \cdot L^{-1}$ in the effluent
Harbin (China), young leachate, [82]	7.5	15993	-	350-400	Anaerobic, lab-scale (4.2 dm^3)	Continuous mode	Plastic cylinders, 40% of reactor volume	Diameter-7 mm; length-9 mm; specific surface area-900 $m^2 \cdot m^{-3}$	35	1.5	10.66	12 and 91% for $N-NH_4^+$ and COD removal, respectively
Thessaloniki (Greece), mature leachate, [40]	7.5	5000	0.2	1800	Aerobic/anoxic, lab-scale (8 dm^3)	Sequencing batch mode	Plastic cubes, 50% of reactor volume	Size-1 cm	-	20	-	95, 65 and 90% for $N-NH_4^+$, COD and BOD_5 removal, respectively

The filter zone in the HBF reactor, in addition to its physical role for biomass retention, has some biological activity contributing to COD reduction in a zone, which, in a classical UASB reactor, biomass is scarce [75]. Nedwell and Reynolds [76] reported steady-state COD removal efficiencies of 81-97% up to 3.75 g COD·L^{-1}·d^{-1}, under methanogenic digestion of a landfill leachate at 30 ºC. Kettunen et al. [77] observed lower COD reductions (52-60%) with a young leachate load of 1.0-1.2 g COD·L^{-1}·d^{-1} but at an unusual temperature digestion, 11 ºC.

2.1.2.2. Moving-Bed Reactors

2.1.2.2.1. Rotating Biological Contactors

The treatment of landfill leachate has also been carried out in rotating biological contactors (RBCs) [36,78,79]. For instance, until 1985, Japan had 1323 plants with RBC arrangements, of which 10% were utilized for bioremediation of leachates generated in municipal landfills [61].

In a RBC unit, the pollutants contained in the leachate are removed by the biofilm that is established on the entire surface area of the support material (typically flat or corrugated discs), which continually rotates [80]. RBCs have been applied for removal of ammonium and organic substances from landfill leachate with high performance (Table 9). If the support material is completely submerged in the leachate to be treated and the reactor tightly closed to avoid air entrance, they can also be used for leachate denitrification [81].

2.1.2.2.2. Suspended Carrier Biofilm Reactors

The suspended carrier biofilm system (SCBR) is a completely mixed reactor where the biomass is grown on small carrier elements that have a density close to the water density and are kept in movement along with a water stream inside a reactor [40,82]. The fluidization inside a reactor can be caused by aeration in an aerobic reactor and by a mechanical stirrer in an anaerobic or anoxic reactor. The treatability of landfill leachate by this process is given in Table 10.

The use of granular activated sludge (GAC) as porous material in the suspended carrier biofilm systems has also been proved effective for the removal of organic matter (even recalcitrant) and of nitrogen from landfill leachate [40]. The main drawback of this process is the large amount of residual suspended solids, which have to subsequently be treated and separated, and the amount of activated carbon addition, which increases the operational costs.

Eldyasti et al. [83] studied an efficient pilot-scale liquid-solid circulating fluidized bed process for the treatment of landfill leachate, employing aerobic and anoxic conditions and lava rock particles as support material. This reactor achieved COD, nitrogen, and phosphorus removal efficiencies of 85%, 80%, and 70%, respectively at a carbon to nitrogen ratio of 3 and nutrients loading rates of 2.15 g COD·L^{-1}·d^{-1}, 0.70 g N·L^{-1}·d^{-1} and 0.014 g P·L^{-1}·d^{-1}.

2.2. Physico-Chemical Systems

The advantages of physico-chemical systems in general include immediate start-up, easy automation, insensivity to temperature changes and, in most cases, simplicity of the materials and of the plant required. Commonly, however, these advantages are outweighed by the drawbacks: the operation costs and the secondary pollution (such as the large quantities of waste sludge generated by the addition of flocculants). As a result, physico-chemical methods are mainly suitable for the pre- or post-treatment of landfill leachate to complement biological degradation techniques [6]. In this role, they are especially useful in the treatment of mature leachate (containing recalcitrant substances) and for the elimination of specific pollutants such as ammonium, heavy metals, suspended solids, colloidal particles and color. Conventional physico-chemical systems include coagulation-flocculation, chemical precipitation, air stripping, adsorption and flotation. Pressure-driven membrane technologies and advanced oxidation processes have emerged as the most promising options for leachate treatment [4,87].

2.2.1. Conventional Physico-Chemical Technologies

2.2.1.1. Coagulation-Flocculation

Coagulation-flocculation is a relative simple technique that has been employed for the removal of suspended solids (SS), colloidal particles, color, non-biodegradable organic matter and heavy metals from landfill leachate [88,89]. The coagulation process destabilizes colloidal particles by the addition of a coagulant. To increase the particle size, coagulation is usually followed by aggregation of the unstable particles into bulky flocks thereby promoting easier settling. The type and concentration of coagulant, pH, velocity gradient, rapid and slow mixing, settling time and flocculant addition play major roles in increasing the probability of the settling of colloidal particles [17,88]. Iron and aluminum salts have been widely used as coagulants as shown in Table 11. Iron salts have proven to be more efficient than aluminm salts [90]. Occasionally, it is observed that the addition of flocculants does not improve COD, color and turbidity removal. However, settling conditions are substantially improved by the formation of flocks that settle better and more quickly [89].

Some drawbacks of the coagulation-flocculation system include high operational cost due to high chemical consumption, the generation of sludge, the sensitivity of the process to pH and in certain cases, the increase on the concentration of aluminum and iron in the resulting effluent [89]. This is why it has been proposed mainly as a pre-treatment method for young leachate, prior to a biological or membrane step, reducing leachate turbidity and counteracting the problem of membrane fouling [62]. It is also often used as a post-treatment technique for mature leachate, in order to reduce non-biodegradable organic matter [17].

Recently, Mariam et al. [87] compared conventional chemical coagulation with electro-coagulation, using aluminum electrodes, for the treatment of leachate. At the optimum reaction time, TOC and turbidity removals by the electro-coagulation process were 67% and 80%, respectively, while at the optimum dosage of $Al_2(SO_4)_3$, TOC and turbidity removals by the classic process were only 10% and 65%, respectively. Orkun and Kuleyin [94] as well as Bouhezila et al. [95] also reported promising results considering electro-coagulation. The latter authors, for instance, using aluminum electrodes, obtained removal efficiencies of 70% and 24% for COD and total nitrogen removal, respectively, after 30 minutes of leachate treatment with a current density of 500 A/m^2, an inter-electrode distance of 2.8 cm and a stirring speed of 150 rpm.

2.2.1.2. Chemical Precipitation

During chemical precipitation, dissolved ions in the solution are transferred to the insoluble solid phase via chemical reactions. Struvite (magnesium ammonium phosphate (MAP)) or lime are usually employed as the precipitant, depending on the target of the removal: $N\text{-}NH_4^+$ or heavy metals [88].

Li and Zhao [96] investigated the removal of ammonium ions from landfill leachate by applying struvite precipitation and observed that the $N\text{-}NH_4^+$ content could be reduced from 5618 $mg{\cdot}L^{-1}$ to 112 $mg{\cdot}L^{-1}$ in 15 min with a Mg^{2+}:NH_4^+:PO_4^{3-} molar ratio of 1:1:1 and a pH between 8.5 and 9.

Thereafter, a number of reports on the application of struvite precipitation as a leachate pre- or post-treatment have been published, exploring influence of factors such as pH, molar ratio of Mg^{2+}:NH_4^+:PO_4^{3-} or the initial concentration of the reagents (Table 12).

Drawbacks of struvite precipitation include the sensitivity of the process employed to pH, the low effectiveness for COD removal and the high dose of precipitant required associated to a high operative cost. In fact, leachates are characterized by low concentrations of magnesium and phosphorus and external sources of these compounds are demanded. Nevertheless, in the case of mature leachates, known by high ammonium concentrations and very low BOD_5/COD ratios, the process could potentially compete even with the cheapest existing method (i.e. biological removal) since in the latter an amount of carbon from an expensive external source is required [99].

Another concern associated to struvite precipitation is the generation of sludge and the need for further disposal. However, if the leachate does not contain any heavy metals, struvite can be used as a valuable fertilizer since it is a slow release source of nitrogen, magnesium and phosphorus [100].

2.2.1.3. Air Stripping

Air stripping has been widely used for removing high concentrations of $N\text{-}NH_4^+$ from landfill leachate. In this process, $N\text{-}NH_4^+$ is transferred from the leachate into the air and is then absorbed from the air into a strong acid such as HCl or H_2SO_4. The air stripping rate is expected to be somewhat lower with low strength leachate than with concentrated leachate [101].

The process is further dependent of pH and temperature. Raising the pH (above 7), $N\text{-}NH_4^+$ turns into nitrogen-ammonia ($N\text{-}NH_3$). Since $N\text{-}NH_3$ is more easily removed by air stripping, high pH values must be employed [65,92]. Additionally, as the solubility of nitrogen-ammonia increases at low ambient temperatures, high temperatures favor the air stripping efficiency [65].

A synthesis of recent works on treatment of landfill leachate using air stripping is presented in Table 13.

Some of the drawbacks associated to air stripping are foaming, which imposes the use of bigger stripping reactors and the generation of carbonate precipitate that may cause severe operation and maintenance problems [100].

Moreover, the high operational pH needs further neutralization, which increases leachate salinity and operational costs [66].

Table 11. Landfill leachate treatment by coagulation-flocculation

Landfill site and age, reference	Leachate feeding characteristics				Coagulation-flocculation operation					Performance
	pH	COD ($mg \cdot L^{-1}$)	BOD_5/COD	$N\text{-}NH_4^+$ ($mg \cdot L^{-1}$)	Scale	Reaction mode	pH	Coagulant type and dose	Flocculant type and dose	
Gaziantep (Turkey), young leachate, [91]	8	2431	0.21	1379	Lab-scale (1 L)	Rapid mixing-20 min, 100 rpm, flocculant addition; slow mixing-40 min, 40 rpm; settle-60 min	8	4000 mg $Al_2(SO_4)_3 \cdot L^{-1}$	8 mg anionic polyelectrolyte- SF-380$\cdot L^{-1}$	≈40, 12, 40, 55 and 3% for COD, Cu, Fe, Mn and Ni removal, respectively
	8	2431	0.21	1379	Lab-scale (1 L)	Rapid mixing-20 min, 100 rpm, flocculant addition; slow mixing-40 min, 40 rpm; settle-60 min	8	4000 mg $Al_2(SO_4)_3 \cdot L^{-1}$ + 16000 mg $Ca(OH)_2 \cdot L^{-1}$	8 mg anionic polyelectrolyte- SF-380$\cdot L^{-1}$	≈40, 85, 47, 95 and 7% for COD, Cu, Fe, Mn and Ni removal, respectively
Chongqing (China), mature leachate, [92]	7.9	3000-4500	0.09-0.22	1000-1750	Lab-scale (1 L)	Rapid mixing-2 min, 250 rpm, flocculant addition; slow mixing-15 min, 80 rpm; settle-60 min	4, 5,6	600 mg $Fe_2(SO_4)_3 \cdot L^{-1}$	1mL polyacrylamide (0.1% v/v) per liter of leachate	28, 32 and 24% COD removal for pH=4, 5 and 6, respectively
	7.9	3000-4500	0.09-0.22	1000-1750	Lab-scale (1 L)	Rapid mixing-2 min, 250 rpm, flocculant addition; slow mixing-15 min, 80 rpm; settle-60 min	5	600, 800, 1000 mg $Fe_2(SO_4)_3 \cdot L^{-1}$	1mL polyacrylamide (0.1% v/v) per liter of leachate	32, 36 and 30% COD removal for 600, 800 and 1000 mg $Fe_2(SO_4)_3 \cdot L^{-1}$, respectively
Mexico city (Mexico), mature leachate, [93]	8-8.5	4580	<0.01	-	Lab-scale (1 L)	Rapid mixing-30 sec, 100 rpm; slow mixing-30 min, <50 rpm; settle-30 min	6	3000 mg $Fe_2(SO_4)_3 \cdot L^{-1}$	-	65 and 97% for COD and color removal, respectively
	8-8.5	4580	<0.01	-	Lab-scale	Rapid mixing-30 sec, 100 rpm; slow mixing-10 min, <50 rpm; settle-30 m	6	3000 mg $Fe_2(SO_4)_3 \cdot L^{-1}$	-	58.2 and 97.1% for COD and color removal, respectively

Table 11. (Continued)

Landfill site and age, reference	Leachate feeding characteristics				Coagulation-flocculation operation					Performance
	pH	COD ($mg \cdot L^{-1}$)	BOD_5/ COD	$N\text{-}NH_4^+$ ($mg \cdot L^{-1}$)	Scale	Reaction mode	pH	Coagulant type and dose	Flocculant type and dose	
Asturias (Spain), young leachate, [62]	8.3	18352	0.6	2445	Lab-scale (1 L)	Rapid mixing-3 min, 180 rpm; slow mixing-17 min, 40 rpm; settle-30 min	3.8	1200 mg $FeCl_3 \cdot L^{-1}$	-	28, 78 and 90% for COD, color and turbidity removal, respectively
	8.3	18352	0.6	2445	Lab-scale (1 L)	Rapid mixing-3 min, 180 rpm; slow mixing-17 min, 40 rpm; settle-30 min	6	5000 mg $Al_2(SO_4)_3 \cdot L^{-1}$	-	27, 84 and 93% for COD, color and turbidity removal, respectively
	8.3	18352	0.6	2445	Lab-scale (1 L)	Rapid mixing-3 min, 180 rpm; slow mixing-17 min, 40 rpm; settle-30 min	6.5	4000 mg aluminum polychloride $\cdot L^{-1}$	-	35, 91 and 90% for COD, color and turbidity removal, respectively
Asturias (Spain), mature leachate, [62]	8.3	4814	0.14	2950	Lab-scale (1 L)	Rapid mixing-3 min, 180 rpm; slow mixing-17 min, 40 rpm; settle-30 min	5.2	1700 mg $FeCl_3 \cdot L^{-1}$	-	73, 98 and 100% for COD, color and turbidity removal, respectively
	8.3	4814	0.14	2950	Lab-scale (1 L)	Rapid mixing-3 min, 180 rpm; slow mixing-17 min, 40 rpm; settle-30 min	6	3200 mg $Al_2(SO_4)_3 \cdot L^{-1}$	-	60, 95 and 92% for COD, color and turbidity removal, respectively
	8.3	4814	0.14	2950	Lab-scale	Rapid mixing-3 min, 180 rpm; slow mixing-17 min, 40 rpm; settle-30 min	7	6000 mg aluminum polychloride $\cdot L^{-1}$	-	62, 97 and 98% for COD, color and turbidity removal, res

Table 12. Landfill leachate treatment by chemical precipitation

Landfill site and age, reference	Leachate feeding characteristics					Chemical precipitation operation			Performance
	pH	COD ($mg \cdot L^{-1}$)	BOD_5/COD	$N\text{-}NH_4^+$ ($mg \cdot L^{-1}$)	Scale	Reaction mode	pH	Precipitant type and molar ratio	
Gaziantep (Turkey), young leachate, [91]	8	37024	0.4	2430	Lab-scale (1 L)	Rapid mixing-20 min, 100 rpm; slow mixing-40 min, 40 rpm; settle-60 min	9.5-10	$Ca(OH)_2$	59, 67, 76 and 29% for Cu, Fe, Mn and Ni removal, respectively
Komurcuoda (Turkey), young leachate, biologically pre-treated, [97]	6.2-8.4	20700	0.6	2330	Lab-scale (0.5 L)	Rapid mixing of magnesium and phosphate-1 min; pH adjustment; settle-30 min	7.5	$Mg^{2+}:NH_4^+:PO_4^{3-} = 1:1:1$	98 and 20% for $N\text{-}NH_4^+$ and COD removal, respectively
Nanjing (China), mature leachate, [98]	8.1	7200	0.2	1960	Lab-scale	Rapid mixing of magnesium and phosphate-30 min; pH adjustment; settle-30 min	8.5, 9.5, 10.5	$Mg^{2+}:NH_4^+:PO_4^{3-} = 1:1:1$	45, 79 and 75% $N\text{-}NH_4^+$ removal for pH 8.5, 9.5 and 10.5, respectively
	8.1	7200	0.2	1960	Lab-scale	Rapid mixing of magnesium and phosphate-30 min; pH adjustment; settle-30 min	9.5	$Mg^{2+}:NH_4^+:PO_4^{3-} = 1:1:0.9$; $Mg^{2+}:NH_4^+:PO_4^{3-} = 1:1:1.1$	69 and 83% $N\text{-}NH_4^+$ removal for $Mg^{2+}:NH_4^+:PO_4^{3-}=1:1:0.9$; and $Mg^{2+}:NH_4^+:PO_4^{3-} = 1:1:1.1$, respectively
Apulia (Italy), mature leachate, [99]	8.5	9700	0.15	2600	Lab-scale (1 L)	Rapid mixing of magnesium and phosphate; pH adjustment; slow mixing-30 min; settle-60 min	9	$Mg^{2+}:NH_4^+:PO_4^{3-} = 1:1:1$; $Mg^{2+}:NH_4^+:PO_4^{3-} = 1.5:1:1$; $Mg^{2+}:NH_4^+:PO_4^{3-} = 2.5:1:1.1$	67, 87 and 98% $N\text{-}NH_4^+$ removal for $Mg^{2+}:NH_4^+:PO_4^{3-}=1:1:1$; $Mg^{2+}:NH_4^+:PO_4^{3-}=1.5:1:1$ and $Mg^{2+}:NH_4^+:PO_4^{3-}=2.5:1:1.1$
Istambul (Turkey), medium-age leachate, [63]	-	-	-	1167	Lab-scale (0.5 L)	Rapid mixing of magnesium and phosphate-3 min; pH adjustment; slow mixing-10 min; settle-30 min	9	$Mg^{2+}:NH_4^+:PO_4^{3-} = 1:1.2:1.2$	97, 26 and 23% for $N\text{-}NH_4^+$, COD and color removal, respectively

Table 13. Landfill leachate treatment by air stripping

Landfill site and age, reference	Leachate feeding characteristics				Scale	Air stripping operation					Performance
	pH	COD ($mg \cdot L^{-1}$)	BOD_5/COD	$N\text{-}NH_4^+$ ($mg \cdot L^{-1}$)		Air source	T (°C)	Chemical for pH adjustment, dose	pH	Time (h)	
Komurcuoda (Turkey), young leachate, [97]	6.2-8.4	20700	0.6	2330	Lab-scale	Air flow	15; 20	$Ca(OH)_2$, 13 $g \cdot L^{-1}$	11	12	87 and 15; 94 and 15% for $N\text{-}NH_4^+$ and COD removal, respectively, at 15 and 20 °C
Dyarbakir (Turkey), young leachate, [65]	6.5-7.5	8500-19200	0.4-0.7	1100-2150	Lab-scale	Air flow rate- 0 $L \cdot h^{-1}$, stirring- 400 rpm	rt	-	9, 10, 11, 12	6	15, 26, 28 and 30% $N\text{-}NH_4^+$ removal for pH 9, 10, 11 and 12, respectively
	6.5-7.5	8500-19200	0.4-0.7	1100-2150	Lab-scale	Air flow rate- 2 $L \cdot h^{-1}$, stirring- 400 rpm	rt	-	9, 10, 11, 12	6	48, 76, 91 and 93% $N\text{-}NH_4^+$ removal for pH 9, 10, 11 and 12, respectively
Asturias (Spain), young leachate, [62]	8.3	7624	-	1750	Lab-scale	Stirring	rt	$Ca(OH)_2$, 18.6 $g \cdot L^{-1}$	13	24	86 and 23% for $N\text{-}NH_4^+$ and COD removal, respectively
Asturias (Spain), young leachate, biologically pre-treated, [62]	6.9	3484		2156	Lab-scale	Stirring	rt	$Ca(OH)_2$, 10 $g \cdot L^{-1}$	12	24	94 and 46% for $N\text{-}NH_4^+$ and COD removal, respectively
Chongquing (China), mature leachate, [92]	7.9-8.5	3000-4500	0.09-0.22	1000-1750	Lab-scale (10 L)	Air flow rate- 900 $L \cdot h^{-1}$	rt	-	11	18	97% $N\text{-}NH_4^+$ removal

2.2.1.4. Adsorption

Adsorption is a mass transfer process by which a substance is transferred from the gas or liquid phase of a mixture to the surface of a solid adsorbent to which becomes bound via physical and/or chemical interactions [88,102]. Due to its inherent physical properties, large surface area, microporous structure, high adsorption capacity and surface reactivity, adsorption using granular activated carbon (GAC) in columns or powdered activated carbon (PAC) is a well-recognized means of leachate treatment [88,102,103]. In general, the application of activated carbon adsorption (GAC or PAC) is effective for the removal of color, heavy metals, biodegradable and non-biodegradable organic compounds from landfill leachate [1,62]. It is easy to operate and has low energy demand, however the frequent need to regenerate the column or the equivalent replacement of powdered activated carbon makes this option expensive [6,104]. Furthermore, activated carbon generally does not have sufficient adsorption capacity for the removal of $N\text{-}NH_4^+$ [103]. So, over the last few years, this method has been used mainly for polishing the mature leachate, for treatment of leachate from biological units or alongside with biological treatment as was previously mentioned. Accordingly, other materials, such as zeolite, coconut shell charcoal, bone meal, iron fines or municipal waste incinerator bottom ash have been chemically modified and tested as low-cost absorbents for landfill leachate treatment.

Table 14 presents some lab-scale experiments reporting landfill leachate treatment via adsorption.

Kurniawan and Lo [104] investigated the combination of granular activated carbon adsorption with H_2O_2 in a low biodegradable leachate. This system achieved substantially higher removal (82% COD, $N\text{-}NH_4^+$ 59%) than the GAC adsorption (58% COD) or H_2O_2 oxidation (33% COD, $N\text{-}NH_4^+$ 4.9%) systems at optimized experimental conditions. The addition of a Fe^{2+} dose at 1.8 $mg{\cdot}L^{-1}$ further improved the removal of refractory compounds by the integrated treatment from 82 to 89% COD removal.

2.2.1.5. Flotation and Dissolved Air Flotation

Flotation and especially dissolved air flotation (DAF), which rely on the collision of air bubbles with the suspended particles, offer advantages over precipitation, adsorption and membrane filtration processes in wastewater treatment. Advantages include better effluent quality, rapid start-up, high rate operation, and a thicker sludge [107]. However, very few studies have so far been devoted to the application of flotation for landfill leachate treatment.

Zouboulis et al. [108] employed flotation in a column, as a post-treatment stage for removing residual humic acids (recalcitrant compounds) from simulated landfill leachates. Under optimized conditions, the treatment performance was very efficient reaching almost 99%, indicating that flotation can serve as a possible alternative technology for the removal of humic acids. Recently, Palaniandy et al. [109] investigated the feasibility of treating a mature landfill leachate with DAF and proved that the coagulation process enhanced the removal of color, COD and turbidity. In fact, in the case of DAF without coagulation, low percent removals of 36%, 33%, and 32% were observed for color, COD, and turbidity, respectively. In turn, with coagulation followed by DAF, the removals were 70%, 79% and 42% for color, COD, and turbidity, respectively.

Table 14. Landfill leachate treatment by adsorption

Landfill site and age, reference	Leachate feeding characteristics				Adsorption operation					Performance
	pH	COD ($mg \cdot L^{-1}$)	BOD_5/COD	$N\text{-}NH_4^+$ ($mg \cdot L^{-1}$)	Scale	Operation mode	Adsorbent type and dose	Adsorbent characteristics	Time (h)	
Chania (Greece), mature leachate previously coagulated, [105]	4.6	500	-	-	Lab-scale	Continuous stirring-100 rpm	Coconut shell: -GAC, 1.25 g YAO M200 W20$\cdot L^{-1}$; -PAC, 1.25 g YAO 30x60$\cdot L^{-1}$	GAC: size-0.6 mm, BET surface area-1250 $m^2 \cdot g^{-1}$; PAC: size-0.15 mm, BET surface area-1250 $m^2 \cdot g^{-1}$	2	25 and 35% COD removal for GAC and PAC, respectively
Hong Kong (China), mature leachate, [104]	8.0	8000	0.08	2595	Lab-scale (0.25 L)	Continuous stirring-200 rpm	Coconut shell: GAC, 15 $g \cdot L^{-1}$	Particle diameter-0.5 mm; BET surface area-900-1100 $m^2 \cdot g^{-1}$	1	58% COD removal
Slovenia, mixed leachate, [106]	8.2	1396	0.01	388	Lab-scale	Column, 0.12 $L \cdot h^{-1}$	Zeolite: Clinoptiolite	-	1.67	100, 32, 48 and 50% for $N\text{-}NH_4^+$, COD, Fe and Mn removal, respectively
Asturias (Spain), young leachate, biologically pre-treated, [62]	6.9	3484	-	2156	Lab-scale	Continuous stirring	Coal: -GAC, 10 g Organosorb 10 ME$\cdot L^{-1}$; -GAC, 10g Filtracarb CC65/1240$\cdot L^{-1}$	-GAC: size-0.4-1.7 mm, BET surface area-1020 $m^2 \cdot g^{-1}$; -GAC: size-0.4-1.7 mm, BET surface area-1050 $m^2 \cdot g^{-1}$	5	48 and 31; 43 and 27% for COD and color removal, for GAC Organosorb and GAC Filtracarb, respectively
	6.9	3484	-	2156	Lab-scale	Continuous stirring	Coal: -GAC, 20 g Organosorb 10 MB L^{-1}	-GAC: size-0.4-1.7 mm, BET surface area-1020 $m^2 \cdot g^{-1}$	5	63 and 45% for COD and color removal, respectively

2.2.2. Pressure-Driven Membrane Technologies

Pressure-driven membrane technologies (microfiltration, ultrafiltration, nanofiltration, reverse osmosis) can be used to remove a wide range of components, ranging from suspended solids (microfiltration) to small organic compounds and ions (reverse osmosis). In these processes, a pressure exerted on the solution at one side of the membrane serves as a driving force to separate the solution into a permeate and a retentate. The permeate is the treated effluent, whereas the retentate is a concentrated solution that must be disposed of or treated by other methods [110].

Microfiltration, ultrafiltration, nanofiltration and reverse osmosis have been found to be highly effective in the treatment of landfill leachate [111–113]. The selection of the appropriate membrane for leachate treatment depends on a number of factors such as the nature and concentration of materials present in the leachate and pH.

2.2.2.1. Microfiltration and Ultrafiltration

Microfiltration (MF) and ultrafiltration (UF) membranes have a porous structure that retains components by a sieving mechanism. In microfiltration, the pore size of the membrane is provided by the manufacturers and serves as a reference for the size of the retained particles. For ultrafiltration, the concept of molecular weight cut-off of the membrane is often used. Rejection increases with the reduction of membrane cut-off [110].

MF and UF membranes are mainly used as pre-treatment stage for another membrane process (nanofiltration or reverse osmosis) or in combination with chemical processes. The former are appropriate to remove suspended solids, colloids and bacteria while the latter are effective in eliminating large dissolved molecules. Landfill leachate purification by MF and UF membranes is shown in Table 15.

Recently, several hybrid systems including MF and UF membranes have been studied. For instance, Ince et al. [114] applied a microfiltration-powdered activated carbon unit to a biologically treated landfill leachate and when 8 g $PAC{\cdot}L^{-1}$ was used a COD removal above 50% was achieved. This result was better than the achieved with a nanofiltration membrane. Likewise, Pirbazari et al. [67] used an ultrafiltration-powdered activated carbon system and observed TOC removals of 95-97% for a previously coagulated leachate.

UF membranes have further been successfully employed in membrane bioreactors as detailed in Table 7.

2.2.2.2. Nanofiltration

Nanofiltration (NF) membranes have smaller pores than MF and UF membranes, which make them suitable for the removal of color, biodegradable and non-biodegradable organic micropollutants with a molecular weight above 200 Da. Furthermore, NF membranes also have a surface charge allowing the retention of ionic species and even the retention of ions with a size below the pore size of the membrane.

Due to its unique properties, NF is able to remove recalcitrant organic compounds and heavy metals, being considered one of the best techniques for landfill leachate treatment [110]. Nanofiltration membranes, compared to reverse osmosis membranes, have many advantages such as, lower operational pressure, high flux, high rejection of polyvalent ions, relatively low investment, operational and maintenance costs [114]. All these advantages increased the use of nanofiltration technologies in the last years (Table 16).

2.2.2.3. Reverse Osmosis

In reverse osmosis (RO), permeation is slower and rejection is not a result of sieving but of a solution-diffusion mechanism. The low permeability of reverse osmosis membranes requires high pressures and, consequently, relatively high-energy consumption [110].

RO membranes have been applied to remove the organic and non-organic fraction of young and mature landfill leachates with very good performances, both at lab [115,116] and full scales [111–113] (Table 16).

Among the reverse osmosis membrane modules applied for leachate treatment, the commonly used modules are the tubular membrane modules [111] and the disc tube module [113]. Recently, a treatment system equipped with the newly developed open channel spiral wound modules for leachate purification was studied [116].

In spite of being promising technologies in the treatment of leachate, pressure-driven membrane processes, particularly reverse osmosis, have obvious drawbacks, which remain unsolved: extensive application is inhibited due to the expensive facilities and European patents, membrane fouling frequently reduces membrane lifetime and decreases productivity, and the large quantity of unusable residual concentrate generated needs further treatment [47].

2.2.3. Chemical Oxidation

Chemical oxidation has being considered for its potential in converting harmful organics into innocuous substances without the production of concentrated residues, which require further disposal, often as hazardous waste. Chemical oxidation of landfill leachate based on chlorine (Cl_2), potassium permanganate ($KMnO_4$) or calcium hydrochlorine ($Ca(ClO)_2$) resulted in COD removals between 20 and 50% [1,6].

Due to its ability to eliminate color, reduce the organic load and improve the biodegradability of recalcitrant contaminants, advanced oxidation processes (AOPs) have increasingly been studied and applied in the treatment of mature leachate [4]. The main purpose of AOPs is to enhance chemical oxidation efficiency by increasing the production of highly free radicals, mainly the hydroxyl radical ($^{\bullet}OH$), a non-selective, very reactive oxidant. These radicals can be produced in reactions without irradiation involving the presence of strong oxidants such as O_3, H_2O_2 and/or a catalyst, or with irradiation employing the energy of ultraviolet's (UV), ultrasounds (US) or electron beams (EB) associated or not to strong oxidants and/or a catalyst. AOPs are affected by many factors such as the amount of constituents in the solution, concentration of oxidants, reaction time, pH, reaction by-products and temperature.

2.2.3.1. AOPs without Irradiation

Among AOPs, the Fenton process, which is based on the electron transfer between hydrogen peroxide (H_2O_2) and ferrous ion (Fe^{2+}) in acidic medium, seems to be the best compromise because the process is technologically simple, there is no mass transfer limitation, and both hydrogen peroxide and iron are cheap and non-toxic [92]. Therefore, the Fenton process has been widely applied on the remediation of young and mature landfill leachate directly, as post- or pre-treatment [121]. For instance, Di Iaconi et al. [38] reported that, when mature leachate was pre-treated by a biological process, the Fenton process achieved an additional COD removal of 85%, while Kochany and Lipczynska-Kochany [27] applied the Fenton process as pre-treatment and observed 66% of SCOD removal and an increase in the leachate BOD_5/COD ratio from 0.63 to 0.88.

Table 15. Landfill leachate treatment using MF or UF

Landfill site and age, reference	Leachate feeding characteristics				Membrane filtration operation						Performance
	pH	COD ($mg \cdot L^{-1}$)	BOD_5/COD	$N\text{-}NH_4^+$ ($mg \cdot L^{-1}$)	Scale, kind of process	Type of membrane, model and module type	Pore size (μm), surface area (m^2), cut-off (kDa)	Velocity ($m \cdot s^{-1}$), flux ($L \cdot m^{-2}$)	Operated pressure (bar)	T (°C)	
Lubna (Poland), mature leachate, [117]	7.3	2300	0.3	12.6	Pilot-scale, MF	Polymeric, Accurel PP, S6/2, tubular	0.2-0.6, 0.11, -	4.1-4.3, 35-40	-	20	26, 30 and 86% for COD, BOD_5 and TSS removal, respectively
Wuhan (China), mature leachate, [118]	6.3	20015	0.16	369	Pilot-scale, UF	Polymeric, YM3, tubular	-, -, 3	-	6	-	28 and 83% for $N\text{-}NH_4^+$ and COD removal, respectively
Lipówka (Poland), mature leachate, biologically pre-treated, [115]	8.6	1660	-	-	Pilot-scale, UF	Polymeric, PSf-15, tubular	-, 0.025, 300	2.5, -	3	25	49 and 100% for COD and SS removal, respectively
France, mature leachate, [119]	8.6	1339	0.06	364	Pilot-scale, UF	Ceramic, KERASEP, tubular	-, -, 15 and 5	-	-	20	47 and 63% COD removal for 15 and 5 kDa, respectively

Table 16. Landfill leachate treatment using NF and RO

Landfill site and age, reference	Leachate feeding characteristics				Membrane filtration operation						Performance
	pH	COD ($mg \cdot L^{-1}$)	BOD_5/COD	$N-NH_4^+$ ($mg \cdot L^{-1}$)	Scale, kind of process	Type of membrane, model and module type	Surface area (cm^2), cut-off (Da)	Velocity ($m \cdot s^{-1}$), flux ($L \cdot h^{-1}$)	Operated pressure (bar)	T (°C)	
Istanbul (Turkey), mature leachate biologically pre-treated, [114]	9.3	2070	-	2180	Lab-scale, NF	Polymeric, FM NP030, -	80, 400	1.1, -	20	25	37 and 41% for $N-NH_4^+$ and COD removal, respectively
Wollongong (Australia), mature leachate, [87]	7.3	635.8 (TOC)	-	918.3 (TN)	Lab-scale, NF	Polymeric, NF 270, -	21.2, 300	-, -	5	-	10 and 90% for TN and TOC removal, respectively
Gujarat (India), mature leachate, [120]	5	56521	0.33	196	Pilot-scale, NF	Polymeric, NF-300, flat sheet	150, 300	-, 900	20	-	52, 94, 67, 86, 84, 98, 95, and 93% for $N-NH_4^+$, COD, BOD_5, TDS, color, Cr^{3+}, Ni^{2+} and Cu^{2+} removal, respectively
Lipówka (Poland), mature leachate physically and biologically pre-treated, [115]	6.5	846	-	-	Lab-scale, RO	Polymeric, OC-SS, flat sheet	155, -	2.5, -	27.6	25	93% COD removal
Hanover (Germany), -, [116]	6-6.5	3100	-	1000	Full-scale, RO	Polymeric, FT-30, open channel spiral wound	25.6 m^2, -	-, 8000	20-40	-	99; 100; 100; 99; 100 and 100% for $N-NH_4^+$, COD, TSS, Cl^-, Fe^{2+} and Cu^{2+} removal, respectively

Regardless its high COD removal efficiency and environmental friendliness, Fenton treatment of wastewater results in the production of ferric hydroxide sludge, which requires an additional separation process and disposal. In order to deal with this problem, it has been suggested that the conventional Fenton process could be modified by the combined application of electricity, i.e. electro-Fenton [122–124]. The electro-Fenton process, in which either or both of H_2O_2 and Fe^{2+}can be generated electrochemically *in situ*, is an indirect electrochemical oxidation that employs $^{\bullet}OH$ produced by the Fenton reaction to oxidize organic compounds [121]. Since more $^{\bullet}OH$ radicals are produced, the oxidation of the organics is enhanced [124]. In a 43 min electro-Fenton study, Mohajeri et al. [124] achieved 94% of COD and 96% of color removal of a mature leachate at pH 3 and H_2O_2/Fe^{2+} molar ratio 1, while current density was 49 $mA{\cdot}cm^{-2}$.

In spite of its advantages, conventional Fenton and Fenton-based processes require low pH and a modification of this parameter is necessary for further treatment or disposal, and deal with H_2O_2, which is very aggressive and can cause corrosion. Furthermore few ammonium is oxidized [121].

Molecular ozone is a strong oxidant having high reactivity and selectivity towards organic pollutants such as humic substances [20]. Wang et al. [125] achieved removal efficiencies of 70, 90 and 67% for COD, color and ammonium, respectively, and an increase in the BOD_5/COD ratio from 0.04 to 0.22, applying ozone directly to a mature leachate, at the natural pH of the leachate (8.3) and consuming 12.5 g $O_3{\cdot}L^{-1}$. Occasionally, as a single process, ozonation has not been very effective. Due to the complexity of the leachate, high doses are required and the reaction takes a long time [93]. Therefore, in the last years, ozonation has increasingly been studied and used for leachate remediation as a pre-treatment or as a polishing step. The use of ozone at high pH (O_3/OH^-) or the combination with H_2O_2 (O_3/H_2O_2) have also become attractive options to oxidize the complex leachate mixtures since they favor the production of the hydroxyl radical, a oxidant even stronger than that of ozone [126]. For instance, Hagman et al. [30] reported that when the same ozone dose was applied, ozone alone ensured a COD removal of about 22%, while the combination of O_3 and H_2O_2 (1 $g{\cdot}L^{-1}$) increased COD removal up to 50%.

The main limitation of using ozone or ozone-based advanced processes lies in the high-energy consumption.

Electrochemical oxidation has been used for landfill leachate treatment over the past 10 years and, under appropriate conditions, has been proved to remove almost all ammonium, most COD and color. During the electro-oxidation of landfill leachate, reduction of pollutants appears to be primarily due to indirect oxidation, utilizing chlorine or hypochlorite, metals mediators such as Ag^{2+}, hydrogen peroxide or ozone, formed by anodic oxidation and originally existing or added in the leachate. Direct anodic oxidation may also occur to some extent, destroying pollutants adsorbed on the anode surface [127]. In order to increase the efficiency of the method and optimize energy consumption, the influence of pre-treatment methods, pH, current density, leachate composition, electrolytes added and anode materials has been studied.

With mature leachate Moraes and Bertazzoli [128] found removal efficiencies of 73, 86 and 49% for COD, color and nitrogen-ammonium, respectively, by employing a electrochemical reactor using TiO_2 anode and Ti cathode at 1160 $A{\cdot}m^{-2}$ with a flow rate of 2000 $L{\cdot}h^{-1}$ during 3 h while Cabeza et al. [129] used boron doped diamond as anode and stainless steel as cathode at 900 $A{\cdot}m^{-2}$ with a flow rate of 660 $L{\cdot}h^{-1}$ and after 6 h got 100%

both of COD and nitrogen-ammonium removal, respectively. These authors also reported that half of the initial ammonium nitrogen was oxidized to $N\text{-}NO_3^-$.

Wide application of electro-oxidation in landfill leachate treatment is limited by its high-energy consumption and potential production of chlorinated organics. Especially because of its expensive operating costs compared with other available technologies, electro-oxidation will be favored as a finishing step in a combined process, as reported by Zhao et al. [130] and Zhang et al. [50], or an auxiliary unit in emergency situations, instead of a full treatment for landfill leachate.

Wet air oxidation is defined as an oxidation technology carried out in the liquid phase under moderately elevated temperature and pressure. The oxidizing agent is oxygen usually added as pressurized air or pure oxygen. The presence of a catalyst or oxidation promoters has resulted in higher oxidation efficiencies.

Few experimental data are available on wet air oxidation of leachate [131]. Rivas et al. [132] applied wet air oxidation using HSO_5^- as promoter to a mature leachate and at pH 8.8, T=250 °C, P=51 bar and after 2.5 h of reaction verified 60% of COD depletion. This result was better than the one achieved with the unpromoted wet air oxidation system or even using hydrogen peroxide as promoter.

Although wet air oxidation offers some advantages such as a small plant for operations and its ability to deal with varying flow rates and composition of the effluent, this process is not cost-effective for leachate treatment with a COD concentration of less than 5000 $mg \cdot L^{-1}$ [4]. Table 17 lists the performance of some AOPs in the treatment of landfill leachate.

2.2.3.2. AOPs with Irradiation

Photochemical technology such as UV irradiation has been applied for leachate treatment, mainly in combination with other strong oxidants.

Wu et al. [133] observed that among O_3, UV/H_2O_2 and UV/O_3 processes, the latter was found to be the most effective in enhancing the biodegradability and eliminating the color of a mature leachate.

Shu et al. [134] applied a thin gap annular UV/H_2O_2 photo reactor for the treatment of landfill leachate. At the maximum UV dosage and 0.233 mol $H_2O_2 \cdot L^{-1}$ the authors achieved 72 and 65% of color and COD removal efficiencies, respectively, in 300 min.

Concerning the $UV/H_2O_2/O_3$ process, Qureshi et al. [135] reported that this combination slightly favored the biodegradability, the removal of organic compounds and color in comparison with the UV/O_3 and UV/H_2O_2 systems, as can be seen in Table 18.

The combination of UV irradiation with H_2O_2 and Fe^{2+} ($UV/H_2O_2/Fe^{2+}$), the so-called photo-Fenton process, is a promising AOP for the treatment of landfill leachate, as UV light may promote photo decarboxylation of ferric carboxylates and reduce ferric to ferrous iron yielding additional hydroxyl radicals by photolysis. Furthermore, the amount of catalytic iron required, and consequently the volume of sludge produced could be strongly reduced in comparison to the conventional Fenton. Due to the fact that this process can be driven by low energy photons, it can also be achieved using solar irradiation [15,136].

Table 17. Landfill leachate treatment using AOPs without irradiation

Landfill site and age, reference	Leachate feeding characteristics				AOP without irradiation operation					Performance
	pH	COD ($mg \cdot L^{-1}$)	BOD_5/COD	$N\text{-}NH_4^+$ ($mg \cdot L^{-1}$)	Scale, kind of process	Oxidants concentration, reaction details	pH	Time (h)	T (ºC)	
Italy, mature leachate, [137]	8.2	10540	0.2	5210	Lab-scale, Fenton	H_2O_2=97 $mmol \cdot L^{-1}$, Fe_2^+=4.924 $mmol \cdot L^{-1}$, molar H_2O_2/Fe_2^+=19.7	3	2	rt	25% COD removal, BOD_5/COD=0.5
Jiangmen (China), mature leachate biologically pre-treated, [138]	5-8	600-700	0.01	<3	Lab-scale, Fenton	H_2O_2=7.2 $mmol \cdot L^{-1}$, Fe_2^+=6 $mmol \cdot L^{-1}$, molar H_2O_2/Fe_2^+=1.2	5	3	rt	47% COD removal
Madrid (Spain), young leachate, [15]	8.3	6119	0.02	1965	Lab-scale, Fenton	H_2O_2=0.15 $mol \cdot L^{-1}$, Fe_2^+=0.1 $mol \cdot L^{-1}$, molar H_2O_2/Fe_2^+=1.5	2.5	1	25	80% COD removal
Wuhan (China), young leachate, [122]	6.4	5000	-	650	Lab-scale, electro-Fenton	H_2O_2=0.34 $mol \cdot L^{-1}$, Fe_2^+=0.028 $mol \cdot L^{-1}$, molar H_2O_2/Fe_2^+=12, Ti/RuO_2-IrO_2 anode and cathode plates, electrode distance=2.1 cm, I=3 A	3	1.25	rt	81% COD removal
Sivas (Turkey), young leachate, [123]	8.4	2350	-	310	Lab-scale, electro-Fenton	H_2O_2=58.8 $mmol \cdot L^{-1}$, cast-iron anode and cathode plates, electrode distance=1.3 cm, I=3 A	3	0.33	rt	28, 72, 90 and 87% for $N\text{-}NH_4^+$, COD, color and $P\text{-}PO_4^{3-}$ removal, respectively
Rio de Janeiro (Brazil), mature leachate physico-chemically pre-treated, [139]	4.5	3945	0.05	-	Pilot-scale, O_3	O_3 consumed=7 $g \cdot L^{-1}$ leachate	4.5	-	-	54% COD removal, BOD_5/COD=0.35

Table 17. (Continued)

Landfill site and age, reference	Leachate feeding characteristics				AOP without irradiation operation					Performance
	pH	COD ($mg \cdot L^{-1}$)	BOD_5/COD	$N\text{-}NH_4^+$ ($mg \cdot L^{-1}$)	Scale, kind of process	Oxidants concentration, reaction details	pH	Time (h)	T (°C)	
Estonia, young leachate, [140]	8.1	14000	0.44	-	Lab-scale, O_3	O_3 dose=33.6 $g \cdot L^{-1}$ leachate, O_3 consumed=14 $g \cdot L^{-1}$ leachate	8.1	4	20	29% COD removal BOD_5/COD=0.41
	8.1	14000	0.44	-	Lab-scale, O_3/OH^-	O_3 dose=33.6 $g \cdot L^{-1}$ leachate, O_3 consumed=18 $g \cdot L^{-1}$ leachate	11	4	20	41% COD removal, BOD_5/COD=0.5
Tunis (Tunisia), mature leachate, [126]	8.7	5230	0.01	-	Lab-scale, O_3	O_3 dose=4.3 $g \cdot L^{-1}$ leachate, O_3 consumed=3.5 $g \cdot g^{-1}$ COD	8.7	0.67	20	19% COD removal, BOD_5/COD=0.19
					Lab-scale, O_3/H_2O_2	H_2O_2=58.8 $mmol \cdot L^{-1}$, O_3 dose=4.3 $g \cdot L^{-1}$ leachate, O_3 consumed=1.5 $g \cdot g^{-1}$ COD				48% COD removal, BOD_5/COD=0.64
Portugal, mature leachate, [141]	3.5	743	0.01	714	Lab-scale, O_3/OH^-	O_3 consumed=4.89 $g \cdot L^{-1}$ leachate	7	1	20	13, 30% for $N\text{-}NH_4^+$ and COD removal, resp; BOD_5/COD=0.08
					Lab-scale, $O_3/OH^-/H_2O_2$	H_2O_2=17.6 $mmol \cdot L^{-1}$, O_3 consumed=5.40 $g \cdot L^{-1}$ leachate	7	1	20	13, 30% for $N\text{-}NH_4^+$ and COD removal, resp; BOD_5/COD=0.16
Penang (Malaysia), mature leachate, [142]	8.9	1414	0.04	-	Lab-scale, electro-oxidation	Na_2SO_4=7 $mmol \cdot L^{-1}$, graphite anode and cathode plates, electrode distance=1.5 cm, I=799 $A \cdot m^{-2}$	8.9	4	rt	68, 70 and 84% for COD, BOD_5 and color removal, respectively
Italy, mature leachate, [131]	5-7	2000	-	-	Lab-scale, wet air oxidation	O_2 from pressurized air, P=35 bar, H_2O_2=0.88 $mol \cdot L^{-1}$	5-7	2	227	55% COD removal

Table 18. Landfill leachate treatment using AOPs with irradiation

Landfill site and age, reference	Leachate feeding characteristics					AOP with irradiation operation				Performance
	pH	COD ($mg \cdot L^{-1}$)	BOD_5/COD	$N\text{-}NH_4^+$ ($mg \cdot L^{-1}$)	Scale, kind of process	Oxidants concentration, reaction details	pH	Time (h)	T (°C)	
Taegu (Korea), mature leachate, [135]	7.8	25950	0.11	1854	Lab-scale, UV/O_3	O_3 dose=70.3 $mg \cdot min^{-1}$, O_3 consumed=28.9 $mg \cdot min^{-1}$, 20 cm synthetic quartz tube UV lamp, I=0.69 $W \cdot cm^{-2}$, 1500 W	7.8	5	20	63 and 88% for TOC and color removal, respectively, BOD_5/COD=0.35
	7.8	25950	0.11	1854	Lab-scale, UV/H_2O_2	H_2O_2=0.2 * COD, 20 cm synthetic quartz tube UV lamp, I=0.69 $W \cdot cm^{-2}$, 1500 W	3	5	20	79 and 96% for TOC and color removal, respectively, BOD_5/COD=0.37
	7.8	25950	0.11	1854	Lab-scale, $UV/H_2O_2/O_3$	H_2O_2=0.2 * COD, O_3 dose=50.3 $mg \cdot min^{-1}$, O_3 consumed=28.1 $mg \cdot min^{-1}$, 20 cm synthetic quartz tube UV lamp, I=0.69 $W \cdot cm^{-2}$, 1500 W	7.8	5	20	88 and 92% for TOC and color removal, respectively, BOD_5/COD=0.39
Curitiba (Brazil), mature leachate, [13]	8.4	5200	0.13	-	Lab-scale, UV/H_2O_2	H_2O_2=0.1 $mol \cdot L^{-1}$, quartz bulb UV lamp, I=3.2*10^{-3} $Einstein \cdot L^{-1} \cdot s^{-1}$, 125 W	8.4	1	-	56 and 97% for COD and TOC removal, respectively, BOD_5/COD=0.4
	8.4	5200	0.13	-	Lab-scale, $UV/H_2O_2/Fe^{2+}$	H_2O_2=0.06 $mol \cdot L^{-1}$, Fe^{2+}=0.18 $mmol \cdot L^{-1}$, quartz bulb UV lamp, I=3.2*10^{-3} $Einstein \cdot L^{-1} \cdot s^{-1}$, 125 W	2.8	1	-	58 and 90% for COD and TOC removal, respectively, BOD_5/COD=0.4
Cantabria (Spain), mature leachate, [136]	7.9	3824	0.18	2017	Lab-scale, $UV/H_2O_2/Fe^{2+}$	H_2O_2=0.44 $mol \cdot L^{-1}$, Fe^{2+}=55.6 $mmol \cdot L^{-1}$, TQ 150 UV lamp, I=8.8*10^{-5} $Einstein \cdot s^{-1}$, 150 W	3-3.5	1	-	78% COD removal
Beijing (China), mature leachate biologically pre-treated, [143]	7.8	560	0.07	700	Lab-scale, UV electrochemical oxidation	Ti/$TiRuO_2$ anode and stainless steel cathode, glass quartz tube UV lamp, I=671 $A \cdot m^{-2}$, 20 W	7.8	2.5	rt	94.5, 74.1 and 41.6% for $N\text{-}NH_4^+$, COD and TOC removal, respectively, BOD_5/COD=0.6

Table 19. Landfill leachate treatment using combined systems

Landfill site and age, reference	Combined system	Scale	Leachate feeding characteristics			Performance
			COD (mg·L^{-1})	BOD_5/COD	$N\text{-}NH_4^+$ (mg·L^{-1})	
Chen-Shi-Li (Taiwan), mature leachate, [133]	Coagulation-flocculation + ozonation + UV irradiation	Lab-scale	6500	0.06	5500	85% COD removal, BOD_5/COD=0.5
Apulia (Italy), mature leachate, [38]	Struvite + aerobic sequencing batch granular biofilter + Fenton	Lab-scale	24400	-	3190	99 and 97% for $N\text{-}NH_4^+$ and COD removal, respectively
Estonia, young leachate, [140]	Coagulation-flocculation + Fenton	Lab-scale	14000	0.44	-	70% COD removal, BOD_5/COD=0.65
Hong Kong, mature leachate, [147]	Ozonation + adsorption	Lab-scale	8000	0.09	2620	92 and 86% for $N\text{-}NH_4^+$, and COD removal, respectively, BOD_5/COD=0.47
Wollongong (Australia), -, [87]	Coagulation-flocculation + NF	Lab-scale	636 (TOC)	-	-	92% TOC removal
Diyarbakir (Turkey), young leachate, [65]	Coagulation-flocculation + air stripping + aerobic/anoxic MBR + RO	Lab-scale	8500-19200	>0.3	1100-2150	100% COD removal
Chongqing (China), mature leachate, [92]	Air stripping + Fenton + SBR + coagulation-flocculation	Lab-scale	4150	0.18	1169	98 and 93% for $N\text{-}NH_4^+$ and COD removal, respectively, BOD_5/COD=0.41
Shiraz (Iran), young leachate, [56]	AD+ AD + AS	Lab-scale	55351	0.81	1460	49-65 and 94% for $N\text{-}NH_4^+$ and COD removal, respectively
Kolenfeld (Germany), mature leachate, [116]	AS + coagulation-flocculation + RO	Full-scale	3100	-	1000	99 and 100% for $N\text{-}NH_4^+$ and COD removal, respectively
Jiangmen (China), mature leachate, [148]	SBR + coagulation-flocculation + Fenton + upflow biological aerated filter	Full-scale	6722	0.1	850	99 and 99% for $N\text{-}NH_4^+$, and COD removal, respectively

In a recent study Hermosilla et al. [15] found that the photo-Fenton treatment of landfill leachate yielded the same COD removal result than a conventional Fenton treatment, but consumed 32 times less iron and produced 25 times less sludge volume. An important drawback of the $UV/H_2O_2/Fe^{2+}$ process is the need to operate under acidic conditions.

Photo electrochemical oxidation has been proven to be a better method for the treatment of complex matrixes, in comparison with individual electrolysis and photocatalysis methods. Tauchert et al. [144] reported that, when applied to untreated mature leachate, the photo electrochemical system was significantly hindered on account of the characteristic dark coloration of the samples. In these conditions, the degradation process was essentially electrochemical permitting typical color and COD removal of about 50% and 20%, respectively. When a previous chemical precipitation process was applied aiming at the elimination of colored species (mainly humic substances), the decoloration and COD removal increased to 90% and 60%, respectively.

There has been little investigation into photochemical technologies using UV irradiation in spite of it being highly effective, and its application to leachate treatment is scarce due to its high operational costs. Therefore, the determination of optimum operational conditions is one of the primary concerns for the future development and potential application of these processes.

Ultrasonication is considered as a new possibility in wastewater treatment for several decades. Sonochemical decomposition of pollutants results from the rapid formation, growth, and violent collapse of cavitation bubbles or the reduction and oxidation due to the generation of $^{\bullet}H$ and $^{\bullet}OH$ radicals [145]. There are few reports about the sonochemical treatment of heavily polluted wastewater like landfill leachate. Recently, after 180 minutes of ultrasound irradiation in a mature leachate, with a power input of 150 W and at pH 11, Wang et al. [146] obtained up to 96% of nitrogen-ammonium reduction. Neczaj et al. [145] used effectively ultrasonication on leachate degradation as pre-treatment of a biological process.

Bae et al. [43] applied electron-beam irradiation to a biologically treated landfill leachate and found that this process was very efficient principally in removing humic substances.

Among the AOPs reviewed, ozone-based and Fenton oxidation processes are the most frequently studied and widely applied methods for the treatment of landfill leachate. A limitation of the AOPs is that they are affected by some chemicals, such as carbonates/hydrocarbonates, phosphate and chloride ions, or aliphatic alkyl compounds, since these compounds can interrupt the chain reaction of the generation of hydroxyl radicals [93]. Another common drawback of AOPs is the high operational cost due to the high-energy consumption of devices such as ozonizers, UV lamps and ultrasounds and/or due to high oxidant doses required for complete degradation (mineralization). Although being expensive to operate, AOPs application is inevitable for the treatment of recalcitrant organic pollutants. Therefore, AOPs have been mainly applied for landfill leachate as post- or pre-treatment together with a physico-chemical and/or a biological technique. A combination of two AOPs in an integrated system may also enhance the formation of $^{\bullet}OH$ radicals, improving the removal efficiency. For instance, in the study of single AOPs such as O_3, O_3/H_2O_2, Fenton and several combined treatment schemes for landfill leachate, Goi et al. [140] found that the best results in COD removal were achieved by the application of combined Fenton and ozonation processes.

2.3. Combined Treatment Systems

Although many different treatments can be applied, not one of the individual biological or physico-chemical techniques described in the previous sections is universally applicable or highly effective for landfill leachate treatment. In general, to set up satisfactory removal of pollutants from the leachate, a combination of several treatment systems is applied. The integrated systems ameliorate the drawbacks of individual processes contributing to a higher efficacy of the overall treatment. It must be noted, however, that the combination of treatment processes must be adjusted to each specific landfill leachate. Table 19 shows the performance of different combined systems on landfill leachate treatment. Though the conclusions should be carefully balanced due to different testing conditions (pH, temperature, strength of leachate, seasonal climate, and hydrology site), this comparison is useful to evaluate the overall treatment performance of each technique to assist the decision-making process.

As seem from Table 19, N-NH_4^+ removal was in the range 49–99% and COD removal was in the range 70–100%. Among the combined treatments reviewed above, it is observed that the combination of activated sludge, coagulation-flocculation and reverse osmosis demonstrated outstanding treatment performances in the removal of N-NH_4^+ (99%) and COD (100%).

3. Landfill Leachate Treatment Costs

The adoption of the optimal treatment process in the industrial environment ultimately depends on favorable process economics. Estimating a reliable treatment cost for landfill leachate is difficult due to many cost components such as collection system, pumping equipment and treatment facility. In addition, changes in the quality and quantity of leachate due to seasonal variations also contribute to the inconsistency of treatment costs data. For this reason, information on the treatment cost of landfill leachate is rarely reported.

From the reviewed literature, only Li et al. [47] estimated the operating treatment cost of leachate using a combined system. According to the economical analysis performed, using combined SBR, coagulation-flocculation, Fenton and upflow biological aerated filter processes would cost US$ 2.7 per m^3 of the treated effluent. The cost covered the reagents required and energy consumption. The same authors stated that the cost of advanced treatment employing a membrane is always up to $\$5\text{-}7 \cdot m^{-3}$.

Altinbaş et al. [149] reported that, depending on the type of precipitation and chemicals employed, the treatment cost of struvite precipitation varies between US$ 2 and $\$4 \cdot m^{-3}$. Tizaoui et al. [126] achieved operating costs of US$ 3.1 and 2.3 per kg of COD removed from leachate using O_3 and O_3/H_2O_2 systems, respectively, while Rivas et al. [150] obtained an operating cost of US$ 0.008 per m^3 of the treated leachate and $mg \cdot L^{-1}$ of COD removed using Fenton oxidation. Bouhezila et al. [95] estimated operating costs of $\$0.54 \cdot m^{-3}$ when aluminum electrodes were used in the electro-coagulation process.

It is important to note that, in the case of advanced oxidation processes, fine-tuning of the operating conditions could change radically the operating costs. Since the biological methods are more financially profitable and more environmental friendly, they should be preferred whenever possible.

Although it is unknown the cost component of leachate treatment in the overall cost of managing a landfill, it is thought that treatment of leachate from MSW will absorb more and more, a significant share of the total cost of MSW treatment.

4. LANDFILL LEACHATE TREATMENT IN PORTUGAL

According to the Portuguese Environmental Agency (APA) in 2006, 459 kg per capita of municipal waste were generated in Portugal [151]. Of this, 66% was landfilled, 18% was incinerated, 10% recycled and only 6% was organically recovered.

In 2010 there were 24 MSW management systems in Portugal, of which 12 were multimunicipal systems and 12 were intermunicipal systems (Figure 2). These systems exploited 34 MSW landfills in operation and 15 already closed, which give rise to considerable quantities of landfill leachate.

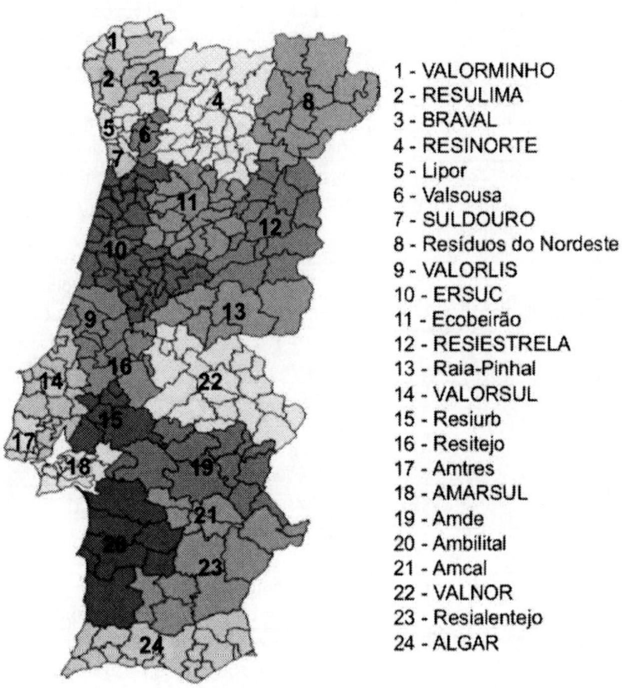

Figure 2. Distribution of waste management systems in Portugal: uppercase letters represent multimunicipal management systems; lowercase letters represent intermunicipal management systems (source: [152]).

Most of the Portuguese MSW landfills began its operation in the early 2000s. The first leachate treatment plants in Portugal were designed similar to the domestic wastewater treatment plants, using the same parameters of scaling and assuming the same degree of efficiency. The analytical control of the treated effluent of these plants during the first years revealed much lower efficiency levels than expected mainly due to an incorrect water balance and the consequent inability to fit the flow, the high pollutant load of landfill leachate and/or its low biodegradability compared with domestic municipal wastewater [153]. Therefore, most of these structures have already changed or improved the sequence of treatment initially deployed. According to the Environment Portuguese Agency and from a data survey carried out by our group, in the first half of 2008, of the 34 municipal solid waste sanitary landfills in operation, 31 had their own leachate treatment or pre-treatment plant while 3 led the produced leachate for the municipal sewer or directly to the wastewater treatment plant. The leachate treatment or pre-treatment plants of the sanitary landfills in operation were based on the systems presented in Table 20.

Table 20. Landfill leachate treatment or pre-treatment plants of the sanitary landfills in operation in Portugal (June 2008)

Management system	Landfill site and operation start	Treatment or pre-treatment system	Final leachate discharge
VALORMINHO	Valença, August 1998	Retention tank + Fast mixing tank (pH adjustment and/or phosphorus addition) + AS + decanting + stabilization tank	Municipal sewer
RESULIMA	Viana do Castelo, December 1998	Stabilization lagoon (and emergency lagoon) + anaerobic lagoon + anoxic tank + aerated lagoon + decanting + Fenton oxidation + decanting	Municipal sewer
BRAVAL	Póvoa de Lanhoso, July 1998	Stabilization lagoon + sand filter + AS + coagulation-flocculation + decanting	Municipal sewer
Amave	Santo Tirso, February 2000	Does not have leachate treatment or pre-treatment plant	Municipal sewer
Lipor	Maia, June 2001	Stabilization lagoon + AS + UF + RO	Surface water
Valsousa	Lousada, November 1998	Entrance structure + stabilization lagoon + AS + decanting + stabilization lagoon	Municipal sewer
	Penafiel, April 1999	Stabilization lagoon + aerated lagoon + stabilization and aerated lagoon (and emergency lagoon)	Municipal sewer
SULDOURO	Vila Nova de Gaia, March 1999	Stabilization lagoon (and emergency lagoon) + fast mixing tank (coagulant addition and pH adjustment) + decanting + AS + decanting + flotation + stabilization lagoon	Municipal sewer
RESAT	Boticas, November 2001	Entrance structure + stabilization lagoon + coagulation-flocculation + decanting	Municipal sewer
Vale do Douro Norte	Vila Real, August 2000	Stabilization and aerated lagoon + coagulation-flocculation + decanting	Surface water - inoperative
Resíduos do Nordeste	Mirandela, September 1997	Fast mixing tank (acid addition, anti-foaming and anti-fouling) + evaporation + condensation + pH adjustment + air stripping + pH adjustment + AS + decanting + stabilization tank	Water resource
REBAT	Celorico de Basto, November 2001	Stabilization lagoon + aerated lagoon + AS + sand filter + cartridge filter + RO	Surface water
RESIDOURO	Lamego, January 2002	Aerated lagoon + stabilization lagoon + sand filter + cartridge filter + RO	Surface water
VALORLIS	Leiria, September 1998	Stabilization lagoon + aerated lagoon + retention lagoon + decanting + constructed wetlands system	Municipal sewer

Management system	Landfill site and operation start	Treatment or pre-treatment system	Final leachate discharge
Ecobeirão	Tondela, May 1999	Two distinct treatment sequences: Stabilization lagoon + AS + decanting + AS +UF + RO +aer. lag Stabilization lagoon + aerated lagoon + decanting + sand filter + cartridge filter + RO + aerated lagoon	Surface water
ERSUC	Aveiro, December 1998	Fast mixing tank (pH adjustment and/or phosphorus addition) + retention lagoon + aerated lagoon + Fenton oxidation + lamellar decanting + aerated tank + sand filter	Municipal sewer
	Coimbra, September 1998	Stabilization lagoon + AS (Carrousel system) + decanting + Fenton oxidation + decanting + pH adjustment	Municipal sewer
	Figueira da Foz, September 1998	Does not have leachate treatment or pre-treatment plant	Municipal sewer
RESIESTRELA	Fundão, November 2001	Stabilization lagoon + AS + UF + RO	Surface water
Raia-Pinhal	Castelo Branco, April 1999	Entrance structure + stabilization and aerated lagoon (and emergency lagoon) + decanting + AS + decanting + sand filter + cartridge filter + RO + stabilization lagoon	Surface water
RESIOESTE	Cadaval, November 2001	Stabilization and aerated lagoon + AS + decanting + sand filter + cartridge filter + RO	Surface water
Ecolezíria	Almeirim, July 2000	Retention lagoon + decanting + AS + decanting + sand filter + cartridge filter + RO	Surface water
Resitejo	Chamusca, May 1999	Retention lagoon + stabilization lagoon + electro-coagulation + decanting + RO	Surface water
VALORSUL	Vila Franca de Xira, June 1998	Entrance structure + fast mixing tank (phosphorus addition) + aerated lagoon + coagulation-flocculation + decanting + pH adjustment	Municipal sewer
AMARSUL	Palmela, January 1997	Entrance structure + aerated lagoon + decanting	Municipal sewer
	Seixal, July 1995	Does not have leachate treatment or pre-treatment plant	Municipal sewer
Gesamb	Évora, January 2002	Stabilization lagoon + sand filter + cartridge filter + RO+ stabilization tank	Surface water
Ambilital	Santiago do Cacém, April 2000	Retention lagoon + entrance structure + aerated lagoon 1 + aerated lagoon 2 + retention lagoon + evaporation lagoon	No discharge - inoperative
Amcal	Cuba, June 1999	Entrance structure + anaerobic lagoon + facultative lagoon + aerated lagoon 1 + aerated lagoon 2 + decanting + constructed wetlands system + stabilization tank	Surface water or no discharge
VALNOR	Avis, May 2000	Entrance structure + stabilization tank + anaerobic lagoon 1 + anaerobic lagoon 2 + aerated lagoon + AS + decanting + coagulation-flocculation + decanting + pH adjustment	Surface water

Table 20. (Continued)

Management system	Landfill site and operation start	Treatment or pre-treatment system	Final leachate discharge
	Abrantes, June 1998	Entrance structure + anaerobic lagoon + aerated lagoon 1 + aerated lagoon 2 + stabilization tank + anaerobic lagoon 1 + anaerobic lagoon 2 + aerated lagoon + AS + decanting + coagulation-flocculation + decanting + pH adjustment	Surface water
Resialentejo	Beja, December 2001	Stabilization lagoon + decanting + sand filter + AS (2 SBR) + coagulation-flocculation + decanting + sand filter + activated carb	Surface water
ALGAR	Portimão, February 1998	Retention lagoon + stabilization and aerated lagoon + sand filter 1 + sand filter 2 + microfilter + RO	Surface water
	Loulé, July 2000	Retention lagoon + stabilization and aerated lagoon + sand filter 1 + sand filter 2 + microfilter + RO	Surface water

NOTE: In the first column, uppercase letters represent multimunicipal management systems; lowercase letters represent intermunicipal management systems.

Of the 31 leachate treatment plants operating in Portugal sanitary landfills, in the first half of 2008, 12 performed the pre-treatment and discharged the pre-treated effluent in the municipal sewer, 18 had treatment systems that allowed the discharge of treated leachate in the aquatic environment and one did not generate any discharge. However, two of these facilities were inoperative for failing to comply with the parameters of discharge in the surface water or due to the inadequate adjustment of the system to the leachate quantity to be treated. It should be noted that in both cases, was under investigation, or already in test phase, a different treatment solution or an optimization of the system already deployed.

Most of leachate treatment plants involve lagoons. With the exception of lagoons, the treatment systems described in Table 20 present great diversity. Some of them are very simple and only remove some organic load because leachate is discharged in the municipal sewer and subsequently is treated in a wastewater treatment plant. This is the case, for example, of the Valsousa system. Other schemes are very complex and robusts allowing the discharge in the aquatic environment, such as the Lipor, REBAT, RESIDOURO, Ecobeirão, RESIESTRELA, Raia-Pinhal, Ecolezíria, Resitejo, Gesamb and ALGAR systems, which integrate in the treatment plant the reverse osmosis process; the Resíduos do Nordeste system, which employs a sophisticated evaporation and condensation process or the Resialentejo system, which utilizes physico-chemical tuning of the final effluent with activated carbon. Naturally, the complexity of the treatment is reflected in its cost, which ranges, for instance from 0.25 € per liter of treated leachate with the AMARSUL system to 6 € per liter of treated leachate with the ALGAR systems (operating costs).

The processing of the sludge produced during treatment has consisted, mainly, on their dehydration followed by disposal in the landfill, which is not environment-friendly and does not meet the requirements stipulated by law. Likewise, the concentrates resulting from the treatment by reverse osmosis, rich in nitrogenous compounds and heavy metals, cannot be MSW landfilled. However, this is a recurring practice in national and European landfills due to the costs associated with their treatment and blanketing.

The current problematic of leachate management in Portugal is intimately connected with the inefficiency of the systems in operation, with subsequent discharge into sewers and water streams of effluents still with high levels of contamination. More specifically, many leachate treatment or pre-treatment plants have been experiencing difficulties in the removal of nitrogenous compounds (mainly ammonium and nitrate).

The removal of nitrogenous compounds is very important since, high concentrations of nitrogen-ammonium ($N\text{-}NH_4^+$) or nitrogen-nitrate ($N\text{-}NO_3^-$), can be toxic to aquatic life, deplete dissolved oxygen (DO) levels, cause eutrophication in receiving water bodies and affect the suitability of wastewater for reuse [154,155]. Furthermore, $N\text{-}NO_3^-$ in the gastrointestinal tract can be reduced to nitrogen-nitrite ($N\text{-}NO_2^-$) causing health problems like methemoglobinemia in infants or form nitrosamines and nitrosamides, potentially human carcinogenic compounds [156,157].

Biological processes (nitrification and denitrification) have been shown to be the most economical and useful approaches among all methods for removing nitrogenous compounds from water and wastewater.

With respect to denitrification, required to reduce nitrate or nitrite to nitrogen gas, all Portuguese leachate treatment plants adopted a pre-anoxic process followed by a nitrification stage with activated sludge and installed a circuit for leachate internal recirculation. The monitoring of the treated leachate has shown that this system is not satisfactory and the concentrations of those parameters exceed the discharge limits. Apparently, only a part of the nitrogen-ammonium is nitrified, and only a part of nitrogen-nitrate is denitrified. For instance, in a leachate treatment plant where the influent had concentrations of about 2500 mg $N\text{-}NH_4^+\cdot L^{-1}$ and 25 mg $N\text{-}NO_3^-\cdot L^{-1}$, after pre-treatment and before being discharged in the municipal sewer the effluent had 435 mg $N\text{-}NH_4^+\cdot L^{-1}$ and 1165 mg $N\text{-}NO_3^-\cdot L^{-1}$. This issue is even more pertinent for the treatment systems that allowed the discharge in the aquatic environment since, in order to avoid negative impacts to the biota or public health, according to the DL 236/98, of August 1 [158], the emission limit value for nitrate for the receiving environment of discharge the effluent is 11.3 mg $N\text{-}NO_3^-\cdot L^{-1}$, for ammonium is 7.8 mg $N\text{-}NH_4^+\cdot L^{-1}$ and for total nitrogen is 15 mg $N\cdot L^{-1}$.

The introduction of an oxidation step after biological treatment, namely the Fenton's oxidation, allows reducing the concentration of recalcitrant compounds. However, in many cases, it has been observed a significant increase of nitrogen-nitrate concentration.

Reverse osmosis units have also been effective in removing refractory substances from leachate. Regarding total nitrogen and nitrate, values exceeding the discharge limits were detected in many facilities.

Therefore, the improvement of nitrification-denitrification after the oxidation or before the reverse osmosis steps must be considered. To achieve this amendment, adequate dissolved oxygen content must be ensured in the aeration tank. Furthermore, the recirculation flow must be performed at a proper rate. The inclusion of a carbon source, such as methanol, in the anoxic tank, may also be needed, as verified in the Asturias leachate treatment plant [62]. In addition to these measures, a different denitrification reactor can be used.

In order to optimize existing nitrate removal strategies and to find new ways to respond to stricter legislation concerning nitrogen discharge, Cortez et al. [84,159] research has been focused on the denitrification of a Portuguese landfill leachate with high nitrate load in an anoxic rotating biological contactor. In one of the studies, Cortez et al. [84] observed that, under a carbon to nitrogen molar ratio (C/N) of 2, the anoxic RBC achieved $N\text{-}NO_3^-$ removal efficiencies above 95% for concentrations up to 100 mg $N\text{-}NO_3^-\cdot L^{-1}$. The highest observed denitrification rate was 55 mg $N\text{-}NO_3^-\cdot L^{-1}\cdot h^{-1}$ (15 g $N\text{-}NO_3^-\cdot m^{-2}\cdot d^{-1}$) at a nitrate concentration of 560 mg $N\text{-}NO_3^-\cdot L^{-1}$. In another article, according to Cortez et al [159], under a phosphorus-phosphate concentration of 10 mg $P\text{-}PO_4^{3-}\cdot L^{-1}$ and nitrogen-nitrate concentrations above 530 mg $N\text{-}NO_3^-\cdot L^{-1}$ the reactor achieved nitrogen-nitrate removal efficiencies close to 100%, without nitrite or nitrous oxide accumulation. Considering the nitrate concentration applied, nitrate removal efficiencies and the negligible accumulation of intermediates, the anoxic reactor showed to be very efficient, having a great potential in the denitrification of a mature landfill leachate with high nitrate load, using acetate as additional carbon source.

Although the reactor presented a very good denitrification performance, the effluent carbon concentration was still above the legal discharge value. In order to increase the biodegradability of the mature leachate and remove and change the recalcitrant organic matter, ozonation and Fenton's oxidation methods were further investigated [141,160]. These treatments led to a moderate biodegrability improvement and carbon reduction, which indicates the high complexity and refractory nature of the leachate used.

This highlights another major problem in leachate treatment or pre-treatment plants operating in Portugal namely, the removal of recalcitrant carbonaceous compounds.

In this context, Vilar et al. [161,162] have been studying the solar photo-Fenton process. According to these authors, when treating a Portuguese mature leachate, the optimal amount of UV solar energy required for photo-treatment to reach a biodegradable effluent, without iron addition, was 100 $kJ_{UV}\cdot L^{-1}$ (10.5 h of photo-Fenton at a constant solar UV power of 30 $W\cdot m^{-2}$), consuming 180 mM of H_2O_2 when used in excess, which means almost 40% mineralization of the leachate, 82% of polyphenols concentration and 83% of aromatic content. The combination of a solar photo-Fenton process with an attached biomass nitrification-denitrification system was also investigated with promising results.

Fernandes et al. [163] evaluated the electrochemical degradation of a biologically pre-treated leachate from a Portuguese sanitary landfill using a boron-doped diamond anode. After 6 hours of electrolysis at 300 $A\cdot m^{-2}$, COD, TKN, and $N\text{-}NH_4^+$ removals were 27%, 11% and 14%, respectively.

5. Major Challenges and Prospects

In order to fully reduce the negative impact on the environment, optimal leachate treatment is still facing various challenges.

Due to the EU Landfill Directive, which requires that the amount of biodegradable municipal waste deposited at landfill progressively been reduced, over a 15-year period, to only 35% of the total amount produced in 1995, the pre-treatment of MSW, has started to be widely used in Germany and Austria and is quickly developing in Italy and the UK. It is expected that this tendency will be adopted soon not only in the other European countries but also worldwide. The pre-treatment of MSW includes mechanical and biological treatment (MBT). MBT of solid waste improves waste settlement characteristics, reduces clogging of leachate drainage systems and accelerates organic degradation in landfills, shortening the monitoring period of the landfill and minimizing the quantity of landfill gas generated [164,165]. However, the fresh leachate generated from pre-treated MSW, contains large amounts of contaminants and a high COD concentration. As a consequence, its treatment has become a new challenge [57].

The continuous hardening of the discharge standards in most countries, the aging of landfill sites with more and more stabilized leachate and the increasingly practiced leachate recirculation in the landfill with consequent modifications in leachate composition demand different treatment solutions.

Therefore, it can be anticipated that several new landfill leachate treatment alternatives will be proposed in the next several years. We are still far from the end of the landfills and even more distant from the end of leachate production.

CONCLUSION

Particular attention was focused on the Portuguese leachate treatment or pre-treatment plants. Nevertheless, several of the reflections and observations that have been done across this study might be helpful as guidelines.

Furthermore, some challenges were identified and a widespread and great progress in this area can be expected in the future.

During the next years, sanitary landfills will continue to be used as a solid waste management strategy for disposal and elimination of MSW, releasing landfill leachate, which varies a great deal from local to local. The leachate complexity makes difficult to formulate general recommendations of universal validity for its treatment. Each of the presented methods offers inherent advantages and drawbacks.

ACKNOWLEDGMENTS

Susana Cortez fully acknowledges the financial support provided from Fundação para a Ciência e Tecnologia (FCT) through the grant SFRH/BPD/73720/2010.

REFERENCES

[1] J. Wiszniowski, D. Robert, J. Surmacz-Gorska, K. Miksch, J. Weber, Landfill leachate treatment methods: A review, *Environ. Chem. Lett.* 4 (2006) 51–61.

[2] K. Kylefors, L. Andreas, A. Lagerkvist, A comparison of small-scale, pilot-scale and large-scale tests for predicting leaching behaviour of landfilled wastes, *Waste Manage.* 23 (2003) 45–59.

[3] D.K. Wall, C. Zeiss, Municipal landfill biodegradation and settlement, *J. Environ. Eng.* 121 (1995) 214–224.

[4] T.A. Kurniawan, W.H. Lo, G.Y. Chan, Radicals-catalyzed oxidation reactions for degradation of recalcitrant compounds from landfill leachate, *Chem. Eng. J.* 125 (2006) 35–57.

[5] P. Kjeldsen, M.A. Barlaz, A.P. Rooker, A. Baun, A. Ledin, T.H. Christensen, Present and long-term composition of MSW landfill leachate: A review, *Crit. Rev. Env. Sci. Tecnol.* 32 (2002) 297–336.

[6] J.M. Lema, R. Mendez, R. Blazquez, Characteristics of landfill leachates and alternatives for their treatment - A review, *Water Air Soil Poll.* 40 (1988) 223–250.

[7] M. El-Fadel, A.N. Findikakis, J.O. Leckie, Modeling leachate generation and transport in solid waste landfills, E*nviron. Technol.* 18 (1997) 669–686.

[8] A.C. Silva, M. Dezotti, G.L. Sant'Anna, Treatment and detoxification of a sanitary landfill leachate, *Chemosphere*. 55 (2004) 207–214.

[9] D. Kulikowska, E. Klimiuk, The effect of landfill age on municipal leachate composition, *Bioresour. Technol.* 99 (2008) 5981–5985.

[10] Z. Salem, K. Hamouri, R. Djemaa, K. Allia, Evaluation of landfill leachate pollution and treatment, *Desalination.* 220 (2008) 108–114.

[11] U. Welander, T. Henrysson, T. Welander, Biological nitrogen removal from municipal landfill leachate in a pilot scale suspended carrier biofilm process, *Water Res.* 32 (1998) 1564–1570.

[12] P.H. Chen, Assessment of leachates from sanitary landfills: Impact of age, rainfall, and treatment, *Environ. Int.* 22 (1996) 225–237.

[13] J.L. de Morais, P.P. Zamora, Use of advanced oxidation processes to improve the biodegradability of mature landfill leachates, *J. Hazard. Mater.* 123 (2005) 181–186.

[14] M.A. Barlaz, D.M. Schaefer, R.K. Ham, Bacterial population development and chemical characteristics of refuse decomposition in a simulated sanitary landfill, *Appl. Environ. Microbiol.* 55 (1989) 55–65.

[15] D. Hermosilla, M. Cortijo, C.P. Huang, Optimizing the treatment of landfill leachate by conventional Fenton and photo-Fenton processes, *Sci. Total Environ.* 407 (2009) 3473–3481.

[16] S. Huo, B. Xi, H. Yu, L. He, S. Fan, H. Liu, Characteristics of dissolved organic matter (DOM) in leachate with different landfill ages, *J. Environ. Sci.* 20 (2008) 492–498.

[17] A.A. Tatsi, A.I. Zouboulis, K.A. Matis, P. Samaras, Coagulation-flocculation pretreatment of sanitary landfill leachates, *Chemosphere.* 53 (2003) 737–744.

[18] H.A. Aziz, S. Alias, M.N. Adlan, F.A.H. Asaari, S. Zahari, Colour removal from landfill leachate by coagulation and flocculation processes, *Bioresour. Technol.* 98 (2007) 218–220.

[19] S.R. Qasim, W. Chiang, Sanitary landfill leachate - generation, control and treatment, Lancaster, Pennsylvania, USA, Technomic Publishing Company, Inc., 1994.

[20] A. Chaturapruek, C. Visvanathan, K.H. Ahn, Ozonation of membrane bioreactor effluent for landfill leachate treatment, *Environ. Technol.* 26 (2005) 65–73.

[21] N.D. Berge, D.R. Reinhart, J. Dietz, T. Townsend, In situ ammonia removal in bioreactor landfill leachate, *Waste Manage.* 26 (2006) 334–343.

[22] L.M. Shao, P.J. He, G.J. Li, In situ nitrogen removal from leachate by bioreactor landfill with limited aeration, *Waste Manage.* 28 (2008) 1000–1007.

[23] S.K. Marttinen, R.H. Kettunen, J.A. Rintala, Occurrence and removal of organic pollutants in sewages and landfill leachates, *Sci. Total Environ.* 301 (2003) 1–12.

[24] R.H. Kadlec, L.A. Zmarthie, Wetland treatment of leachate from a closed landfill, *Ecol. Eng.* 36 (2010) 946–957.

[25] D.R. Birchler, M.W. Milke, A.L. Marks, R.G. Luthy, Landfill leachate treatment by evaporation, *J. Environ. Eng.* 120 (1994) 1109–1131.

[26] J. Rodriguez, L. Castrillon, E. Maranon, H. Sastre, E. Fernandez, Removal of non-biodegradable organic matter from landfill leachates by adsorption, *Water Res.* 38 (2004) 3297–3303.

[27] J. Kochany, E. Lipczynska-Kochany, Utilization of landfill leachate parameters for pretreatment by Fenton reaction and struvite precipitation--A comparative study, *J. Hazard. Mater.* 166 (2009) 248–254.

[28] D.P. Smith, Submerged filter biotreatment of hazardous leachate in aerobic, anaerobic, and anaerobic/aerobic systems, Hazard. *Waste Hazard. Mater.* 12 (1995) 167–183.

[29] H. Alvarez-Vazquez, B. Jefferson, S.J. Judd, Membrane bioreactors vs conventional biological treatment of landfill leachate: a brief review, *J. Chem. Technol. Biotechnol.* 79 (2004) 1043–1049.

[30] M. Hagman, E. Heander, J. la C. Jansen, Advanced oxidation of refractory organics in leachate – potential methods and evaluation of biodegradability of the remaining substrate, *Environ. Technol.* 29 (2008) 941–946.

[31] E. Neczaj, E. Okoniewska, M. Kacprzak, Treatment of landfill leachate by sequencing batch reactor, *Desalination.* 185 (2005) 357–362.

[32] Z. Liang, J. Liu, Landfill leachate treatment with a novel process: Anaerobic ammonium oxidation (Anammox) combined with soil infiltration system, *J. Hazard. Mater.* 151 (2008) 202–212.

[33] C.-C. Wang, P.-H. Lee, M. Kumar, Y.-T. Huang, S. Sung, J.-G. Lin, Simultaneous partial nitrification, anaerobic ammonium oxidation and denitrification (SNAD) in a full-scale landfill-leachate treatment plant, *J. Hazard. Mater.* 175 (2010) 622–628.

[34] J.P.Y. Jokela, R.H. Kettunen, K.M. Sormunen, J.A. Rintala, Biological nitrogen removal from municipal landfill leachate: low-cost nitrification in biofilters and laboratory scale in-situ denitrification, *Water Res.* 36 (2002) 4079–4087.

[35] S. Renou, J.G. Givaudan, S. Poulain, F. Dirassouyan, P. Moulin, Landfill leachate treatment: review and opportunity, *J. Hazard. Mater.* 150 (2008) 468–493.

[36] D. Kulikowska, T. Jóźwiak, P. Kowal, S. Ciesielski, Municipal landfill leachate nitrification in RBC biofilm - Process efficiency and molecular analysis of microbial structure, *Bioresour Technol.* 101 (2010) 3400–3405.

[37] J. Tsilogeorgis, A. Zouboulis, P. Samaras, D. Zamboulis, Application of a membrane sequencing batch reactor for landfill leachate treatment, *Desalination.* 221 (2008) 483–493.

[38] C. Di Iaconi, R. Ramadori, A. Lopez, Combined biological and chemical degradation for treating a mature municipal landfill leachate, *Biochem. Eng. J.* 31 (2006) 118–124.

[39] J. Wiszniowski, J. Surmacz-Gorska, D. Robert, J.V. Weber, The effect of landfill leachate composition on organics and nitrogen removal in an activated sludge system with bentonite additive, *J. Environ. Manage.* 85 (2006) 59–68.

[40] M.X. Loukidou, A.I. Zouboulis, Comparison of two biological treatment processes using attached-growth biomass for sanitary landfill leachate treatment, *Environ. Pollut.* 111 (2001) 273–281.

[41] D.A. Dzombak, K.M. Lagnese, D.B. Spengel, R.G. Luthy, Comparison of activated sludge and RBC treatment of leachate from a solid waste landfill, in: *Proceedings of the*

Water Quality Management of Landfills Conference, Alexandria, Virginia, Water Pollution Control Federation, 1990: pp. 4–39–4–58.

[42] D.M. Shiskowski, D.S. Mavinic, Biological treatment of a high ammonia leachate: influence of external carbon during initial startup, *Water Res.* 32 (1998) 2533–2541.

[43] B.-U. Bae, E.-S. Jung, Y.-R. Kim, H.-S. Shin, Treatment of landfill leachate using activated sludge process and electron-beam radiation, *Water Res.* 33 (1999) 2669–2673.

[44] Ellouze, M., Aloui, F., Sayadi, S., Performance of biological treatment of high-level ammonia landfill leachate, *Environ. Technol.* 29 (2008) 1169–1178.

[45] A.I. Zouboulis, M.X. Loukidou, K. Christodoulou, Enzymatic treatment of sanitary landfill leachate, *Chemosphere.* 44 (2001) 1103–1108.

[46] A. Uygur, F. Kargi, Biological nutrient removal from pre-treated landfill leachate in a sequencing batch reactor, *J. Environ. Manage.* 71 (2004) 9–14.

[47] H. Li, S. Zhou, Y. Sun, P. Feng, J. Li, Advanced treatment of landfill leachate by a new combination process in a full-scale plant, *J. Hazard. Mater.* 172 (2009) 408–415.

[48] E. Klimiuk, D. Kulikowska, Effectiveness of organics and nitrogen removal from municipal landfill leachate in single- and two-stage SBR systems., *Pol. J. Environ. Stud.* 13 (2004) 525–532.

[49] Z.-Y. Xu, G.-M. Zeng, Z.-H. Yang, Y. Xiao, M. Cao, H.-S. Sun, et al., Biological treatment of landfill leachate with the integration of partial nitrification, anaerobic ammonium oxidation and heterotrophic denitrification, *Bioresour. Technol.* 101 (2010) 79–86.

[50] H. Zhang, X. Ran, X. Wu, D. Zhang, Evaluation of electro-oxidation of biologically treated landfill leachate using response surface methodology, *J Hazard Mater.* 188 (2011) 261–268.

[51] D. Frascari, F. Bronzini, G. Giordano, G. Tedioli, M. Nocentini, Long-term characterization, lagoon treatment and migration potential of landfill leachate: a case study in an active Italian landfill, *Chemosphere.* 54 (2004) 335–343.

[52] N. Schwarzenbeck, K. Leonhard, P. Wilderer, Treatment of landfill leachate - high tech or low tech? A case study, *Water Sci. Technol.* 48 (2004) 277–284.

[53] M.K. Mehmood, E. Adetutu, D.B. Nedwell, A.S. Ball, In situ microbial treatment of landfill leachate using aerated lagoons, *Bioresour. Technol.* 100 (2009) 2741–2744.

[54] R. Mendez, J.M. Lema, R. Blazquez, M. Pan, C. Forjan, Characterization, digestibility and anaerobic treatment of leachates from old and young landfills, *Water Sci. Technol.* 21 (1989) 145–155.

[55] C.-Y. Lin, Anaerobic digestion of landfill leachate, *Water SA.* 17 (1991) 301–306.

[56] S. Kheradmand, A. Karimi-Jashni, M. Sartaj, Treatment of municipal landfill leachate using a combined anaerobic digester and activated sludge system, *Waste Manage.* 30 (2010) 1025–1031.

[57] J. Liu, J. Zhong, Y. Wang, Q. Liu, G. Qian, L. Zhong, et al., Effective bio-treatment of fresh leachate from pretreated municipal solid waste in an expanded granular sludge bed bioreactor, *Bioresour. Technol.* 101 (2010) 1447–1452.

[58] R.H. Kettunen, J.A. Rintala, Performance of an on-site UASB reactor treating leachate at low temperature, *Water Res.* 32 (1998) 537–546.

[59] K.J. Kennedy, E.M. Lentz, Treatment of landfill leachate using sequencing batch and continuous flow upflow anaerobic sludge blanket (UASB) reactors, *Water Res.* 34 (2000) 3640–3656.

[60] O.N. Agdag, D.T. Sponza, Anaerobic/aerobic treatment of municipal landfill leachate in sequential two-stage up-flow anaerobic sludge blanket reactor (UASB)/completely stirred tank reactor (CSTR) systems, *Process Biochem.* 40 (2005) 895–902.

[61] E. Castillo, M. Vergara, Y. Moreno, Landfill leachate treatment using a rotating biological contactor and an upward-flow anaerobic sludge bed reactor, *Waste Manage.* 27 (2007) 720–726.

[62] L. Castrillón, Y. Fernández-Nava, M. Ulmanu, I. Anger, E. Marañón, Physico-chemical and biological treatment of MSW landfill leachate, *Waste Manage.* 30 (2010) 228–235.

[63] E. Akkaya, A. Demir, D. Karadag, G. Varank, M.S. Bilgili, B. Ozkaya, Post-treatment of anaerobically treated medium-age landfill leachate, *Environ Prog Sustain Energy*. 29 (2010) 78–84.

[64] J. Bohdziewicz, E. Neczaj, A. Kwarciak, Landfill leachate treatment by means of anaerobic membrane bioreactor, *Desalination.* 221 (2008) 559–565.

[65] H. Hasar, S.A. Unsal, U. Ipek, S. Karatas, O. CInar, C. Yaman, et al., Stripping/flocculation/membrane bioreactor/reverse osmosis treatment of municipal landfill leachate, *J. Hazard. Mater.* 171 (2009) 309–317.

[66] R. Canziani, V. Emondi, M. Garavaglia, F. Malpei, E. Pasinetti, G. Buttiglieri, Effect of oxygen concentration on biological nitrification and microbial kinetics in a cross-flow membrane bioreactor (MBR) and moving-bed biofilm reactor (MBBR) treating old landfill leachate, *J. Membrane Sci.* 286 (2006) 202–212.

[67] M. Pirbazari, V. Ravindran, B.N. Badriyha, S.-H. Kim, Hybrid membrane filtration process for leachate treatment, *Water Res.* 30 (1996) 2691–2706.

[68] A. Gálvez, L. Giusti, M. Zamorano, A.F. Ramos-Ridao, Stability and efficiency of biofilms for landfill leachate treatment, *Bioresour. Technol.* 100 (2009) 4895–4898.

[69] A. Zayen, S. Mnif, F. Aloui, F. Fki, S. Loukil, M. Bouaziz, et al., Anaerobic membrane bioreactor for the treatment of leachates from Jebel Chakir discharge in Tunisia, *J. Hazard. Mater.* 177 (2010) 918–923.

[70] C. Visvanathan, M.K. Choudhary, M.T. Montalbo, V. Jegatheesan, Landfill leachate treatment using thermophilic membrane bioreactor, *Desalination.* 204 (2007) 8–16.

[71] K. Knox, Leachate treatment with nitrification of ammonia, *Water Res.* 19 (1985) 895–904.

[72] R. Gourdon, C. Comel, P. Vermande, J. Veron, Fractionation of the organic matter of a landfill leachate before and after aerobic or anaerobic biological treatment, *Water Res.* 23 (1989) 167–173.

[73] T. Stephenson, S.J.T. Pollard, E. Cartmell, Feasibility of biological aerated filters (bafs) for treating landfill leachate, *Environ. Technol.* 25 (2004) 349.

[74] A. Gálvez, M. Zamorano, E. Hontoria, A. Ramos, Treatment of landfill leachate with aerated and non-aerated submerged biofilters, *J. Environ. Sci. Health.,* Part A. 41 (2006) 1129–1144.

[75] A. Tilche, S.M. Vieira, Discussion report on reactor design of anaerobic filters and sludge bed reactors, *Water Sci. Technol.* 24 (1991) 193–206.

[76] D.B. Nedwell, P.J. Reynolds, Treatment of landfill leachate by methanogenic and sulphate-reducing digestion, *Water Res.* 30 (1996) 21–28.

[77] R.H. Kettunen, T.H. Hoilijoki, J.A. Rintala, Anaerobic and sequential anaerobic-aerobic treatments of municipal landfill leachate at low temperatures, *Bioresour. Technol.* 58 (1996) 31–40.
[78] G. Cema, J. Wiszniowski, S. Zabczynski, E. Zablocka-Godlewska, A. Raszka, J. Surmacz-Gorska, Biological nitrogen removal from landfill leachate by deammonification assisted by heterotrophic denitrification in a rotating biological contactor (RBC), *Water Sci. Technol.* 55 (2007) 35–42.
[79] K. Egli, F. Bosshard, C. Werlen, P. Lais, H. Siegrist, A.J.B. Zehnder, et al., Microbial composition and structure of a rotating biological contactor biofilm treating ammonium-rich wastewater without organic carbon, *Microbial Ecol.* 45 (2003) 419–432.
[80] S. Cortez, P. Teixeira, R. Oliveira, M. Mota, Rotating biological contactors: a review on main factors affecting performance, *Rev Environ Sci Biotechnol.* 7 (2008) 155–172.
[81] D.B. Spengel, D.A. Dzombak, Treatment of a landfill leachate with rotating biological contactors: bench-scale experiments, *Res. J. Water Pollut. C. Fed.* 63 (1991) 971–981.
[82] S. Chen, D. Sun, J.-S. Chung, Simultaneous removal of COD and ammonium from landfill leachate using an anaerobic-aerobic moving-bed biofilm reactor system, *Waste Manage.* 28 (2008) 339–346.
[83] A. Eldyasti, N. Chowdhury, G. Nakhla, J. Zhu, Biological nutrient removal from leachate using a pilot liquid-solid circulating fluidized bed bioreactor (LSCFB), *J. Hazard. Mater.* 181 (2010) 289–297.
[84] S. Cortez, P. Teixeira, R. Oliveira, M. Mota, Denitrification of a landfill leachate with high nitrate concentration in an anoxic rotating biological contactor, *Biodegradation.* 22 (2010) 661–671.
[85] U. Welander, T. Henrysson, T. Welander, Nitrification of landfill leachate using suspended-carrier biofilm technology, *Water Res.* 31 (1997) 2351–2355.
[86] K. Isaka, S. Yoshie, T. Sumino, Y. Inamori, S. Tsuneda, Nitrification of landfill leachate using immobilized nitrifying bacteria at low temperatures, *Biochem. Eng.* J. 37 (2007) 49–55.
[87] T. Mariam, L.D. Nghiem, Landfill leachate treatment using hybrid coagulation-nanofiltration processes, *Desalination.* 250 (2010) 677–681.
[88] T.A. Kurniawan, W.H. Lo, G.Y. Chan, Physico-chemical treatments for removal of recalcitrant contaminants from landfill leachate, *J. Hazard. Mater.* 129 (2006) 80–100.
[89] E. Marañón, L. Castrillón, Y. Fernández-Nava, A. Fernández-Méndez, A. Fernández-Sánchez, Coagulation-flocculation as a pretreatment process at a landfill leachate nitrification-denitrification plant, *J. Hazard. Mater.* 156 (2008) 538–544.
[90] M. Vedrenne, R. Vasquez-Medrano, D. Prato-Garcia, B.A. Frontana-Uribe, J.G. Ibanez, Characterization and detoxification of a mature landfill leachate using a combined coagulation–flocculation/photo Fenton treatment, *J. Hazard. Mater.* 205–206 (2012) 208–215.
[91] F. Cecen, G. Gursoy, Characterization of landfill leachates and studies on heavy metal removal, *J. Environ. Monitor.* 2 (2000) 436–442.
[92] J.-S. Guo, A.A. Abbas, Y.-P. Chen, Z.-P. Liu, F. Fang, P. Chen, Treatment of landfill leachate using a combined stripping, Fenton, SBR, and coagulation process, *J. Hazard. Mater.* 178 (2010) 699–705.

[93] I. Monje-Ramirez, M.T.O. de Velásquez, Removal and transformation of recalcitrant organic matter from stabilized saline landfill leachates by coagulation-ozonation coupling processes, *Water Res.* 38 (2004) 2359–2367.

[94] M.O. Orkun, A. Kuleyin, Treatment performance evaluation of chemical oxygen demand from landfill leachate by electro-coagulation and electro-fenton technique, *Environ Prog Sustain Energy.* 31 (2012) 59–67.

[95] F. Bouhezila, M. Hariti, H. Lounici, N. Mameri, Treatment of the OUED SMAR town landfill leachate by an electrochemical reactor, *Desalination.* 280 (2011) 347–353.

[96] X.Z. Li, Q.L. Zhao, Efficiency of biological treatment affected by high strength of ammonium-nitrogen in leachate and chemical precipitation of ammonium-nitrogen as pretreatment, *Chemosphere.* 44 (2001) 37–43.

[97] B. Calli, B. Mertoglu, B. Inanc, Landfill leachate management in Istanbul: applications and alternatives, *Chemosphere.* 59 (2005) 819–829.

[98] T. Zhang, L. Ding, H. Ren, Pretreatment of ammonium removal from landfill leachate by chemical precipitation, *J. Hazard. Mater.* 166 (2009) 911–915.

[99] C. Di Iaconi, M. Pagano, R. Ramadori, A. Lopez, Nitrogen recovery from a stabilized municipal landfill leachate, *Bioresour. Technol.* 101 (2010) 1732–1736.

[100] X.Z. Li, Q.L. Zhao, X.D. Hao, Ammonium removal from landfill leachate by chemical precipitation, *Waste Manage.* 19 (1999) 409–415.

[101] S.K. Marttinen, R.H. Kettunen, K.M. Sormunen, R.M. Soimasuo, J.A. Rintala, Screening of physical-chemical methods for removal of organic material, nitrogen and toxicity from low strength landfill leachates, *Chemosphere.* 46 (2002) 851–858.

[102] K.Y. Foo, B.H. Hameed, An overview of landfill leachate treatment via activated carbon adsorption process, *J. Hazard. Mater.* 171 (2009) 54–60.

[103] A.A. Halim, H.A. Aziz, M.A.M. Johari, K.S. Ariffin, M.N. Adlan, Ammoniacal nitrogen and COD removal from semi-aerobic landfill leachate using a composite adsorbent: Fixed bed column adsorption performance, *J. Hazard. Mater.* 175 (2010) 960–964.

[104] T.A. Kurniawan, W.H. Lo, Removal of refractory compounds from stabilized landfill leachate using an integrated H2O2 oxidation and granular activated carbon (GAC) adsorption treatment, *Water Res.* 43 (2009) 4079–4091.

[105] C. Papastavrou, D. Mantzavinos, E. Diamadopoulos, A comparative treatment of stabilized landfill leachate: Coagulation and activated carbon adsorption vs. electrochemical oxidation, *Environ. Technol.* 30 (2009) 1547.

[106] M. Cotman, A.Z. Gotvajn, Comparison of different physico-chemical methods for the removal of toxicants from landfill leachate, *J. Hazard. Mater.* 178 (2010) 298–305.

[107] J. Rubio, E. Carissimi, J.J. Rosa, Flotation in water and wastewater treatment and reuse: recent trends in Brazil, *Int. J. Environ. Pollut.* 30 (2007) 197 – 212.

[108] A.I. Zouboulis, W. Jun, I.A. Katsoyiannis, Removal of humic acids by flotation, *Colloids Surf.,* A. 231 (2003) 181–193.

[109] P. Palaniandy, M.N. Adlan, H.A. Aziz, M.F. Murshed, Application of dissolved air flotation (DAF) in semi-aerobic leachate treatment, *Chem. Eng. J.* 157 (2010) 316–322.

[110] B.V.D. Bruggen, C. Vandecasteele, T.V. Gestel, W. Doyen, R. Leysen, A review of pressure-driven membrane processes in wastewater treatment and drinking water production, *Environ. Prog.* 22 (2003) 46–56.

[111] K. Linde, A. Jönsson, R. Wimmerstedt, Treatment of three types of landfill leachate with reverse osmosis, *Desalination.* 101 (1995) 21–30.

[112] W.Y. Ahn, M.S. Kang, S.K. Yim, K.H. Choi, Advanced landfill leachate treatment using an integrated membrane process, *Desalination.* 149 (2002) 109–114.

[113] K. Ushikoshi, T. Kobayashi, K. Uematsu, A. Toji, D. Kojima, K. Matsumoto, Leachate treatment by the reverse osmosis system, *Desalination.* 150 (2002) 121–129.

[114] M. Ince, E. Senturk, G. Onkal Engin, B. Keskinler, Further treatment of landfill leachate by nanofiltration and microfiltration-PAC hybrid process, *Desalination.* 255 (2010) 52–60.

[115] J. Bohdziewicz, M. Bodzek, J. Górska, Application of pressure-driven membrane techniques to biological treatment of landfill leachate, *Process Biochem.* 36 (2001) 641–646.

[116] F. Li, K. Wichmann, W. Heine, Treatment of the methanogenic landfill leachate with thin open channel reverse osmosis membrane modules, *Waste Manage.* 29 (2009) 960–964.

[117] W. Piatkiewicz, E. Biemacka, T. Suchecka, A polish study: treating landfill leachate with membranes, *Filtr. Sep.* 38 (2001) 22–23.

[118] K.W. Pi, L.X. Gao, M.X. Fan, W.Q. Gong, D.J. Wan, Two-stage biodegradation coupled with ultrafiltration for treatment of municipal landfill leachate, *Process Saf. Environ. Prot.* 87 (2009) 336–342.

[119] S. Renou, S. Poulain, J.G. Givaudan, P. Moulin, Amelioration of ultrafiltration process by lime treatment: Case of landfill leachate, *Desalination.* 249 (2009) 72–82.

[120] L.B. Chaudhari, Z.V.P. Murthy, Treatment of landfill leachates by nanofiltration, *J. Environ. Manage.* 91 (2010) 1209–1217.

[121] Y. Deng, J.D. Englehardt, Treatment of landfill leachate by the Fenton process, *Water Res.* 40 (2006) 3683–3694.

[122] H. Zhang, D. Zhang, J. Zhou, Removal of COD from landfill leachate by electro-Fenton method, *J. Hazard. Mater.* 135 (2006) 106–111.

[123] E. Atmaca, Treatment of landfill leachate by using electro-Fenton method, *J. Hazard. Mater.* 163 (2009) 109–114.

[124] S. Mohajeri, H.A. Aziz, M.H. Isa, M.A. Zahed, M.N. Adlan, Statistical optimization of process parameters for landfill leachate treatment using electro-Fenton technique, *J. Hazard. Mater.* 176 (2010) 749–758.

[125] F. Wang, M.G. El-Din, D.W. Smith, Oxidation of aged raw landfill leachate with O3 only and O3/H2O2: treatment efficiency and molecular size distribution analysis, *Ozone Sci. Eng.* 26 (2004) 287–298.

[126] C. Tizaoui, L. Bouselmi, L. Mansouri, A. Ghrabi, Landfill leachate treatment with ozone and ozone/hydrogen peroxide systems, *J. Hazard. Mater.* 140 (2007) 316–324.

[127] Y. Deng, J.D. Englehardt, Electrochemical oxidation for landfill leachate treatment, *Waste Manage.* 27 (2007) 380–388.

[128] P.B. Moraes, R. Bertazzoli, Electrodegradation of landfill leachate in a flow electrochemical reactor, *Chemosphere.* 58 (2005) 41–46.

[129] A. Cabeza, O. Primo, A. Urtiaga, I. Ortiz, Definition of a clean process for the treatment of landfill leachates integration of electrooxidation and ion exchange technologies, *Sep. Sci. Technol.* 42 (2007) 1585–1596.

[130] G. Zhao, Y. Pang, L. Liu, J. Gao, B. Lv, Highly efficient and energy-saving sectional treatment of landfill leachate with a synergistic system of biochemical treatment and electrochemical oxidation on a boron-doped diamond electrode, *J. Hazard. Mater.* 179 (2010) 1078–1083.

[131] D. Goi, G. Di Giorgio, I. Cimarosti, B. Lesa, G. Rossi, G. Dolcetti, Treatment of landfill leachate by H2O2 promoted wet air oxidation: COD-AOX reduction, biodegradability enhancement and comparison with a Fenton-type oxidation, *Chem. Biochem. Eng. Q.* 23 (2009) 343–349.

[132] F.J. Rivas, F.J. Beltrán, F. Carvalho, P.M. Alvarez, Oxone-promoted wet air oxidation of landfill leachates, *Ind. Eng. Chem. Res.* 44 (2005) 749–758.

[133] J.J. Wu, C.C. Wu, H.W. Ma, C.C. Chang, Treatment of landfill leachate by ozone-based advanced oxidation processes, *Chemosphere.* 54 (2004) 997–1003.

[134] H.-Y. Shu, H.-J. Fan, M.-C. Chang, W.-P. Hsieh, Treatment of MSW landfill leachate by a thin gap annular UV/H2O2 photoreactor with multi-UV lamps, *J. Hazard. Mater.* 129 (2006) 73–79.

[135] T.I. Qureshi, H.-T. Kim, Y.-J. Kim, UV-catalytic treatment of municipal solid-waste landfill leachate with hydrogen peroxide and ozone oxidation, *Chin. J. Chem. Eng.* 10 (2002) 444–449.

[136] O. Primo, M.J. Rivero, I. Ortiz, Photo-Fenton process as an efficient alternative to the treatment of landfill leachates, *J. Hazard. Mater.* 153 (2008) 834–842.

[137] A. López, M. Pagano, A. Volpe, A.C. Di Pinto, Fenton's pre-treatment of mature landfill leachate, *Chemosphere.* 54 (2004) 1005–1010.

[138] X. Wang, S. Chen, X. Gu, K. Wang, Pilot study on the advanced treatment of landfill leachate using a combined coagulation, Fenton oxidation and biological aerated filter process, *Waste Manage.* 29 (2009) 1354–1358.

[139] D.M. Bila, A.F. Montalvão, A.C. Silva, M. Dezotti, Ozonation of a landfill leachate: evaluation of toxicity removal and biodegradability improvement, *J Hazard Mater.* 117 (2005) 235–242.

[140] A. Goi, Y. Veressinina, M. Trapido, Combination of ozonation and the Fenton processes for landfill leachate treatment: evaluation of treatment efficiency, *Ozone Sci. Eng.* 31 (2009) 28–36.

[141] S. Cortez, P. Teixeira, R. Oliveira, M. Mota, Ozonation as polishing treatment of mature landfill leachate, *J. Hazard. Mater.* 182 (2010) 730–734.

[142] M.J.K. Bashir, M.H. Isa, S.R.M. Kutty, Z.B. Awang, H.A. Aziz, S. Mohajeri, et al., Landfill leachate treatment by electrochemical oxidation, *Waste Manage.* 29 (2009) 2534–2541.

[143] X. Zhao, J. Qu, H. Liu, C. Wang, S. Xiao, R. Liu, et al., Photoelectrochemical treatment of landfill leachate in a continuous flow reactor, *Bioresour. Technol.* 101 (2010) 865–869.

[144] E. Tauchert, S. Schneider, J.L. de Morais, P. Peralta-Zamora, Photochemically-assisted electrochemical degradation of landfill leachate, *Chemosphere.* 64 (2006) 1458–1463.

[145] E. Neczaj, M. Kacprzak, J. Lach, E. Okoniewska, Effect of sonication on combined treatment of landfill leachate and domestic sewage in SBR reactor, *Desalination.* 204 (2007) 227–233.

[146] S. Wang, X. Wu, Y. Wang, Q. Li, M. Tao, Removal of organic matter and ammonia nitrogen from landfill leachate by ultrasound, *Ultrason. Sonochem.* 15 (2008) 933–937.

[147] T.A. Kurniawan, W.H. Lo, G.Y. Chan, Degradation of recalcitrant compounds from stabilized landfill leachate using a combination of ozone-GAC adsorption treatment, *J. Hazard. Mater.* 137 (2006) 443–455.

[148] Y. Wu, S. Zhou, D. Chen, R. Zhao, H. Li, Y. Lin, Transformation of metals speciation in a combined landfill leachate treatment, *Sci Total Environ.* 409 (2011) 1613–1620.

[149] M. Altinbaş, C. Yangin, I. Ozturk, Struvite precipitation from anaerobically treated municipal and landfill wastewaters, *Water Sci. Technol.* 46 (2002) 271–278.

[150] F.J. Rivas, F. Beltrán, O. Gimeno, F. Carvalho, Fenton-like oxidation of landfill leachate, *J. Environ. Sci. Health.,* Part A. 38 (2003) 371.

[151] APA, Caracterização da situação dos Resíduos Urbanos em Portugal continental em 2006 - Resumo, (2008).

[152] APA, Sistemas de Gestão de Resíduos Sólidos Urbanos - Situação actual, (2010).

[153] J.Q. Levy, C. Santana, Funcionamento das estações de tratamento de águas lixiviantes e acções para a sua beneficiação, (2004).

[154] M. Kumar, J.-G. Lin, Co-existence of anammox and denitrification for simultaneous nitrogen and carbon removal--Strategies and issues, *J. Hazard. Mater.* 178 (2010) 1–9.

[155] D. Paredes, P. Kuschk, T.S.A. Mbwette, F. Stange, R.A. Muller, H. Koser, New aspects of microbial nitrogen transformations in the context of wastewater treatment - a review, *Engineering in Life Sciences.* 7 (2007) 13–25.

[156] S. Aslan, H. Cakici, Biological denitrification of drinking water in a slow sand filter, *Journal of Hazardous Materials.* 148 (2007) 253–258.

[157] S. Ghafari, M. Hasan, M.K. Aroua, Bio-electrochemical removal of nitrate from water and wastewater--A review, *Bioresource Technology.* 99 (2008) 3965–3974.

[158] Ministério do Ambiente, Decreto-Lei n.º 236/1998, Diário Da República. I Série-A (1998) 3676–3722.

[159] S. Cortez, P. Teixeira, R. Oliveira, M. Mota, Mature landfill leachate treatment by denitrification and ozonation, *Process Biochem.* 46 (2011) 148–153.

[160] S. Cortez, P. Teixeira, R. Oliveira, M. Mota, Evaluation of Fenton and ozone-based advanced oxidation processes as mature landfill leachate pre-treatments, *J. Environ. Manage.* 92 (2011) 749–755.

[161] V.J.P. Vilar, E.M.R. Rocha, F.S. Mota, A. Fonseca, I. Saraiva, R.A.R. Boaventura, Treatment of a sanitary landfill leachate using combined solar photo-Fenton and biological immobilized biomass reactor at a pilot scale, *Water Res.* 45 (2011) 2647–2658.

[162] V.J.P. Vilar, S.M.S. Capelo, T.F.C.V. Silva, R.A.R. Boaventura, Solar photo-Fenton as a pre-oxidation step for biological treatment of landfill leachate in a pilot plant with CPCs, *Catal Today.* 161 (2011) 228–234.

[163] A. Fernandes, M.J. Pacheco, L. Ciríaco, A. Lopes, Anodic oxidation of a biologically treated leachate on a boron-doped diamond anode, *J Hazard Mater.* 199–200 (2012) 82–87.

[164] D.Q. Zhang, S.K. Tan, R.M. Gersberg, Municipal solid waste management in China: Status, problems and challenges, *J. Environ. Manage.* 91 (2010) 1623–1633.

[165] H.D. Robinson, K. Knox, B.D. Bone, A. Picken, Leachate quality from landfilled MBT waste, *Waste Manage.* 25 (2005) 383–391.

In: Landfills
Editors: G. B. C. Cabral and B. A. E. Botelho

ISBN: 978-1-62257-121-5

Chapter 3

GROUNDWATER PROTECTION BY PERMEABLE ADSORBING BARRIERS AT SOLID WASTE LANDFILLS

I. Bortone*[1*]*, S. Chianese*[1]*, A. Di Nardo*[1]*, M. Di Natale*[1]*,
***A. Erto*[#2] *and D. Musmarra*[1]**

[1]CIRIAM, Centro Interdipartimentale di Ricerca in Ingegneria Ambientale,
Dipartimento di Ingegneria Civile, Seconda Università di Napoli, Italy
[2]Università di Napoli Federico, Italy

ABSTRACT

The accidental discharge of leachate from solid waste landfills often leads to groundwater contamination both by inorganic and organic compounds.

A major contribution to groundwater pollution at solid waste landfills comes from chlorinated organic compounds such as Tetrachloroethylene (PCE) and Trichloroethylene (TCE). They are among the most dangerous pollutants, due to their high toxicity and persistence grade mainly originating from plastic material degradation due to rainfall water leaching. The environmental fate of these species strictly depends on their main physical and chemical properties; they are denser and less viscous than water, thus migrating from superficial water and contaminated soils to groundwater. Consequently, the remediation of polluted sites must be designed accordingly.

Several remediation technologies have been developed for aquifers contaminated by organic and inorganic compounds; they range from ex-situ treatments, coupled with the *pump and treat* technique, to in situ treatments, performed with Permeable Reactive Barriers (PRB).

In a PRB-based treatment, the barrier is commonly built with reactive materials whose hydraulic conductivity is higher than that of the surrounding soils, so that the contaminated groundwater, moving under natural hydraulic gradient, is forced to pass through the barrier without any external energy input. Nowadays, Pump and Treat (PT) technology is widely used for groundwater remediation. This technology has good

* CIRIAM, Centro Interdipartimentale di Ricerca in Ingegneria Ambientale, Dipartimento di Ingegneria Civile, Seconda Università di Napoli, via Roma 29-81031 Aversa (C. E.), Italy.

Dipartimento di Ingegneria Chimica, Università di Napoli Federico II, P. leTecchio, 80-80125 Naples, Italy.

remediation efficiency but has high operational costs. For this reason, there is a growing interest for PRB installations as an effective alternative to Pump and Treat remediation methods, thanks to its low operating and maintenance costs. The mechanism of action of a PRB depends on the reactive material chosen to build the barrier. Chlorinated organic compounds can be effectively removed from polluted water and wastewater also by adsorption, a process that combines good efficiency with a very simple process configuration. The pollutant is trapped into the barrier made of adsorbing material and does not precipitate. This particular PRB made of adsorptive material is named Permeable Adsorbing Barrier (PAB). The chapter, after a short description of the main remediation technologies, describes a specific procedure for the design of a PAB for the remediation of an aquifer contaminated by PCE and TCE. As case study, the design of a PAB at a solid waste landfill site in Giugliano in Campania (area North of Napoli, Italy) is presented. In this site, there are many solid waste landfills where, over the past 20 years, about eight million tons of urban and special wastes have been deposited and contamination by different compounds has been detected.

INTRODUCTION

In this paragraph a short summary of the main remediation technologies for groundwater and soil remediation is presented, emphasis on Permeable Reactive Barriers (PRB).

Technologies for Groundwater and Soil Remediation

Two main approaches to remediation can be usually adopted for the decontamination of soil and groundwater from pollutants of different nature:

- In-situ;
- Ex-situ.

The remediation with an *in-situ* technology consists in the application of a technique aimed at the degradation, the removal or the containment of pollutants, without the extraction of contaminated material from the subsurface. On the contrary, remediation with an *ex-situ* technology consists in two phases: first, the extraction of polluted media from the subsurface and, then, the treatment of the contaminated materials directly on the site (on-site *ex-situ* remediation technologies) or off the site of decontamination (off-site *ex-situ* remediation technologies). Both *in-situ* and *ex-situ* technologies can be performed with the application of biological treatments or physicochemical treatments. A synthetic description of the most important techniques for the decontamination of polluted groundwater and soils, is reported in the following:

a) physicochemical treatments:

a) 1) in-situ.

- *Permeable Reactive Barriers/Permeable Adsorbing Barriers* (Chemical Reduction and/or Adsorption): these techniques require a wall of reactive/adsorbing media put in the subsurface through which, under natural gradient, a groundwater flow passes and the polluted plume is captured. Contaminants are transformed into environmentally tolerable forms or adsorbed by the material to achieve remediation concentration goals, at check points downstream the natural groundwater flow [US EPA, 1997a; US EPA, 2002c; Costello, 2003; ITRC, 2005a; ESTCP, 2011a; ITRC, 2011a];

- *Soil Vapor Extraction*: volatile and semivolatile organic pollutants are removed from the soil by imposing a vacuum to generate a controlled flow. This technology is often implemented in conjunction with Air Sparging.[US EPA, 1997b; US EPA, 1997c; American Academy of Environmental Engineers, 1998; Wisconsin Department of Natural Resources, 2003; US EPA, 2004c; US EPA, 2006a; US EPA, 2006b; California Environmental Protection Agency, 2010; ESTCP, 2011a];
- *Air Sparging*: volatile and semivolatile organic contaminants are extracted from the unsaturated zone by stripping the vapour phase. Air or oxygen is injected into the polluted aquifer in order to volatilize the contaminants.[US EPA, 1997b; American Academy of Environmental Engineers, 1998; Leeson et al., 2002; Wisconsin Department of Natural Resources, 2003];
- *Electrokinetics*: a low-intensity electric direct current is applied through the contaminated soil by electrodes The cathode draws positive-charge organic compounds, metal ions and ammonium ions; the anode draws negative-charge organic compounds, chloride, nitrate, cyanide and fluoride [ITRC, 1997a; US EPA, 2002c; Saichek and Reddy, 2005; US EPA, 2006b];
- *Flushing Technologies* (Cosolvent/Alcohol Flooding, Surfactant Flushing): the polluted soil zone is flooded with large volumes of water, often integrated with surfactants, co-solvents, or different chemicals. Then the water and the chemicals used are recuperated with the flushed contaminants [American Academy of Environmental Engineers, 1993; ITRC, 2000; US EPA, 2002c; ITRC, 2003a; ITRC, 2004; US EPA, 2006b; US EPA, 2009a];
- *Thermal Processes* (Steam Injection, Electrical Heating, Radio Frequency Heating, Thermal Conduction, Dynamic Underground Stripping): the heat is transferred to the contaminated soil and/or groundwater, in situ, in several different ways and with the combination of many different techniques. Thanks to the heat, organic pollutants are cracked or volatilized, their mobility grows and the vapour-phase pollutants can be captured by wells and cleaned up in an ex situ treatment unit. Thermal Processes are mainly used for both Dense and Light Non Aqueous Phase Liquids (DNAPLs or LNAPLs) removal [US EPA, 1997a; US EPA, 1997b; US EPA, 1997c; US DOE, 2000; ITRC, 2004; US EPA. 2004a; US EPA, 2006b; Powell et al., 2007; US EPA, 2007a; US EPA, 2009a; ESTCP, 2010; ESTCP, 2011a];
- *Chemical Oxidation*: volatile and semivolatile hazardous pollutants are chemically converted to non-hazardous or less toxic elements, that are inert or less mobile. This remediation technology is also applied to DNAPL source zones and to the dissolved-phase chemicals emanating from the source zones. The oxidizing agents most usually adopted to treat hazardous pollutants in soil and

groundwater are: hydrogen peroxide, catalyzed hydrogen peroxide, potassium permanganate, sodium permanganate, sodium persulfate and ozone [ESTCP, 1999; ITRC, 2000; ITRC, 2004; US EPA, 2004c; ITRC, 2005b; US EPA, 2006b; US EPA, 2006c; SERDP, 2006; SERDP, 2007a; SERDP, 2007b; Tsitonaki et al., 2008; US EPA, 2009a; ESTCP, 2011a; SERDP, 2011];

- *Solidification/Stabilization*: solidification process consists in the encapsulation of contaminants by decreasing the exchange surface and/or by covering the polluted media with low-permeability materials; this technique involves mechanical processes or chemical reactions between contaminant materials and solidifying reagents, such as cement, kiln dust or lime/fly ash, and addition of water where required. Stabilization consists in the reduction of mobility or solubility of polluted materials through chemical reactions. Solidification and stabilization are applied to areas of sludge or soils; they can be also adopted as *ex-situ* techniques with the excavation and the subsequent remediation of solids in the treatment units [US EPA, 1986; US EPA, 1989a; US EPA, 1993; US EPA, 2000a; US EPA, 2002c; U.K. Environment Agency, 2004a; U.K. Environment Agency, 2004b; Paria and Yuet, 2006; US EPA, 2007a; ITRC, 2011b];

a) 2) ex-situ:

- *Air Stripping*: a hot air flow is forced through contaminated groundwater or surface water to extract pollutants; liquid contaminants evaporate and the gas is conveyed in an unit treatment and cleaned up. This technology is typically adopted to clean up groundwater as part of the pump and treat technique and it is mainly used to remove Volatile Organic Compounds (VOCs). [US EPA, 1989b; US DOE, 1995; Hand et al., 1999; Stocking et al., 2000; US Army Corps of Engineers, 2001; US DOE, 2001; Adams et al., 2002;US EPA, 2004c; Pour et al., 2005; The California MTBE Research Partnership, 2006; US EPA, 2007a];
- *Pump and Treat*: this technique is composed of two different treatment phases. A pump process is adopted to extract contaminated groundwater and a treat process is used to clean up the removed water. The most used treatments are: adsorption with activated carbon, air stripping and chemical oxidation [US EPA, 1990; US EPA, 1996a; US EPA, 1996b; US EPA, 1997d; US EPA, 2002a; US EPA, 2002b; US EPA, 2005a; US EPA, 2007b; US EPA, 2007c; US EPA, 2008a; US EPA, 2009b; ESTCP, 2011a];
- *Soil Washing*: a water-based system is adopted to remove pollutants adsorbed on fine soil particles, previously separated from bulk soil. The wash water may be enhanced with a basic leaching agent, surfactant or chelating agent or by adjustment of pH to improve extraction of organic contaminants and heavy metals. Soils and wash water are mixed in a treatment unit and separated after the treatment [US EPA, 1991; American Academy of Environmental Engineers, 1993; ITRC, 1997b; ITRC, 1997c; US EPA, 1997e; US EPA, 2002c; Levisauskas et al., 2004; Jang et al., 2005; Pearl et al., 2006; US EPA, 2007a; CL: AIRE, 2007; Dermont et al., 2008; US EPA, 2011a];

- *Thermal Processes* (*Thermal Desorption, Hot Gas Decontamination, Plasma High-Temperature Recovery, Pyrolysis, Thermal Off-Gas Treatment, Vitrification*): pollutants are destroyed or separated from the contaminated materials through the exposition to elevated temperatures. The ex situ technologies involve excavation of soils. The remediation of polluted media takes place in apposite treatment units such as treatment cells and combustion chambers. This type of processes needs shorter treatment time and is characterized by uniformity of action thanks to selection, homogenization and continuous mixing of polluted materials. Thermal desorption and hot gas decontamination are separation technologies; pyrolysis and thermal treatment destroy the pollutants; vitrification destroys or separates organics and immobilizes some inorganic contaminants [US EPA, 1997e; US EPA, 1997f; US DOE, 1998; NFESC, 1998; US NRL, 1999; US EPA, 2002c; US EPA, 2004b; US EPA, 2004c; US EPA, 2007a];

b) biological treatments:

b) 1) in situ:

- *Bioremediation*: organic pollutants in soil, sludge, and solids are degraded by microorganisms, that use contaminants as a food and energy source; this operation requires the control of the pH and the temperature. Bioremediation processes can be aerobic or anaerobic: aerobic processes require oxygen to take place and the end products are carbon dioxide and water; anaerobic processes do not require oxygen and the end products can be methane, sulphide, elemental sulphur, hydrogen and ammonia. Different microorganisms adapted for deterioration of specific contaminants can be used on purpose. Bioremediation can be also used as an *ex-situ* biological treatment; this requires the excavation of contaminated area. [US EPA, 1997b; US EPA, 2000b; US EPA, 2001a; ITRC, 2002; ITRC, 2004; US EPA, 2004c; Moretti, 2005; California Department of Toxic Substances Control, 2006; ICSS, 2006; ITRC, 2008a; Hazen, 2009a; Hazen, 2009b; US EPA, 2009a; Hazen, 2010; US EPA, 2010a; ESTCP, 2011a; ESTCP, 2011b];
- *Bioventing and Biosparging*: are forms of in situ bioremediation. For this operation, extraction wells are used to blow air through the soil to improve the growth of microorganisms degrading organic compounds [US EPA, 1995a; US EPA, 1995b; US EPA, 2001a; ESTCP, 2003; Wisconsin Department of Natural Resources, 2003; US EPA, 2004c; ICSS, 2006; Lambert et al., 2009; ESTCP, 2011a];
- *Monitored Natural Attenuation*: in some cases the pollution in soil and groundwater can be eliminated or attenuated by natural processes. Even if these processes naturally take place in most contaminated sites, suitable conditions are desirable or the decontamination will not be sufficiently rapid or complete. These conditions are usually monitored to be sure that natural attenuation is effective [US EPA, 1999; US EPA, 2002d; US EPA, 2004c; US EPA, 2004d; PNNL,

2006; SRNL, 2006; ITRC, 2007; SRNL, 2007; ITRC, 2008b; SERDP, 2008; US EPA, 2011b];

- *Phytotechnologies* (Phytoremediation): these are defined as technologies that apply vegetation to remove and destroy, to control and immobilize, or both, pollutants present in soil, sediment, surface water and groundwater [ITRC, 1997d; US EPA, 2000c; US EPA, 2001b; US EPA, 2002c; ITRC, 2003b; Kamath et al., 2004; McClain et al., 2004; US EPA, 2005b; US EPA, 2006b; US EPA, 2007d; ITRC, 2009; US EPA, 2010b; ESTCP, 2011a];

b) 2) ex-situ:

- *Bioreactor landfills* : degradation and stabilization of organic materials take place through the addiction of liquid and air to improve microbiological growth in dedicated treatment units. There are three bioreactor configurations: aerobic, anaerobic and hybrid [Costello, 2003; US EPA, 2005c; ITRC, 2006; FNADE-ADEME, 2007; US EPA, 2007e; US EPA, 2008b; Townsend et al., 2008].

PRB for Groundwater Remediation

Permeable Reactive Barriers (PRBs) are *in situ* technology used to remediate contaminated groundwater. It is a cost effective and passive system composed of a reactive medium removing the contaminants in the groundwater as the plume passes through it and to achieve the restoration of the groundwater. The separation of pollutants from the aquifer takes place by conversion to less hazardous compounds, using reactive medium (Permeable Reactive Barriers, PRB) and/or by adsorption on porous active medium (Permeable Adsorbing Barriers, PAB).

In a PRB/PAB treatment, the barrier is commonly built with reactive materials whose hydraulic conductivity is higher than that of the surrounding soils, so that the contaminated groundwater is forced to pass through the barrier itself, moving under natural hydraulic gradient [Di Natale F. et al., 2007; Erto et al., 2010a; Gibert et al., 2011; Kacimov et al., 2011; Liu et al., 2011]. The mechanism and the effectiveness of action of a PAB depends on the choice of the reactive material used to build the barrier and its reactivity with the target pollutant/pollutants [Di Natale M. et al., 2007; Di Nardo et al., 2010b; Erto et al., 2010a; Gibert et al., 2011; Miller et al., 2011]. This technology represents a good alternative to classic groundwater remediation methods, for its low operational and maintenance costs, often characterized by very high volumes and low pollutant concentrations. Usually, PRB/PABs are used to remove heavy metals [Eljamal et al., 2011; Han et al., 2011; Jun et al., 2009; Ouellet-Plamondon et al., 2011; Park et al., 2011; Zhang et al., 2011; Calabrò et al., 2012], Light and Dense Non-Aqueous Phase Liquids (LNAPLs and DNAPLs), chemical fertilizers [Hosseini et al., 2011; Suzuki et al., 2012], and for the prevention of aquifer pollution [Bartzas and Komnitsas, 2010].

The design of a PRB/PAB for the remediation of a polluted aquifer, for which a hydraulic, geotechnical and contaminant characterization has been previously performed, is very complex and it mainly consists in the definition of the barrier material and location, orientation and dimensions [Di Natale M. et al., 2007; Di Natale et al., 2008a; Erto et al.,

2011a]. Design choices are very important and a good compromise between technical and economic issues has to be found.

The remediation of a polluted site is a very expensive process hence a careful design based on the knowledge of both specific site hydrology and contaminant plume characteristics is necessary [Di Nardo et al., 2010a; Erto et al., 2011a]. In recent years, computer codes have been developed to describe groundwater and contaminant transport in order to determine the optimal remediation strategy. Traditionally, several optimization strategies applied to Pump and Treat technology (PT) have been developed in order to minimize operational costs, using either time-invariant [Wang and Ahlfeld, 1994; Guan and Aral, 1996; McKinney and Lin, 1996] or time-varying optimization techniques [Minsker and Shoemaker, 1998]. They are based on linear and non-linear programming techniques and are mainly aimed at finding the best strategies to identify wells number and location, and to define optimal pumping rates at the wells.

Other simulation-based optimization procedures for groundwater remediation rely on genetic algorithm [Hsiao and Chang, 2002], fuzzy logic [He et al., 2008a] and statistical inference techniques [He et al., 2008b; He et al., 2009] and are also aimed at identifying the optimal strategies among a number of alternatives [Lu et al., 2007]. These studies, mainly referred to PT and surfactant-enhanced flushing remediation techniques, demonstrate that simulation and optimization are effective tools for supporting decision-makers in groundwater remediation design. Many commercial computer codes can be used to support PRB/PAB design, but no specific dedicated software has been developed yet. Furthermore, the modelling scales of transport and adsorbing processes are very different. In fact, the thickness of a barrier is very small as compared to the size of the surrounding aquifer and to the distance from the pollution source, hence the computational time of contaminant transport is much longer than the permanence time of the contaminant inside the barrier. For these reasons, simultaneous modelling at a small and large scale, in time and space, cannot be approached by direct computation in a unique simulation framework and separate simulations are often necessary [Erto et al., 2011a]. Therefore, in PRB/PAB design, also a less-than-exact optimization procedure can be helpful to support designers. In particular, in this chapter an original heuristic approach to optimize the design of a PAB made of granular activated carbon is proposed.

Description of PAB Modelling

In this paragraph the basic principles of PAB are described. A Permeable Reactive Barrier consists of a wall of reactive material (such as granular iron or other typically reduced metal, lime, electron donor-releasing compounds, electron acceptor-releasing compounds or adsorbing material, such as activated carbon, metal oxide, etc.), whose hydraulic conductivity is higher than that of the surrounding soils, installed in the path of a plume of contaminated groundwater (Figure 2.1).

Figure 2.1. Permeable Reactive Barrier (PRB).

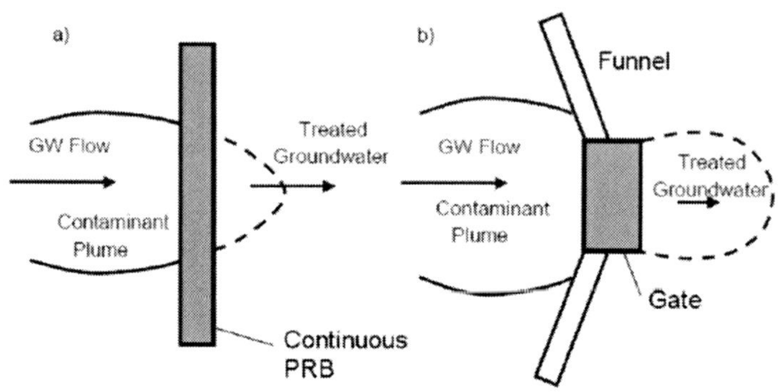

Figure 2.2. PRB configurations: a) Continuous barrier; b) Funnel and gate.

As the groundwater flows through this zone, without any external energy input, the contaminants are degraded to innocuous components through chemical and/or biological reactions, captured on the adsorbing media, or chemically altered so that they form insoluble precipitates. PRB can be installed at the downgradient edge of the plume as a containment system and/or immediately downgradient of the source area to prevent further migration of elevated levels of contaminant mass. Several configurations of in situ PRB systems are feasible, and the applicability of this technology depends on the geologic and hydrogeologic conditions, as well as on the contaminant distribution in the vertical and horizontal directions at a specific site.

Two different configurations can be used in full-scale implementations of reactive barriers:

1. The continuous trench; and
2. The funnel and gate (Figure 2.2),although other designs are being researched and evaluated [US EPA, 1998].

The continuous PRB is the simplest configuration, in which the barrier and, subsequently, the reactive media extend across the entire width of the contaminant plume (Figure 2.2a). This system can be installed by digging a trench and filling it with permeable material to create an artificially permeable environment.

a)

b)

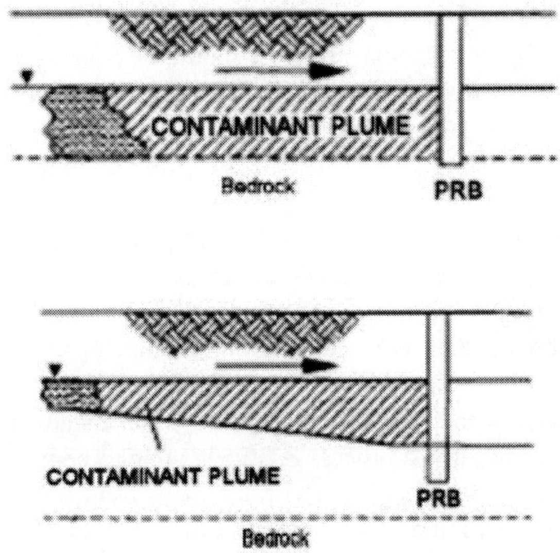

Figure 2. 3. PRB configurations: a) penetrating PRB system and b) hanging PRB.

The funnel and gate configuration (Figure 2.2b), instead, consists of an impermeable funnel, typically made up of interlocking sheet pilings or slurry walls, emplaced to enclose and direct the flow of contaminated water toward a gate or gates containing the permeable zone of reactive material. This latter can be implemented using various configurations, depending upon the plume width and depth and the type of mass transfer reaction required. The options for the funnel (cut-off walls) include sheet pile walls or slurry walls and options for the gate include permeable non-reactive materials such as pea gravel (for air sparging) or oyster shells (for biodegradation) and reactive materials such as activated carbon or zero-valence iron. Funnelling or directing the groundwater toward a treatment gate may increase the natural groundwater flow velocity several times. Funnel and gate designs need to extend beyond the extent of the plume to ensure that all the contaminated groundwater is captured an treated. The goal of a continuous barrier or a funnel and gate is to ensure that contaminant flow beneath, around, or above the system does not occur.

PRB systems can be constructed through the entire thickness of an aquifer, if the contamination extends across the full depth of an aquifer, or, when the contaminant plume is shallow, penetration of the system may be required only to the depth of contamination. The above two configurations are known as a fully penetrating PRB system and a hanging PRB system, respectively (Figure 2.3).

The variety of treatment materials available for use in PRB applications is an important issue because it allows the customization of PRB design, based on specific site characteristics.

The choice of suitable reactive media depends on many factors, of which the most important is the chemical composition of the contaminant.

The most used chemical or physical processes are degradation, precipitation and sorption. The degradation processes consist of chemical or biological (aerobic or anaerobic) reactions that lead to the decay of contaminant to less dangerous compounds. These reduction reactions may induce solid precipitation that can wear the barrier and seriously affect its removal efficiency, diminishing its porosity and conductivity [Mackenzie et al., 1999; Kamolpornwijit et al., 2003; Li et al., 2006]. Moreover, for some compounds, dechlorination reactions are very slow at high water flow-rates, hence barriers should be very thick in order to extend the contaminant residence time in the system [Moon et al., 2005]. An example of chemical degradation can be oxidation of zero valent iron Fe^0. This latter is the most widely used reactive material in PRBs owing to its success in treating common organic and inorganic contaminants in groundwater, such as chlorinated volatile organic compounds (VOCs) [US EPA, 1998; Vogan et al., 1999; Farrell et al., 2000; D'Andrea et al., 2005; Moon et al., 2005; Plagentz et al., 2006; Higgins and Olson, 2009; Jun et al., 2009].

Precipitation is the transformation of contaminants to the insoluble solid forms that are captured in the reactive media. For example, the reduction of some metals that are precipitated in the form of hydroxides or sulfites can be favored by an increase of pH.

Sorption is the removal contaminant from polluted water by adsorption or ion exchange, using specific reactive media. It is a process relatively simple (compared to others), and can achieve satisfactory removal efficiencies. The most often used media is granular activated carbon, zeolites and peat for removal of organic compounds and heavy metals [Czurda e Haus, 2002; Lorbeer et al., 2002; Ake et al., 2003; Komnitsas et al., 2006; Di Natale M. et al., 2007; Di Natale F.et al., 2007; Di Natale et al., 2008a; Di Natale et al., 2008b; Di Natale et al., 2009; Di Natale et al., 2011].

The main advantages of PRB technology are:

- Availability of the site for other uses or activities after installation of a PRB;
- Passive in-situ remediation of groundwater without external energy source;
- Capacity to treat chlorinated hydrocarbons to very low or non-detect levels;
- Lower influence to the groundwater regime;
- More cost efficacy than pump-and-treat systems for long-term unattended operation;
- Low acquisition and operational costs.

Conversely, the main disadvantages of PRB are:

- Unknown long-term effects from chemical and/or biological precipitate formation;
- Construction complications deriving from subsurface utilities and/or above ground structures;
- Limitationto depths of less than 50 feet using current construction technologies;
- Necessity to remove or replace the PRB reactive media when exhaust.

Fate and Transport of Contaminants in Porous Media

In this paragraph some basic hydrogeological concepts, and the basic equations of groundwater motion are introduced, paying particular attention to transport equation in porous media.

The complex system of solids and voids, filled with fluids, that is the aquifer, can be simplified as a porous media [Bear, 1979]. The balance laws for flow are obtained formally through integration from the microscopic level (the level of observation of a fluid particle) to the macroscopic level (the level of observation of the porous medium). Even if problems could be described and solved at the microscopic level, measurements cannot be taken at that level (i.e., at a point within the void space), in order to validate the model. To overcome this difficulty, the porous medium domain is visualized as a continuum with fluid or solid matrix variables defined at every point.

The transition from microscopic to macroscopic studies of transport phenomena is achieved via the introduction of a Representative Elementary Volume (REV) of the porous medium domain (Figure 2.4).

A REV is defined as the smallest volume in the domain over which the averages of its fluid and solid properties are independent of small perturbations in its size; each domain can then be identified as a portion of space occupied by a number of fluid phases and a solid phase (i.e., the solid matrix).

The rate of water flow through a porous media depends on the properties of the water, the properties of the porous media, and the gradient of the hydraulic head, as represented by Darcy's law, which can be written as:

$$q_i = -K_{ij} \cdot \frac{\partial h}{\partial x_j} \qquad i, j = 1,2,3 \tag{2. 1}$$

where q_i is the specific discharge, [LT^{-1}]; K_{ij} is the hydraulic conductivity of the porous medium (a second-order tensor), [LT^{-1}]; and h is the hydraulic head, [L].

The specific discharge has units of velocity and is also known as the Darcy velocity (v). It should be noted that the Darcy velocity is an artificial velocity as it corresponds to the total quantity of water flowing divided by the cross-sectional area. The average velocity, v_w of the water can therefore be calculated as the Darcy velocity divided by the effective porosity, n_e:

$$v_w = \frac{v}{n_e} \tag{2. 2}$$

The equations used to describe the transient flow of a compressible fluid in a non-homogeneous anisotropic aquifer, in groundwater applications, is formulated by applying the law of conservation of mass, which explains that the mass quantity leaving or entering a specified volume of aquifer during a given time interval is equal to the change in the amount of that quantity stored in the volume. Consequently the general form of the equation derived by combining Darcy's law with the continuity equation leads to:

$$\frac{\partial}{\partial x_j}\left(K_{ij} \frac{\partial h}{\partial x_j} \right) = S_s \frac{\partial h}{\partial t} + Q \tag{2. 3}$$

where i and j are the principal coordinate directions (equal to 1,2,3), K_{ij} is the hydraulic conductivity, h is the aquifer head, S_s the specific storage, Q the local sources or sinks per unit volume and t is the time.

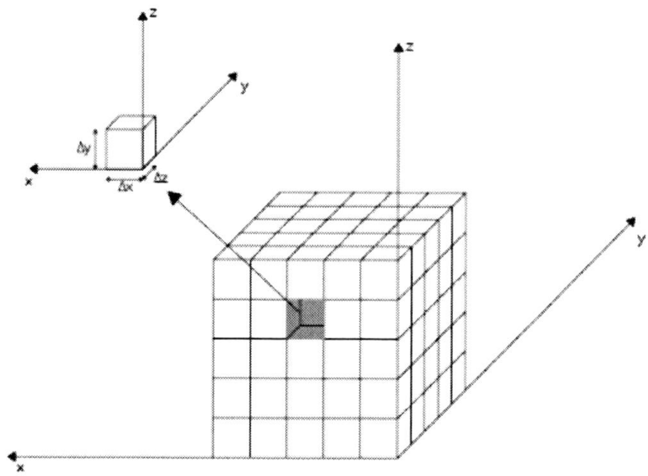

Figure 2. 4. Representative Elementary Volume (REV).

The specific storage, or specific storage coefficient S_s in equation (2. 3) represents the amount of water released from storage or stored per unit of volume in an aquifer, per unit of change in head. Darcy's law gives three equations for a flow in three directions. The flow equation yields the fourth equation for the four unknowns: the three components of groundwater flux and the head.

The transport of solutes in the saturated zone is governed by the Advection-Dispersion Equation (ADE) which for a porous medium with uniform porosity distribution is formulated as follows:

$$\frac{\partial C}{\partial t} = \frac{\partial}{\partial x_i}\left(D_{ij} \frac{\partial C}{\partial x_j} \right) - \frac{\partial}{\partial x_i}(v_i C) + \frac{q_s}{n_s} C_s + \sum_{k=1}^{N} R_k \qquad (2.4)$$

where:

C is the concentration of contaminants dissolved in groundwater, [ML^{-3}];

t is time,[T];

x_i is the distance along the respective Cartesian coordinate axis,[L];

D_{ij} is the hydrodynamic dispersion coefficient, [L^2T^{-1}];

v_j is the seepage or linear pore water velocity, [LT^{-1}];

q_s is the volumetric flux of water per unit volume of aquifer representing sources (positive) and sinks (negative), [T^{-1}];

C_s is the concentration of the sources or sinks, [ML^{-3}];

n_s is the porosity of the porous medium, dimensionless;

$\sum_{k=1}^{N} R_k$ is a chemical reaction rate, [$ML^{-3}T^{-1}$].

The first term on the right side of eq. II.4 represents the change in concentration due to hydrodynamic dispersion. Dispersion in porous media refers to the spreading of contaminants over a greater region than that would be predicted from the groundwater velocity vectors. As described by Anderson [1984], dispersion is caused by mechanical dispersion, as a result of deviations of actual velocity on a microscale from the average groundwater velocity, and molecular diffusion, as a result of concentration variations. The molecular diffusion effect is generally secondary and negligible as compared to the mechanical dispersion effect, and only becomes important when groundwater velocity is very low. The sum of the mechanical dispersion and the molecular diffusion is termed hydrodynamic dispersion. The hydrodynamic dispersion tensor for isotropic porous media is defined as:

$$D_{ij} = \alpha_{ijef} \frac{v_e v_f}{|v|} + D_m \qquad i,j,e,f = 1,2,3 \tag{2. 5}$$

where α_{ijef} is the dispersivity of the porous medium [L]; v_e and v_f are the components of the flow velocity of the fluid in the e and f directions, respectively [$L\ T^{-1}$]; D_m is the effective coefficient of molecular diffusion [$L^2\ T^{-1}$]; and $|v|$ is the magnitude of the velocity vector [$L\ T^{-1}$] [Scheidegger, 1961; Konikow and Grove, 1977; Bear, 1979; Domenico and Schwartz, 1990]. The dispersivity α_{ijef} is a measure of the heterogeneity present in the aquifer and expresses the microscopic configuration of the solid-liquid interface. A very heterogeneous porous medium has a higher dispersivity than a slightly heterogeneous porous medium. It is a fourth-order tensor and for an isotropic porous medium is related principally to the longitudinal dispersivity α_L, and the transversal dispersivity α_T [Bear and Verruyt, 1987]. Consequently the components of the mechanical dispersion tensor may be expressed as follows:

$$D_{ij} = \begin{bmatrix} \alpha_L \frac{v_x^2}{|v|} + \alpha_T \frac{v_y^2}{|v|} + D_m & (\alpha_L - \alpha_T)\frac{v_x v_y}{|v|} + D_m \\ (\alpha_L - \alpha_T)\frac{v_x v_y}{|v|} + D_m & \alpha_L \frac{v_y^2}{|v|} + \alpha_T \frac{v_x^2}{|v|} + D_m \end{bmatrix} \tag{2. 6}$$

The second term represents advective transport and describes the movement of solutes at the average seepage velocity of the flowing groundwater.

For many practical problems concerning contaminant transport in groundwater, the advection term dominates. In order to measure the degree of advection domination, a dimensionless Peclet number is used, that is defined as:

$$Pe = \frac{|v|\delta}{D} \tag{2. 7}$$

where δ is a characteristic length, commonly taken as the grid cell width, [L] and D is the dispersion coefficient, $[L^2T^{-1}]$.

The third term represents the effects of solute mass dissolved in water entering the simulated domain through sources or injections, or solute mass dissolved in water leaving the simulated domain through sinks.

The fourth term lumps all of the chemical, geochemical, and biological reactions that cause transfer of mass between the liquid and solid phases or conversion of dissolved chemical species from one form to another.

The chemical attenuation of inorganic chemicals can occur by sorption/desorption, precipitation/dissolution, or oxidation/reduction; organic chemical can adsorb or degrade by microbiological processes.

Considering the third term as negligible the equation (2. 4) can be written as:

$$\frac{\partial C}{\partial t} = \frac{\partial}{\partial x_i}\left(D_{ij}\frac{\partial C}{\partial x_j}\right) - \frac{\partial}{\partial x_i}(v_i C) + \sum_{k=1}^{N} R_k \qquad (2.\ 8)$$

In the case of a Permeable Adsorbing Barrier in which adsorption phenomena take place, the $\sum_{k=1}^{N} R_k$ term of general equation (2. 4) is represented by adsorbing phenomena occurring into the barrier as, under isothermal conditions, groundwater flow can be considered as steady and no volumetric sinks (leakage, infiltration, evaporation, etc.) occur. This sink term can be expressed as follows:

$$\sum_{k=1}^{N} R_k = \frac{\rho_b}{n_b}\frac{\partial \omega}{\partial t} = K_c a\left[C - C^*(\omega)\right] \qquad (2.\ 9)$$

In equation (2. 9), ω represents the pollutant concentration on solid, ρ_b the dry adsorbing material bulk density, n_b the barrier porosity and $C^*=C^*(\omega)$ derives from the adsorption isotherm and defines the mass transfer driving force in the transport model equations.

Further explanations about adsorption phenomena are provided in the subsequent paragraph.

It is important here to recall that adsorbing material properties and pollutant adsorption isotherms have to be known or they have to be measured in dedicated laboratory tests in order to completely define the eq. 2. 4.

As to initial conditions, the initial pollutant liquid concentrations are determined and the corresponding solid concentrations are assumed to be zero, throughout the entire flow domain.

The boundary conditions are as follows:

$$C = 0 \qquad \begin{Bmatrix} x = 0 & \forall y \forall t \\ y = 0 & \forall x \forall t \\ y = Y & \forall x \forall t \end{Bmatrix} \tag{2. 10}$$

$$\frac{\partial C}{\partial t} + \frac{\partial}{\partial x_i}(v_i C) - \frac{\partial}{\partial x_i}\left(D_{ij} \frac{\partial C}{\partial x_j} \right) = 0 \quad x = X \quad \forall y \forall t$$

where X, in a coordinated system with x-axis parallel to groundwater flux direction, is the distance between the barrier and the western boundary of the polluted domain, and Y is the extension of the domain in y direction.

Adsorption Modelling

Adsorption is a mass transfer phenomenon in which a molecule of a chemical species in a fluid phase (liquid or gas) binds to the surface of a porous solid, thanks to the presence of attractive forces at solid-fluid interface. The species subject to adsorption is commonly referred to as *(ad)sorbate*, while the solid phase is called *(ad)sorbent.* Adsorption directly involves the outer surface of the solid adsorbent and adsorption capacity (ω), as the mass of adsorbate captured by the solid per mass of adsorbent is usually related to the extension of its specific surface. Therefore, the most widely used adsorbent materials are those with high porosity i.e. activated carbon, silica gel, some synthetic polymers, alumina and some natural or waste materials that, if properly treated, can be used for this purpose. The choice of solid material is carried out according to the pollutant to remove. As an example, charcoal and carbonaceous solids with reduced surface charge are more suitable than other sorbents for the adsorption of molecules with low polarity, such as many organic compounds [Stumm and Morgan, 1996].

Adsorption processes are widely used in water and wastewater treatment, as the latest scientific guidelines count it among the BATs (best available techniques) for water depuration [European IPPC Bureau, 2003].

The performance of an adsorption treatment mainly depends on the thermodynamic aspects of solute-solvent-sorbent interactions and on the transport phenomena involving the diffusive-convective transport within the porous media. The equilibrium conditions are the most significant limits for the application of a given sorbent since the uptake of a specific compound may change within orders of magnitude by varying the process parameters, such as concentration, pH, ionic strength, temperature and chemical composition of the aqueous solution [Suzuki, 1990; Do, 1998].

The characterization of the adsorption process can be done either from a thermodynamic point of view, in order to have information about adsorption capacity and its dependence on the main process parameters, or by a kinetic standpoint, focusing on fluid-dynamic characterization, mass transport from fluid to solid phase and, more in general, on the rate of the overall process.

From a thermodynamic point of view, adsorption is a spontaneous process ($\Delta G<0$) and is characterized by a decrease in the entropy of the substance adsorbed and incorporated into the solid ($\Delta S<0$). Since, at constant temperature and pressure, the following relationship is valid:

$$\Delta G = \Delta H - T\Delta S \tag{2. 11}$$

It is easy to deduce that adsorption is a exothermic process, i.e. characterized by decrease of enthalpy, also referred as heat of adsorption, ($\Delta H<0$) and, therefore, favored by low temperatures.

On the basis of the nature of the forces which are involved in attraction between adsorbent and adsorbate, it is common to distinguish two different types of adsorption:

- Physical adsorption;
- Chemical adsorption.

In physical adsorption, the contaminant molecule is bound to surface of the adsorbent material by weak electrostatic forces with very low binding energy (heat of adsorption equal to or less than 5 kJ/mole). These forces are always attractive and allow the creation of bonds; commonly, interactions between molecules with permanent dipoles are denominated Dipole-Dipole bonds (Van der Walls forces), while those between temporary dipoles induced by fast electron density fluctuation of a molecule or molecules with uneven charge distribution, are termed dispersion forces or London forces [Stumm and Morgan, 1996]. Usually the process is not dissociative and adsorbate is bound to the surface as a molecule and independently from its geometric shape.

In chemical adsorption the molecule is adsorbed through proper chemical bonds (heat of adsorption order of 20-100 kJ/mole) and, therefore, almost always in a dissociative manner. Unlike physical adsorption, chemical adsorption can be done exclusively in monolayer [Ruthven, 1984]. The link may be ionic or covalent type, and in any case leads to a substantial change of electronic configurations. Moreover, in this case, adsorption is much more selective and in many cases even irreversible.

Adsorption Thermodynamic Models

Thermodynamic study and modeling of adsorption processes are carried out by means of adsorption isotherms representing the relationships between the mass of solute (adsorbate) bound per gram of the adsorbent solid, ω, and the concentration of solute C in the fluid, at constant temperature. In the case of specific interactions between adsorbates and the active sites on solid surface, the most accredited theory is that proposed by Langmuir [1918]. In this case it is believed that the process is modeled by a chemical *pseudo*-reaction between adsorbate and active sites, the latter characterized by the same energy of interactions with the adsorbate molecules.

Moreover, it is assumed that no further interactions between adsorbed molecules and with the solvent are present and that adsorption takes place in monolayer. The Langmuir model is among the most used in the description of real systems, particularly in the case of adsorption of ions from aqueous solutions, precisely because it presupposes the existence of specific interactions between the solute and surface of the adsorbent solid.

Although with somewhat restrictive assumptions, this model allows for an accurate description of the phenomenon when the adsorbed quantity is low and this has no interaction with the solvent. In particular, the Langmuir model is able to adequately describe the case of a single layer of molecules adsorbed on the surface of the solid. When this eventuality cannot be hypothesized, two different models can be considered. In the presence of low concentrations of solute bound, lower than those needed to achieve the full coverage of the solid surface, the adsorption process can be described by a linear relationship between the concentration of the solute in solution and the adsorption capacity, with a similar formula, even conceptually, to Henry's law for absorption. Conversely, when adsorption is hypothesized to take place on more than one layer, it is necessary to consider the existing interactions between the molecules adsorbed on intermediate layers and the solute in solution. The descriptive model for such a phenomenon has been developed by Brunauer, Emmett and Teller [1938] and is usually referred to as BET isotherm.

This equation is based on the assumption that the number of sites is exactly defined and the solute bound to the first layer follows the assumptions of the Langmuir model, while the one bound to the first successive layers adsorb with a mechanism similar to condensation. As a result, the enthalpy of adsorption for the first layer, constant for all sites in, is different than that for the successive layers, which is equal to the enthalpy of liquefaction. The Frumkin isotherm, also known as Fowler-Guggenheim isotherm, is based on the assumptions of the Langmuir model but in his formulation takes into account the interaction between adsorbed molecules [Stumm and Morgan, 1996]. Modeling of adsorption phenomena may also take into account of the existence of a distribution of the interaction energy of the active sites present on the surface of the solid adsorbent. The Langmuir isotherm is representative of a system with a single energy level for all active sites, as descriptive of the ideal case of a reaction between adsorbed species and a specific active site. If we assume a uniform distribution of energies of the active sites on the solid surface, the adsorption isotherm is expressed by the relation of Temkin [1941].

The most common case is the one in which, on the surface of the solid, the lower active sites are the most popular, while increasing their energy level results in a decrease in the number of active sites that can be associated to that level (i.e. a decreasing exponential distribution of energies of the active sites).

In this case, the functional form of the adsorption isotherm is expressed by the equation of Freundlich [1878]. Table 2.1 summarizes the adsorption isotherms most commonly used, the respective fields of application and the meaning of parameters present in each of them. It should be noted that the Freundlich isotherm is reduced to that of Henry if $n=1$. This represents the condition of the same binding energy for all active sites. This condition is identical to that of Langmuir, which also reduces the isotherm Henry if the concentration C tends to zero.

Multicomponent Adsorption Models

Real systems are often characterized by the simultaneous presence of different pollutants, then in the analysis of adsorption it is appropriate to consider multiple chemical species adsorbing on the same surface with different mechanisms, which may also involve a competition to the same active sites [Do, 1998; Erto et al., 2011b; Erto et al., 2012].

The design of adsorption units mainly depends on sorbent adsorption capacity in equilibrium conditions and their performance is influenced by the multicomponent competitive interactions of the compounds simultaneously present in solution.

Table 2. 1. Adsorption isotherms for single compound systems

Isotherm	Parameter	Properties
Langmuir: $\omega^* = \frac{\omega_{MAX} K C^*}{1 + K C^*}$	K = Langmuir equilibrium constant; ω_{MAX} = maximum adsorption capacity (monolayer) [mg/g].	Monolayer. No solute-solvent interactions. Active sitesisoenergetic: $\phi(E) = \delta(E°)$.
Freundlich: $\omega^* = K C^*$	$K = aRT\pi/\sin(n\pi)$; T = temperature [°K]; n = depends on binding energy, $0 < n < 1$.	Exponential distribution of active sites: $\phi(E) = \alpha \exp(-\alpha_2 E)$.
Langmuir - Freundlich: $\omega^* = \frac{\omega_{MAX} K C^{*n}}{1 + K C^{*n}}$	K = Langmuir-Freundlich equilibrium constant; ω_{MAX} = maximum adsorption capacity (monolayer) [mg/g] n = empiricalparameter.	Non-linear distribution of active sites:
Temkin: $\omega^* = p + q \cdot \ln C^*$	p and q = empirical constants.	Uniform distribution of active site: $\phi(E) = \cos t$.
Henry: $\omega^* = H^* C^*$	H^* = Henry constant = $K\omega_{MAX}$.	Infinite dilution of adsorbate.
B.E.T.: $\omega^* = \frac{\omega_{MAX} K C^*}{[1 - (C^*/C_S)(KC^*/C_S)][1 - (C^*/C_S)]}$	K = equilibrium constant; C_S = water solubility of the adsorbate [mg/l].	Validity of Langmuir assumptions for the first layer. Multilayer. Heat of adsorption variable for the other layers.
Frumkin (Flower – Guggenheim): $\theta = \frac{K C^* \exp(\alpha\theta)}{1 + K C^* \exp(\alpha\theta)}$	K = equilibriumconstant; $\theta = \omega^* / \omega_{MAX}$.	Monolayer. Interactions between adsorbed molecules:

Hence, the assessment of the adsorption capacity of a compound in the presence of other solutes as well as the definition of appropriate theoretical models is necessary for a proper design of a cost-effective depuration system.

The interference and competition phenomena arising in these systems can be described, and in some cases also predicted, by specific adsorption models whose mathematical formulation can be more or less complex, depending on the physical adsorption mechanism hypothesized. The equations representing the adsorption models investigated are reported in the following, together with the description of the main assumption and basic hypotheses behind their formulation.

Many of the adsorption models previously described for single compound systems can now be applied to multicomponent adsorption phenomena. In particular it is possible an extension of Henry, Langmuir and Freundlich to such systems.

Henry model turns out to be simply additive, as it suggests the absence of interactions between different chemical species to adsorb. Therefore, in the case of adsorption of different chemical species *N*, the expression of the isotherms of Henry turns out to be [Benjamin, 2002]:

$$\omega = \sum_{k=1}^{N} H_k^* C_k \qquad (2.\ 12)$$

where:

C_k is the concentration of the i-th species [ML^{-3}];

H^*_k, is the Henry constant correspondent.

As to Freundlich isotherm, the typical expression for a system of *N* total species present in solution and in competition with the solute to the same active sites is as follows [Benjamin, 2002]:

$$\omega_A = K_A C_A \left(\sum_{k=1}^{N} a_{Ak} C_A \right)^{\left(n_A^{-1}\right)} \qquad (2\ .13)$$

where:

C_A, is the concentration of species [ML^{-3}];

K_A is the Freundlich coefficient;

a_{Ak}, is a parameter that takes into account for the interactions between the *A* species and the k-th of the other *M* species present in solution.

The Langmuir model for multicomponent adsorption systems is based on the same assumptions as its single-compound counterpart.

Unlike the other two models, Langmuir Multicomponent Model allows to distinguish between different operating conditions [Benjamin, 2002]:

- Adsorption of N species on different active sites, i.e., in the absence of competition:

$$\omega = \sum \omega_k = \sum_{k=1}^{N} \frac{\omega_{MAX} K_k C_k}{1 + K_k C_k} \tag{2. 14}$$

where:

ω_k is the adsorption capacity of the k-th species [-];

ω_{MAX}, is a parameter representing the maximum capacity of the solid adsorption [-];

K_k, is the constant of Langmuir on the k-th species [L^3M^{-1}];

C_k is the equilibrium concentration of the k-th species [ML^{-3}].

With these assumptions, the value of ω_k is equal to the single-component case.

- Adsorption of N chemical species, including the solute A, competing for the same active sites:

$$\omega_A = \omega_{MAX} \frac{K_A C_A}{1 + \sum_{k=1}^{N} K_k C_k} \tag{2. 15}$$

In this case, the total adsorption capacity of the solid towards all the species, is given by:

$$\frac{\omega}{\omega_{MAX}} = \frac{\sum_{k=1}^{N} K_k C_k}{1 + \sum_{k=1}^{N} K_k C_k} \tag{2. 16}$$

In all the above cases, it is worth noticing that the adsorption constants of all compounds ($\omega_{k,MAX}$ and K_k) should be derived individually from single-component adsorption tests, as they represent intrinsic properties of the sorbate-sorbent couple [Do, 1998]. Moreover, the above equations are valid only if ω_{MAX} is a constant value for all the involved compounds as it is an intrinsic property of the adsorbent.

The Ideal Adsorbed Solution Theory (IAST) was originally proposed by Myers and Prausnitz [1965] for gas mixture adsorption and then extended to liquid solutions by Radke and Prausnitz [1972]; it provides a thermodynamically consistent method for predicting multicomponent adsorption isotherms using single-component isotherm data. The IAST is based on the assumption that the adsorbed mixture forms an ideal solution in equilibrium with liquid phase at a constant spreading pressure for each solute (π_k). The spreading pressure of a solute is an intensive thermodynamic variable for adsorption equilibria; it is defined as the difference between the interfacial tension of the pure solvent-solid interface and that of the solution-solid interface [Ruthven, 1984].The liquid-solid equilibrium, in analogy with Raoult's law for vapour –liquid equilibria, is represented by the following equation:

$$C_k = C_k^0(T,P,\pi) \cdot z_k \qquad k = 1..N \tag{2.17}$$

where C_k is the equilibrium liquid concentration of solutes in a multicomponent system, C_k^0 is the single-solute (single-component) liquid concentration in equilibrium with ω_k^0, the single-component adsorption capacity, at same pressure, temperature and spreading pressure of a multicomponent system, and z_k, is the mole fraction on carbon surface, that can be expressed as:

$$z_k = \frac{\omega_k}{\omega_T} \qquad k = 1...N \tag{2.18}$$

$$\omega_T = \sum_{k=1}^{N} \omega_k \tag{2.19}$$

In this equation ω_T is the total surface loading, i.e. the sum of solutes adsorption capacities (ω_k) in a multicomponent system.

The IAST model also assumes:

(i) Identical chemical potentials of adsorbed and liquid phases, resulting in equation (II.17);
(ii) A thermodynamically inert solid;
(iii) A negligible effect of pressure on liquid phase properties and
(iv) A constant adsorption area per mole of solute, for the mixture or the single-solute, at the spreading pressure of the mixture. These properties are expressed by equation (2. 20):

$$\frac{1}{\omega_T} = \sum_k \frac{z_k}{\omega_k^0} \qquad k = 1..N \tag{2. 20}$$

The spreading pressure of the mixture is equal to that of single-compound systems and it is defined, starting from Gibbs adsorption isotherm, by the following equation [Radke and Prausnitz, 1972]:

$$\pi_k\left(C_k^0\right) = \frac{RT}{A} \cdot \int_0^{C_k^0} \frac{\omega_k^0\left(C_k^0\right)}{C_k^0} dC_k^0 \qquad k = 1..N \tag{2. 21}$$

where the integrand is represented by the single-component adsorption isotherm, A is the specific adsorbent area, R is the universal gas constant and T is the absolute temperature.

Multicomponent adsorption isotherms can be predicted using the equation set (2. 17)-(2. 21) by assigning C_k values and calculating the corresponding ω_k values, starting from the single-component adsorption isotherms, that can be known by any adsorption model [Langmuir, Freundlich, etc.].

Kinetic Aspects of Adsorption

Adsorption process can be described also from a kinetic point of view through the study of kinetic and fluid-dynamic parameters, including flow rate, velocity and time of solid-liquid contact in the plant configuration selected for the conduction of the process. The analysis includes an assessment of mass transfer mechanisms and of the diffusivity of pollutant molecules. More generally, it is possible to say that adsorption processes are characterized by complex kinetic phenomena involving elementary step:

a) Extra-particle transport mechanisms:
   1) Fluid-dynamic motion field: mixing of fluid phase, which determines the conditions of transport from particle to fluid;
   2) External particle mass transfer: convective-diffusive phenomena determining mass transfer from the surrounding fluid phase to the surface of the adsorbent particles and vice-versa;
b) Intra-particle transport mechanisms:
   3) Diffusion in macropores/micropores: namely the diffusion of solute molecules inside the pores of adsorbent particles;
   4) Surface diffusion: the migration of solute molecules on the surface of the adsorbent particles that contributes to the transport of the same particle inside;
c) Chemical reaction between solute and active sites.

The mass transport of solutes from the solution to the solid surface can be presented with a series of successive steps, each of which represents a different contribution to the overall resistance:

- External transport: transport through the water layer (film) more directly in contact with the solid particle;
- Internal diffusion: transport within the pores until the solute reaches the active site;
- Adsorption: formation of the bond between the adsorbent solid and the adsorbate.

The last two mechanisms are in series with the first but act in parallel with each other. In any case, the slowest of them represents the limiting step and therefore determines the overall rate of adsorption. Generally, under the turbulent regime, the diffusion inside the pores offers the greatest resistance.

In general, the rate of the process decreases with the increase in the size of the molecule, because the diffusion coefficients decrease. Moreover, when the controlling mechanism is the internal diffusion, the rate adsorption decreases when adsorbent pore size decreases.

Conversely, when the controlling mechanism is the external transport, an increase in the relative velocity fluid-solid determines an increase in the overall adsorption rate.

The adsorption process can be conducted in ex-situ operations, coupled with pump and treat technique if applied to the depuration of groundwater. Most adsorption processes are carried out in fixed-bed columns, i.e. tubular reactors in which an absorbent material is loaded up, and the current liquid containing the contaminant to be removed passes through. As previously reported, adsorption for groundwater depuration can be performed also as in situ operations, performed with Permeable Adsorbing Barriers (PAB).

The operating principle of these barriers is exactly the same as that of the columns. A configuration of this type of reactor does not ensure the achievement of the equilibrium between the solute and the entire treated flow volume; hence the concentration profile along the column is variable over time.

The determination of the concentration of the substance to be absorbed in the fluid flow, at each point of the column and for each runtime is therefore a starting point for understanding the dynamic evolution of an adsorption column or of a PAB.

The kinetic study of adsorption phenomena and the influence of the characteristic parameters of the process is usually performed using a process configuration in pilot scale, e.g., a column with a fixed bed of adsorbing material and through the realization of experimental breakthrough curves. They represent experimental relationships between concentration of the pollutant in the output flow from the column and the time of adsorption. Alternatively, once the transport properties of the system under examination are known (mass transfer coefficients, pollutant diffusion, etc.) and starting from the knowledge of adsorption isotherms, a simulation of the process can be made, with the help of dedicated software.

In both cases, an equation of mass balance of the substance to be adsorbed, flanked by a mass transfer equation on the solid, can be written. The equation of the mass balance, in the volume chosen as control, should take into account the fact that the fluid flow that moves along the column or the PAB, leaving part of the pollutant adsorbed on the solid surface.

For a two-dimension system (x; y), the mass balance equation that takes into account the contribution of the solid the fluid phases and representing the phenomenon of adsorption can be written as:

$$\frac{\partial C}{\partial t}+\frac{\rho_b}{n_b}\frac{\partial \omega}{\partial t}+\frac{\vec{u}\,\nabla C}{n_b}-\nabla\left(D_h \nabla C\right)=0 \qquad (2.\ 22)$$

Equation (2. 22) is a partial derivative differential equation, very similar to eq. 2. 4 written for a generic porous media, in which:

$\bar{u}$ is the unit flux vector, i.e. the apparent speed of the fluid, which is defined as the ratio between the flow input the column (PAB) and its section $[LT^{-1}]$;

C is the concentration of the contaminant in the fluid flow $[ML^{-3}]$;

ρ_b is the apparent density (bulk) of activated carbon $[ML^{-3}]$;

ω is the concentration of pollutant on solid [-];

D_h is the hydrodynamic dispersion coefficient, that includes the pollutant diffusivity and mechanical dispersion in the column (PAB) $[L^2T^{-1}]$;

n_b is the porosity of the adsorbing material [-].

Equation (2. 22) was written with the following assumptions:

It is assumed that there is no concentration gradient neither in the solid particle pores nor within the fluid film surrounding the particles; hence the accumulation of pollutant in the liquid contained in the pores of the adsorbent material is neglected.

Furthermore, It has been hypothesized that the change in the flow speed in the column is not appreciable.

To solve equation (2. 22) it is required to calculate the diffusional mass transport and an experimental relationship that correlates the adsorption capacity of the pollutant with its concentration in the fluid phase (i.e. an adsorption isotherm).

In aqueous solutions the axial dispersion can be described by neglecting molecular diffusion phenomena and considering turbulent mixing only.

The description of the phenomenon must include, as said above, also an equation representing the mass flow per unit volume N, near the surface of the sorbent. This flow can be described by the product of a global transport coefficient, K_c, a specific area, a, and a driving force:

$$N = K_c a\left(C - C^*\right) \tag{2. 23}$$

where:

C^* is the concentration of solute in fluid phase in equilibrium with the adsorption capacity of the solid $[ML^{-3}]$.

The material balance of the solute adsorbed on the solid, having a specific surface a, is:

$$\frac{\rho_b}{n_b}\frac{\partial \omega}{\partial t} = K_c a\left[C - C^*(\omega)\right] \tag{2. 24}$$

where:

ρ_b is the bulk density, which is mass of granular material per unit volume of a bulk-packed column $[ML^{-3}]$;

a is the specific surface area of adsorbent $[L^{-1}]$;

K_c is the global mass transfer coefficient $[T^{-1}]$;

n_b is still the porosity of the adsorbing material [-].

This kinetic adsorption model supposes the knowledge of global mass transfer coefficient K_c, which includes all individual mechanisms of fluid-particle transport and intraparticle diffusion, and the surface reaction kinetic for adsorption of solutes. In general, the latter term can be neglected, it being much higher than the other. Therefore, the global transport coefficient, $K_c a$, can be described as the result of a series of external, fluid-particle, and intraparticle transport phenomena according to the equation:

$$\frac{1}{K_c a} = \frac{1}{K_{est} a_{est}} + \frac{1}{K_{\text{int}} a_{\text{int}}} \tag{2. 25}$$

where:

K_{est} is the coefficient of fluid-particle transport;

K_{int}, is the diffusive intraparticle coefficient;

a_{est}, is the external specific surface particle $a_{est} = \frac{6}{d_p}$, being d_p the solid particle average diameter;

a_{int}, is the specific surface of solid (i.e. B.E.T area).

There are several relationships in the literature for the determination of fluid-particle mass transport, which enables the estimation of the term K_{est}. Typically, these relationships are expressed in terms of number of Sherwood, defined as:

$$Sh = \frac{d_p K_{est}}{D_m} \tag{2. 26}$$

where:

d_p, is the particle diameter [L];

D_m, is the molecular diffusivity of the solute [LT^{-1}].

The Sherwood number is obtained from the adimensionalization of the Navier-Stokes transport equations and it depends on two adimensional groups, the Reynolds and Schmidt numbers. It can be calculated by adimensional relationships available in the literature such as [Perry, 2004]:

$$Sh = \frac{1.09}{\varepsilon} Re^{0.33} Sc^{0.33} \quad 0.0015 < Re < 55 \tag{2. 27}$$

$$Sh = \frac{1.25}{\varepsilon} Re^{0.69} Sc^{0.33} \quad 55 < Re < 1050 \tag{2. 28}$$

$$Sh = 2.0 + 0.6 \cdot Sc^{1/3} \cdot \left(\frac{Re}{\varepsilon} \right)^{1/2} \tag{2. 29}$$

In which the Reynolds number, *Re*, and the Schmidt number, *Sc*, are defined as:

$$Re = \frac{\rho \cdot u \cdot d_p}{\mu} \qquad Sc = \frac{\mu}{D_m \cdot \rho} \tag{2. 30}$$

where:

ρ is the fluid density [ML^{-3}];

u is the fluid velocity [LT^{-1}];

μ is the dynamic viscosity of the fluid [$ML^{-1}T^{-1}$].

The intraparticle transport coefficient, K_{int}, can be expressed as follows [Perry, 2004]:

$$K_{int} = \frac{\varepsilon_p D_p}{d_p} \tag{2. 31}$$

where:

D_p, is the solute diffusivity in the pores, which can be calculated as:

$$D_p = \frac{D_m}{\tau_p} [1 - \lambda_m]^{-2} \left[1 + \frac{9}{8} \lambda_m \ln \lambda_m - 1.539 \lambda_m \right] \tag{2. 32}$$

where:

τ_p, is the tortuosity of the solid, that, for an activated carbon, can be considered equal to 3 [Suzuki, 1990; Perry, 2004];

λ_m, is the ratio of the Stokes-Einstein radius of the solute, r_s, and the pore radius of the solid, r_p:

$$\lambda_m = \frac{r_s}{r_p} \tag{2. 33}$$

The Stokes-Einstein radius is defined as a function of the Boltzmann constant, K^*, the temperature T, the water viscosity, μ, and D, the molecular diffusivity of the solute:

$$r_s = \frac{K^* T}{6\pi \cdot \mu \cdot D_m} \tag{2. 34}$$

In most cases, the fluid-particle mass transfer coefficient turns out to be far lower than its intraparticle counterpart, so, it is fair to say that the phenomenon of mass transfer is limited by the transport in the film fluid surrounding the particle (in particular if the adsorbent particle size is small enough to be the diffusion negligible). This is typical if the pollutant present in solution is an organic compound [Suzuki, 1990].

Finally, it should be noted that equation (2. 24) requires the knowledge of the concentration C^*, which represents the concentration of solute in the fluid phase in equilibrium with the adsorption capacity of solute, ω. Therefore, it is necessary to define a relationship between ω and C^* among those listed in Table 2.1, i.e. identify the most suitable adsorption isotherm.

PAB Design

This section describes the general issues for a PAB design and the heuristic procedure proposed.

The primary physical function of the PAB is to intercept groundwater flux (and plume) and to assure a sufficient resident time inside the reactive media in order to achieve the desired decontamination.

The design of a PAB for a polluted aquifer consists in the definition of the barrier location, orientation and dimensions. A good design is essentially time-invariant because, once the PAB is dimensioned and positioned, the strategy is unchangeable during the entire remediation process. Design choices are very important and a good compromise between technical and economic issues has to be found. A PAB design is very complex and may be time-consuming because it must consider the natural motion of groundwater and pollutant and all the reactive or adsorption processes occurring inside the barrier [Di Natale M. et al., 2007; Di Natale et al., 2008a; Erto et al., 2011a]. The problem cannot be approached by direct calculation and empirical iterative procedure, even supported by numerical simulations, has to be applied.

In particular, barrier properties have to be chosen in advance and then it is necessary to check whether they allow for a thorough pollutant capture during the whole lifetime of the barrier. Safety factors may be incorporated into the input parameters or applied to the design thickness. A typical dimensioning of PAB design incorporates the variability in the input design parameters into a probabilistic model to determine the PAB thickness required to achieve a certain confidence level. Moreover, as a final result, the minimum dimensions of the barrier have to be identified for it to be cost-effective.

Several hydrodynamic codes, such as FeFlow 5.3 by WASY GmbH [Trefry and Muffels, 2007], PMWIN by US Geological Survey [McDonald et al., 1988; Chiang and Kinzelbach, 1996], ChemFlux by SoilVision [Fredlund, 2006], allow one-, two- and three-dimension calculation to simulate groundwater flow, contaminant transport and some generation/consumption pollution processes (dissolution, bio-degradation, adsorption, etc.) but, none of them is specifically dedicated to optimize the PAB design. They can be used to describe the fate of the pollutant crossing the barrier but none of them allows for barrier design optimization. Indeed, these software can only be used for preliminary PAB dimensioning and a *trial and error* procedure is required to define the barrier dimensions and the chemical properties of the barrier adsorbing material.

In order to pre-define PAB properties, the first step of the design procedure consists in the definition of the size of the calculation domain that must be large enough to include the whole pollutant plume and longer than the plume itself along the direction of the natural groundwater flow. Then the calculation grid must be chosen, also considering that larger domains or smaller calculation cells produce more precise results but calculation is more time-consuming. Then the size of the calculation grid must be defined specifically for the barrier, by refining the calculation cells in order to reach a more detailed concentration profile inside the barrier itself.

The second step is to take into account the following general criteria:

- Position: barrier has to be as closest as possible to the pollutant plume;
- Orientation: barrier has to be orthogonal to groundwater flow lines, in order to have the same residence time of fluid inside the PAB and to ensure the minimum dimensions of the barrier itself [Craig et al., 2006];
- Length and height: equal to the corresponding pollutant plume dimensions;
- Thickness: to be chosen considering that the contaminated flow travelling through the barrier should be long enough for adsorption process to take place; therefore the barrier thickness must satisfy the following inequality:

$$\frac{W}{v_b} > (K_c a)^{-1} \tag{3. 1}$$

where K_c represents the mass transfer coefficient for adsorption reactions, a represents the external specific surface area of the adsorbent particles, and v_b is the average groundwater flow velocity through the barrier, which must be previously identified by a complete hydro-geological characterization. In equation (3. 1) the values of a and K_c can be determined by a preliminary adsorbent characterization, often provided by the vendor or by laboratory dedicated tests. Once the geometrical properties of the barrier, together with the adsorbing material have been chosen, the barrier performance must be checked by evaluating the evolution in time of the pollutant plume over the calculation domain considering that the pollutant concentration downstream the barrier (C_W) has to be everywhere lower than the fixed limit value (C_{lim}) for the whole barrier lifetime.

Eventually, as the pollutant concentration at barrier inlet may vary during the barrier working period, the occurrence of desorption phenomena within the barrier must also be taken into consideration. Adsorption continues until the pollutant concentration at the barrier inlet remains lower than the equilibrium values corresponding to the amount of pollutant adsorbed on carbon, but if the pollutant inlet concentration decreases, the pollutant adsorbed may be desorbed from the barrier solid, giving rise to a contaminated plume at the exit of the barrier itself. When desorption occurs, a thicker barrier ensures a slower release of the pollutant adsorbed, avoiding critical outbound concentrations. Therefore, the barrier must be designed both to retain intense concentration peaks and for long-term performances, also considering the occurrence of any desorption phenomena.

Heuristic Procedure for the Optimization of PAB Size

In this chapter a simulation-based heuristic optimization of PAB is illustrated to define the position and dimensions of the barrier and minimize fixed costs that are proportional to its volume.

The heuristic procedure for barrier design starts with the definition of the size of the calculation domain and the calculation grid (i.e. the dimensions of each3D calculation cell), as previously illustrated.

Barrier properties can be defined by evaluating the evolution in time of the pollutant plume in the calculation domain considered. To this aim the domain ($\mathcal{D}$), including the initial position of the pollutant plume, is defined on the basis of the preliminary characterization of aquifer pollution. The boundary of this domain is represented by the positive axes of a coordinated system, the barrier and its projections on the axes (i.e. $\mathcal{A}$ and $\mathcal{C}$ lines) as illustrated in Figure 3.1. A non-contaminated domain ($\mathcal{N}$), defined as the external part of the $\mathcal{D}$ domain, can be simultaneously identified. The definition of these domains and the evaluation of pollutant concentration inside allow to identify the minimum barrier dimensions that assure the respect of regulatory limits in the whole space and for all times. In fact, when an appropriate PAB size has been identified the pollutant concentrations within the $\mathcal{N}$domain must be lower to the regulatory limits for the whole barrier lifetime. Of course, the evaluations of pollutant concentration have to be performed outside the $\mathcal{D}$ domain, i.e. over $\mathcal{A}$, $\mathcal{B}'$ and $\mathcal{C}$ lines. Finally, in Figure 3.1, PAB geometrical properties such as: distance from pollutant plume (*E*), orientation angle (*m*), length (*L*) and thickness (*W*) are reported. In particular, the orientation angle is defined as the angle between the direction of groundwater flux ($\overline{v}$) and the barrier direction ($\overline{b}$).

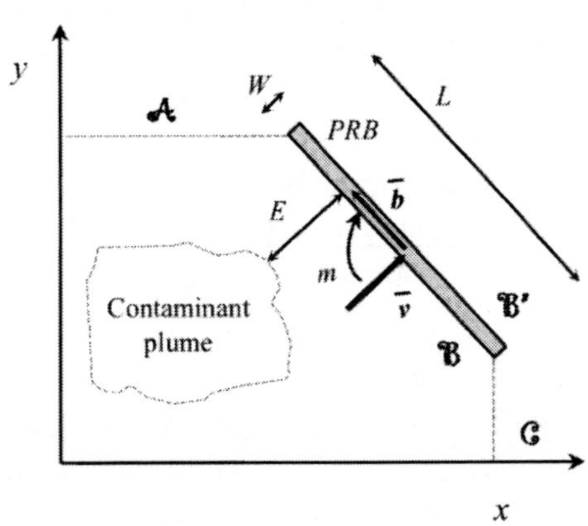

Figure 3.1.Contaminated domain and PAB properties.

The procedure consists in a simulation-based heuristic optimization by coupling two computer codes and using the iterative procedure schematically shown in the flow chart of Figure 3.2 that allows to:

$$minimize\ G = \{E, L, H, W, m\} \quad (3.\ 2)$$

subject to the following constraints:

$$\begin{cases} C_{A,C} < C_{\lim} \\ C_W \leq C_{\lim} \\ m \leq 90° \pm \xi \end{cases} \quad (3.\ 3)$$

The first code is a commercial 3D model flow, PMWIN, that uses a block-centred finite difference scheme for the saturated zone and that includes a groundwater flow model, MODFLOW [McDonald et al., 1988], and a pollutant transport model, MT3D [Zheng, 1990]. In particular, this code solves the numerical integration of equations (2. 3)–(2. 8) with the boundary conditions (2. 10); specifically, MODFLOW toolbox solves Darcy equation (2. 1) and groundwater flow equation (2. 3) while MT3D toolbox solves transport equations (2. 4)–(2. 5). The second code, ADSORP-CODE© (by CIRIAM, Seconda Università di Napoli), developed by the authors, describes the adsorption phenomena involving the pollutant when it passes through the barrier. In particular, ADSORP-CODE solves equations (2. 8) and (2. 9) inside the barrier with their initial and boundary conditions. ADSORP-CODE has been structured as a flexible tool, able to:

1) Consider any kind of adsorption isotherms [Henry, Freundlich, Langmuir, etc.];
2) Refine a grid in space (grid dimensions) and time (calculation time steps) in order to properly describe the local phenomena occurring inside the barrier.

The heuristic procedure for PAB optimal design, as illustrated in Figure 3.2, requires the following input data: computational grid (Δx, Δy, Δz), soil hydraulic conductivity (K_{ij}), soil porosity (n_s), initial hydraulic height (h_0), initial distribution of pollutant concentration (C_0=$C_0(x, y, z)$), hydrodynamic dispersion coefficient (D_{ij}), overall mass transfer coefficient for adsorption reactions (K_c), external specific surface of adsorbent particles (a), dry adsorbing material bulk density (ρ_b), barrier porosity (n_b) and adsorption capacity, the latter is obtained from adsorption isotherm models (ω=$\omega(C)$). In order to start optimizing PAB design, the procedure requires the following barrier geometrical properties: PAB distance from pollutant plume (E), PAB orientation angle (m) and PAB length (L), height (H), thickness (W). A preliminary choice can be made by setting the minimum size of the $\mathcal{D}$ domain, as the one that includes the pollutant plume in its initial position (see Figure 3.1). Once input data and preliminary choices have been defined, the MODFLOW toolbox allows the calculation, in each grid cell, of hydraulic height (h) and groundwater velocity (v) and their evolution with respect to run time. Consequently, the preliminary choice for barrier orientation angle has to be changed until groundwater velocity vector $\bar{v}$ is as orthogonal as possible to PAB direction ($\bar{b}$), with a defined maximum tolerance ($\pm\varepsilon$).

The MT3D toolbox allows for the calculation of pollutant concentrations (C) in the $\mathscr{D}$ domain and their evolution in time. The calculation results, and in particular their value on the $\mathscr{A}$ and $\mathscr{C}$ lines of the boundary, are used to adjust the preliminary choices made for E, L, H and W, considering that the barrier has to be tall and long enough to intercept the whole pollutant plume during its motion under natural hydraulic gradient. Furthermore, ADSORP-CODE is used to calculate the out-flowing pollutant concentration ($C_W(t)$) at each point of the barrier (i.e. on the $\mathscr{B}$' line) and the efficacy of the barrier is evaluated by comparing these concentrations with the limit value (C_{lim}). The preliminary PAB thickness (W) is correct if the pollutant concentration out-flowing the barrier is always and everywhere lower than limit value (C_{lim}), otherwise it must be increased up to $C_W(t)<C_{lim}$. The procedure has to be repeated in order to minimize the dimensioning set $G=\{E, L, H, W, m\}$.

Eventually, results are graphically represented by means of AMBSIT© (by CIRIAM, Seconda Università di Napoli), a GIS (Geographic Information System) application specifically developed to improve contour plots of pollutant concentration.

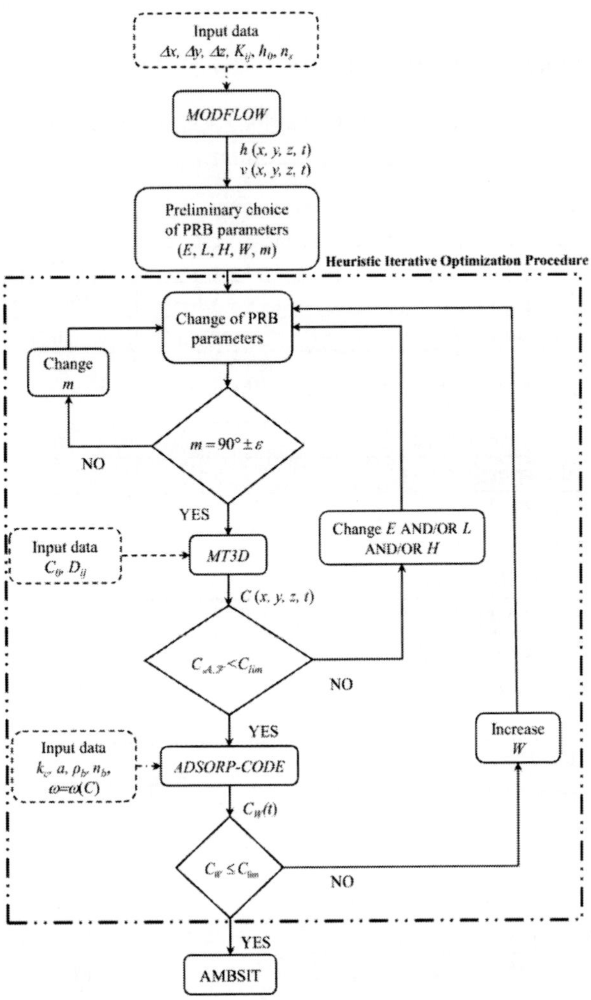

Figure 3.2. PAB design flow chart.

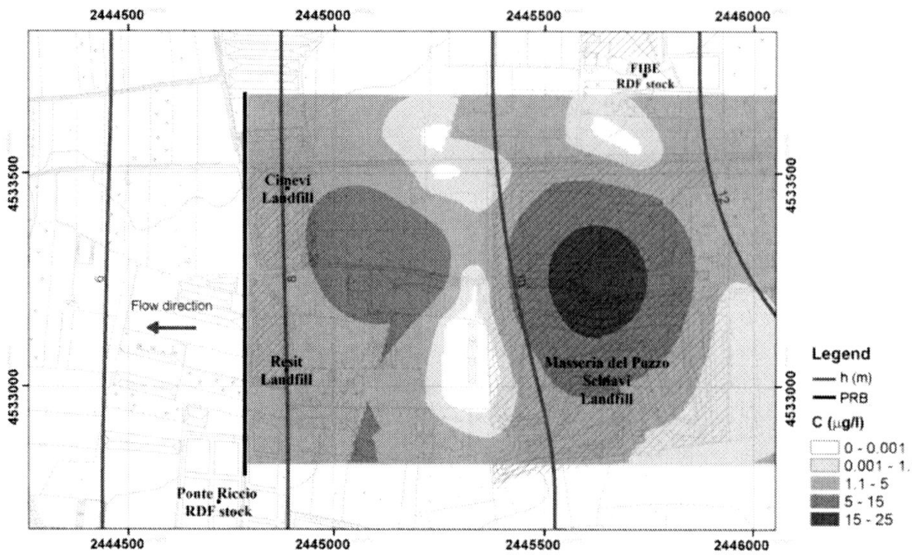

Figure 4.1. PCE iso-concentration for the case study.

Table. 1. Case study: aquifer characteristic, PAB-A properties and numerical model parameters

Aquifer characteristic	
Polluted area total extent, *A*	1.1 km^2
Aquifer bed height, *H*	10 m
Piezometric gradient, *J*	0.01 m m^{-1}
Porosity, n_s	0.25
Dry soil bulk density, ρ_s	1500 kg m^{-3}
Hydraulic conductivity, K_{ij}	$5 \cdot 10^{-5}$ m s^{-1}
Longitudinal dispersivity, α_L	1 m
Transverse dispersivity, α_T	0.01 m
Molecular diffusion coefficient, D_m	10^{-8} m^2 s^{-1}
Numerical model parameters	
Horizontal space step, Δx	6 m
Transversal space step, Δy	6 m
Vertical space step, Δz	10 m
Time step, Δt	60 days
PAB-A properties	
Reactive media	Activated carbon
Horizontal space step, Δx_b	$\Delta x/100$
Transversal space step, Δy_b	Δy
Barrier width, *W*	3 m
Barrier length, *L*	900 m
Distance from contaminant plume, *E*	6 m
Barrier distance from domain boundary, *X*	1266 m
Transversal domain extension, *Y*	996

PAB Application: Case Study

The procedure described in Figure 3.2 has been applied to the design of a PAB for the remediation of a contaminated aquifer near a solid waste landfill in Giugliano in Campania, a town in the metropolitan area North of Napoli (Italy). In this area, many solid waste landfills exist and over the past 20 years, about eight million tons of urban and special waste have been deposited in these landfills [Di Nardo et al., 2010b; Erto et al., 2011a]. A preliminary site characterization, including geotechnical, hydraulic and pollution properties, is available. The groundwater aquifer, located at a depth of 35-40m from the land surface and confined by an aquitard (50m), is contaminated by a large number of pollutants, both inorganic and organic [Di Nardo et al., 2010a; Di Nardo et al., 2010b]; in particular a severe PCE and TCE pollution has been found.

These compounds are considered among the most dangerous organic pollutants [US EPA, 1988] due to their high toxicity and persistence grade and, consequently, they represent a great risk for environment and public health. In EU Directives (1999/13/EC-Solvent Emissions Directive; 2000/60/EC-Water Framework Directive and 2006/118/EC-Groundwater Directive) these compounds are in the list of the most dangerous pollutants [ATSDR, 1997]. TCE was found at 301 of the 1689 National Priorities List sites and 391 of the 1444 ATSDR's (Agency for Toxic Substances and Disease Registry) CEP (Completed Exposure Pathway) sites [ATSDR, 2007].

In the case-study considered, the soil composition (Neapolitan yellow tuff)can be approximated to a single mineral whose hydraulic conductivity is $5*10^{-5}$m/s [Di Nardo et al., 2010a; Di Nardo et al., 2010b]. The adsorption capacity for organic compounds of this material is very low [Di Nardo et al., 2010b; Erto et al., 2010a], and can, realistically, be assumed to be zero throughout the entire flow domain (as reported in the paragraph II.1). The groundwater flow lines are East-West oriented, with piezometric heights ranging between 5 and 12.5m a.s.l, under a piezometric gradient (J) of 0.01m/m. An impermeable layer is present at the bottom of the aquifer, hence a horizontal movement of water in the aquifer with the pollutants can be predicted. The aquifer characteristics are summarized in Table 4.1 [Di Nardo et al., 2010b].

In Figure 4.1, the PCE iso-concentrations in the actual conditions are reported, together with the map of the area, the hydraulic heights, the direction of flow and the position of PAB. Figure 4.1 shows that PCE concentration values change in the area with peaks more than 20 times higher than the Italian regulatory limits for groundwater quality ($C_{\lim}^{PCE}$), set at 1.1μg/l. Thus a defined volume of contaminated groundwater can be identified and an *in-situ* treatment by PAB installation can be planned.

For the sake of simplicity, the total volume of polluted groundwater has been assumed to be constant during the monitoring time of the aquifer and the dissolution of further amounts of pollutant or the dilution of its concentration due to rainfall have been neglected. Therefore, the evolution of PCE concentration profiles is predicted by considering just advective phenomena as well as adsorption phenomena inside the barrier.

In the design of the barrier of this case study, a commercially available non-impregnated granular activated carbon (GAC), Aquacarb 207EA™ (Sutcliffe Carbon) has been considered. This material has a BET surface area (S_{bet}) of 950m^2/g and an average pore diameter (d_{pore}) of around 26Å. Dry bulk density (ρ_b) is 500kg/m^3, porosity (n_b) is 0.4m^3/m^3

and hydraulic conductivity is about 0.001m/s. A thorough solid characterization is reported in Erto et al. (2011b). The adsorption isotherm of PCE-Aquacarb 207EA™at a temperature of 10°C, typical of groundwater, is reported in Figure 4.2. The Langmuir model, whose equation is reported in table II.1, resulted to be the best for PCE adsorption on the selected activated carbon [Erto et al., 2009; Erto et al., 2010b; Erto et al., 2010c]. Specifically, the calculated Langmuir parameters are respectively ω_{max}=913.9mg/g and K=19.830l/mol.

According to the heuristic procedure previously described, several iterations based on different geometrical parameters of the barrier have been performed, by changing W in the range 1 to 4m with a ΔW step of 0.1m, to determine the optimal position and dimensions of the barrier.

The PAB resulted to be a continuous trench penetrating the aquifer at full-depth (50m) up to the aquitard and the PAB optimal parameters resulted to be as follows: E= 6m, m=90°, coincident with North direction, and L= 900m.

Figure 4.3 reports PCE outlet concentration over the run time, at barrier points S, S_1 and S_2 (as indicated in Figure 4.3), with different thickness values of PAB, from W=1m to W=4m.

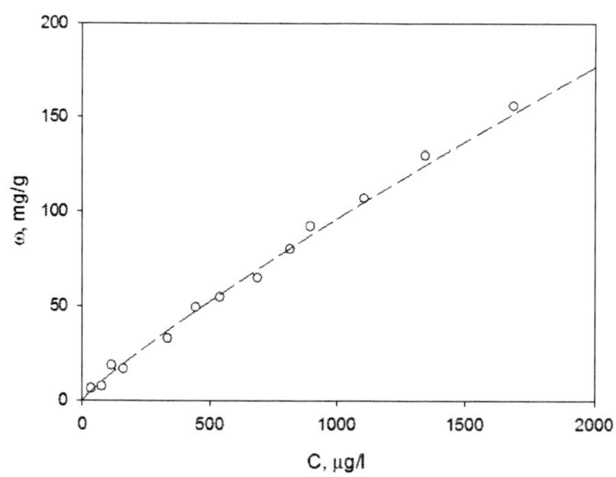

Figure 4. 2. PCE adsorption isotherm ontoAquacarb 207EA™ GAC at a temperature of 10°C.

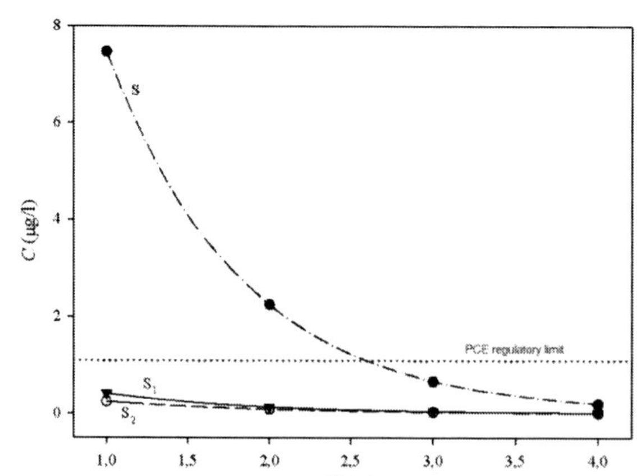

Figure 4. 3. PCE concentrations of PAB at points S, S_1 and $S_2$for different thickness values (from 1m to 4m).

Table 4. 1. PCE max inlet concentrations C_{IN} and corresponding PCE max outlet concentrations C_W of points S, S_1 and S_2 for different PAB thickness values during all simulation time (E=6m, L=900m and m=90°)

PAB point	$maxC_{IN}$	$maxC_{W=1m}$	$maxC_{W=2m}$	$maxC_{W=3m}$	$maxC_{W=4m}$
S	24.47	7.46	2.24	0.67	0.20
S_1	0.67	0.24	0.08	0.03	0.01
S_2	1.27	0.42	0.13	0.04	0.01

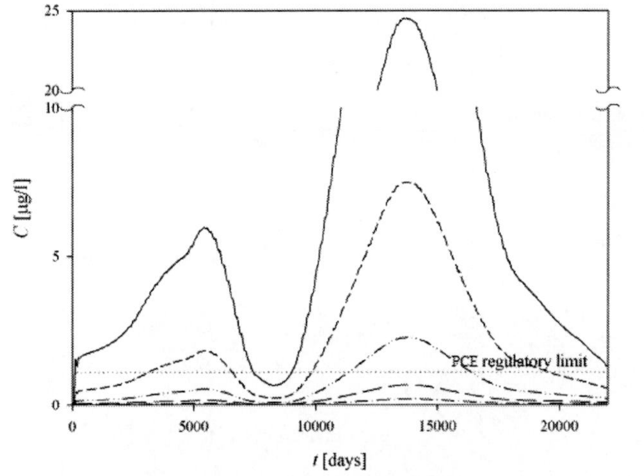

(——) CIN – (- - -) CW=1m–(·· — ··) CW=2m – (— —) CW=3m– (·—·—)CW=4m.

Figure 4. 4. PCE concentrations in the barrier at point S (E=6m, L=900m and m=90°).

Points S_1 and S_2 are the extreme points of the barrier, while point Sis the most critical point, i.e. where the highest PCE concentration is reached during the run time. In Figure 4.3 the regulatory limit for PCE concentration has been included, in order to outline the validity of the heuristic procedure followed for barrier design.

Moreover, the maximum inlet and outlet concentration values at points S, S_1 and S_2 during the simulation time, for the barrier thickness 1, 2, 3 and 4m are reported in Table IV.2. Results in Figure 4.3 and Table IV.2 clearly show that the sizing problem is strongly nonlinear; these results show that thickness must be at least 2.6m in order to guarantee that PCE concentration, out flowing the barrier, is within the regulatory limits. However, in order to consider an appropriate design safety factor, the optimal barrier thickness can be set at 3m.

In Figure 4.4, inlet (C_{IN}) and outlet (C_W) PCE concentration at barrier point S, is reported, over the whole run time, for different barrier thickness values (1, 2, 3 and 4m).

The curves in Figure 4.4 clearly show that the PCE concentration flowing out of the barrier is always lower than the standard limit ($C_{\lim}^{PCE}$) only for barrier thickness values of 3 and 4m, even when inlet PCE concentration decreases (after a run time of about 20 years) and desorption of previously captured PCE may occur. Figure 4.4 also shows that the PAB is effective for PCE peak concentrations that become smaller when flowing through the barrier.

a)

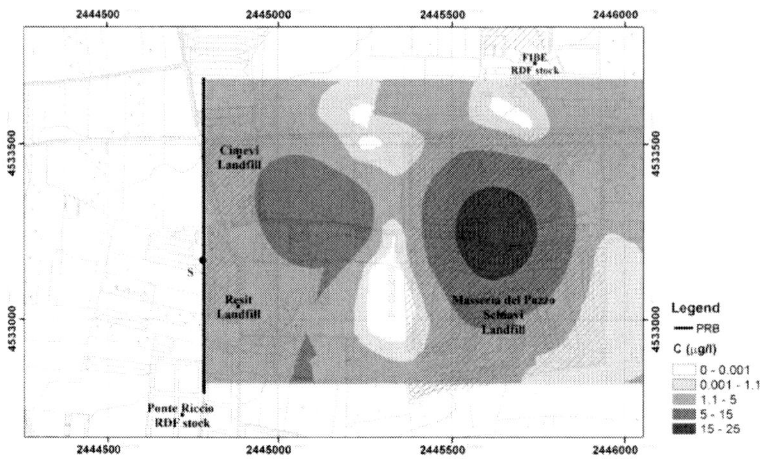

b)

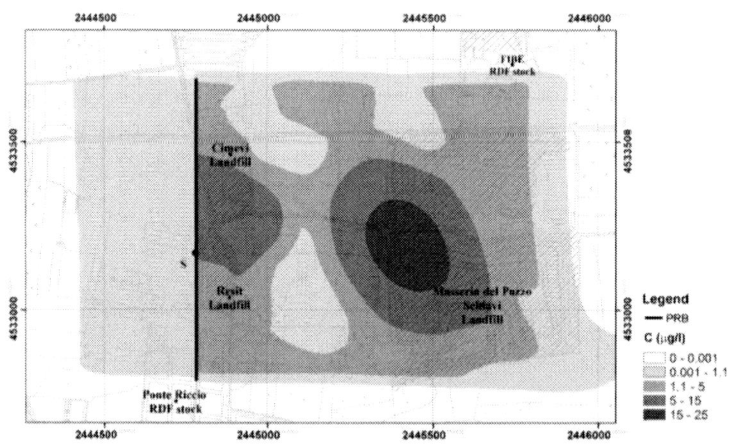

c)

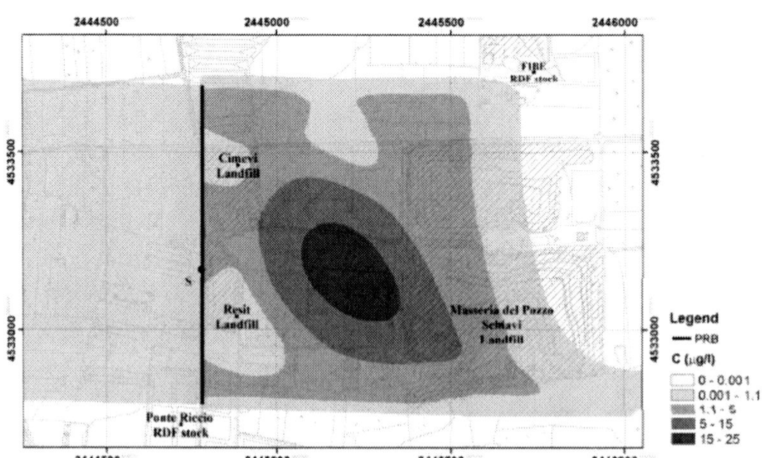

Figure 4.5. (Continued).

d)

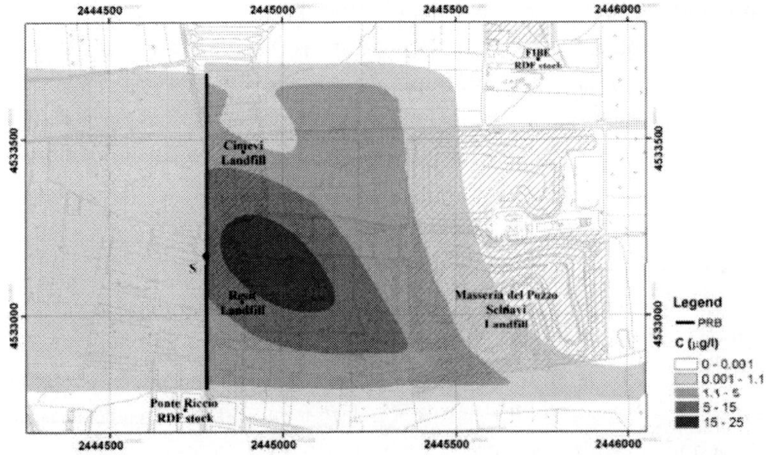

e)

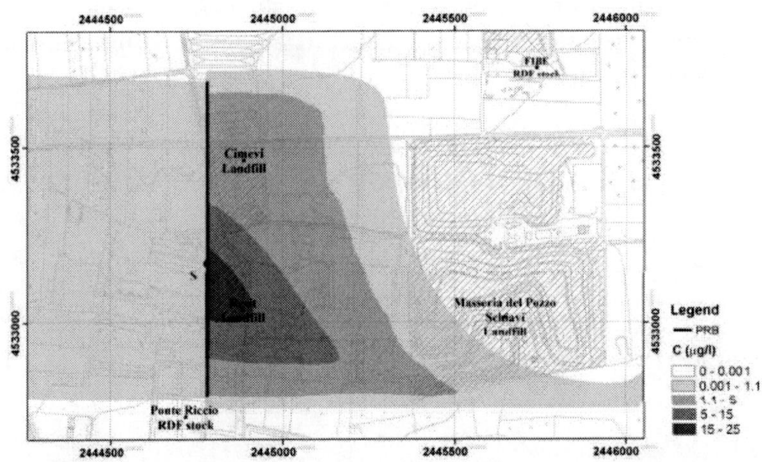

f)

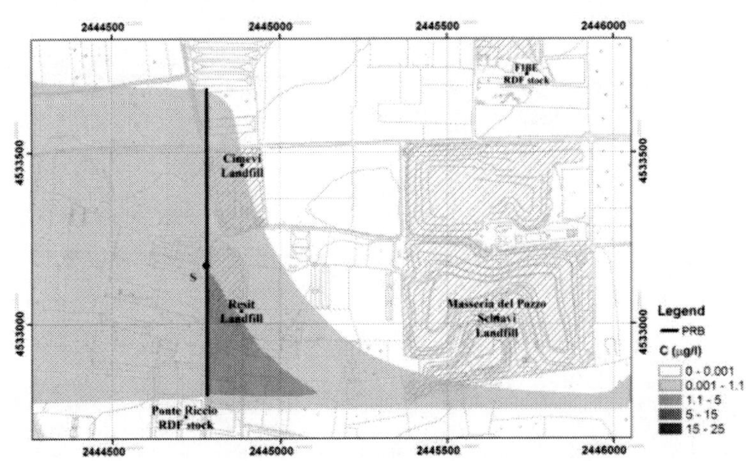

Figure 4.5. PCE iso-concentrations as a function of the working time and for PAB thickness of 3m (E=6m, L=900m and m=90°).

As a final verification and once the optimal barrier parameters in (3. 2) have been defined, the numerical simulations allow to calculate PCE concentration all over the domain as a function of run time in the form of contour plots, as reported in Figure 4.5:

Figure 4.5 shows the evolution of pollutant spots over run time, as snapshots taken every 10 years. As can be observed, pollutant spots move towards the barrier where they are captured; indeed, over a runtime of about 60 years, the PCE concentration flowing out of the barrier (C_W) is always lower than the concentration limit ($C_{\lim}^{PCE}$) set at 1.1μg l^{-1} and the remediation of the aquifer can be considered as complete. Figure 4.5 also shows that run time must be prolonged to 60 years in order to assess barrier performances. This very long time is due to the hydraulic properties of the aquifer considered. Eventually, results in Figure 4.5 showthat the whole pollutant plume passes through the barrier and no overflow is observed at the barrier ends. It is worth noticing that as adsorption takes place, the barrier starts to saturate; therefore, the barrier width has to be large enough to ensure the capture of the maximum PCE concentration. Moreover, when the PCE concentration approaching the barrier decreases with time the desorption of previously captured PCE may occur. Moreover, Figure 4.5 shows that the barrier attenuates the oscillation of pollutant concentration and shifts it forward with respect to run time; furthermore, at the end of the run time the barrier has a residual adsorption capacity that may represent a protection with respect to other pollutant waves.

Table 4. 3. Langmuir parameters for PCE e TCE onto Aquacarb 207EA™ at temperature 10°C and pH 7

Langmuir parameters	PCE	TCE
ω_{MAX} [mg/g]	913.8896	255.4416
K [l/μg]	0.000119	0.00035

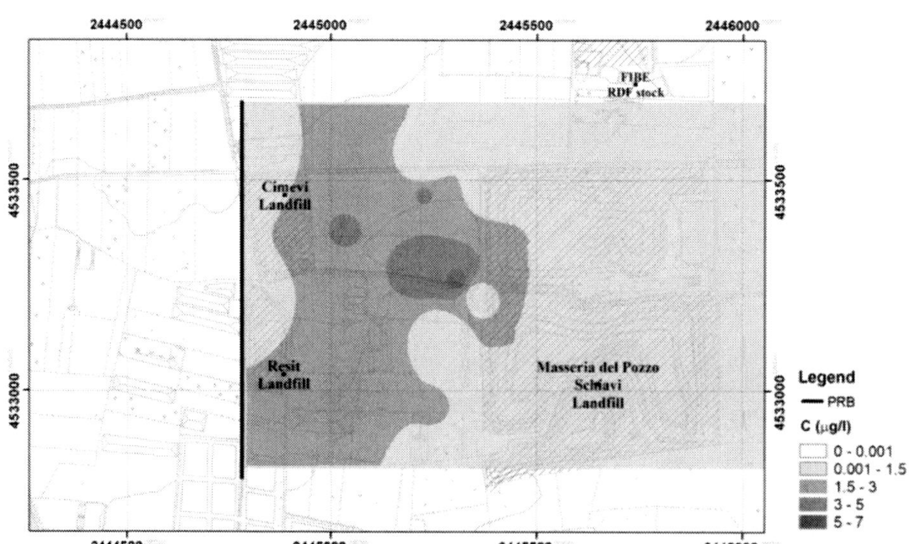

Figure 4.6. TCE iso-concentration for the case study.

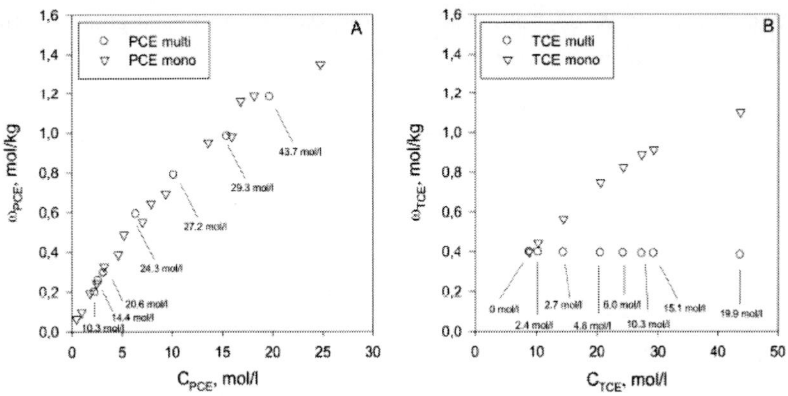

Figure 4.7. Adsorption isotherms for PCE and TCE both in single-component and multicomponent systems onto Aquacarb207EA^{TM}.T= 10°C and pH = 7.

The efficacy and the versatility of the PAB designed for PCE capture have been tested by considering the simultaneous TCE and PCE contamination, present in the studied site. The simultaneous and competitive capture of TCE and PCE has been modeled considering the equations described in the chapter II.2.2.

In particular, in Figure 4.6 TCE iso-concentrations in the initial conditions with the direction of flow and position of the PAB are reported. Figure 4.6 shows that TCE concentration values change in the area with peaks more than 6 times higher than the Italian regulatory limits for groundwater quality ($C_{\lim}^{TCE}$), set at 1.5μg/l.

Literature results [Erto et al., 2011b; Erto et al., 2012] on the adsorption of PCE and TCE in multi-component systems revealed that the PCE adsorption capacity does not depend by the presence of TCE; on the contrary, TCE adsorption is influenced by the presence of the PCE proportionally. For low concentration values of the two pollutants, such as those relating to the concerned aquifer, the multi-component Langmuir model (Eq. 2. 15) satisfactorily interprets the experimental data. In particular, for PCE/TCE multi-component system adsorbed on Aquacarb 207EA^{TM}activated carbon, the experimental adsorption isotherms and the Langmuir parameters, at the temperature of 10°C,are reported respectively in Table 4.3 and Figure 4.7 (Erto et al., 2009; Erto et al., 2012). For the case study considered, in Figure 4.8, the inlet (C_{IN}) and outlet (C_W) PCE and TCE concentrations are reported; specifically the breakthrough curves for the thickness value (W) of 3m, in the points S and P of the barrier, during the whole run time, are showed. The points S and P on the barrier (see Figures 4.5 and 4.9) represent the most critical points during simulations, namely with the higher values of concentrations of PCE and TCE, respectively. The curves show clearly that the outlet concentrations of PCE and TCE are always lower than the respective regulatory limits, even when inbound concentrations decrease and/or desorption phenomena may occur.

In Figure 4.9, instead, the results of numerical simulations reported as iso-concentration curves over time are provided. These figures illustrate the evolution over time of the TCE pollutant plume, starting from the initial condition and after 10, 20, 30, 40 and 60 years, respectively. They point out that the entire PCE plume is captured in a period of about 60 years old, while the TCE plume, being less extended, needs a period of about 25 years.

The model results show that, in the presence of the multi-component adsorption, the PAB dimensioned first to remediate the PCE pollutant is also able to simultaneously treat the TCE, keeping the concentrations of both the compounds below the regulatory limits, equal respectively to 1.1μg/l for the PCE and 1.5μg/l for TCE.

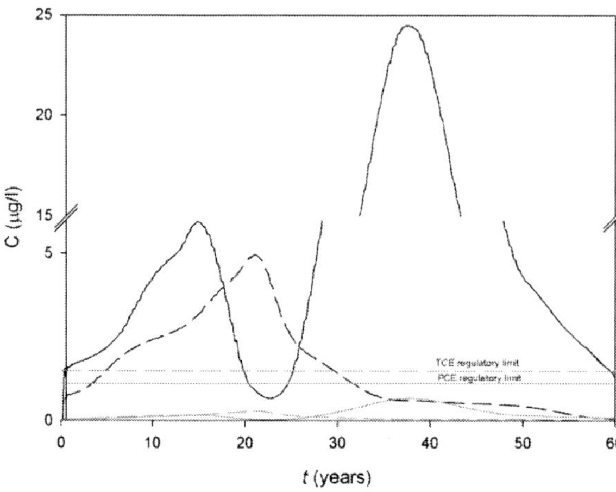

( ) PCE C_{IN} — ( )TCE C_{IN}— ( )PCE$C_{W=3m}$— ( )TCE $C_{W=3m}$

Figure 4. 8. PCE and TCE concentrations in the barrier at point S and P with the thickness W=3m (E=6m, L=900m and m=90°).

a)

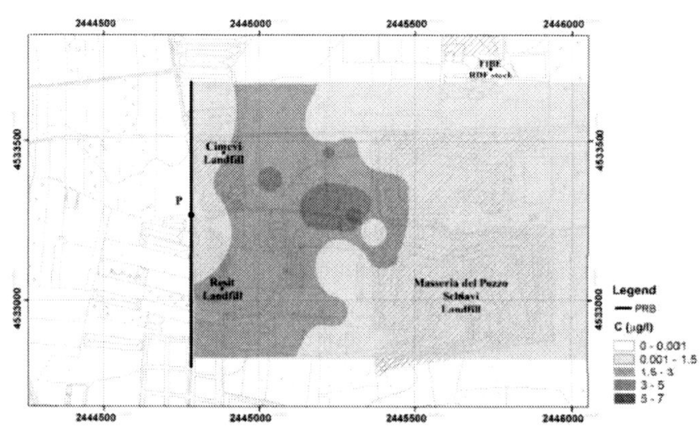

b)

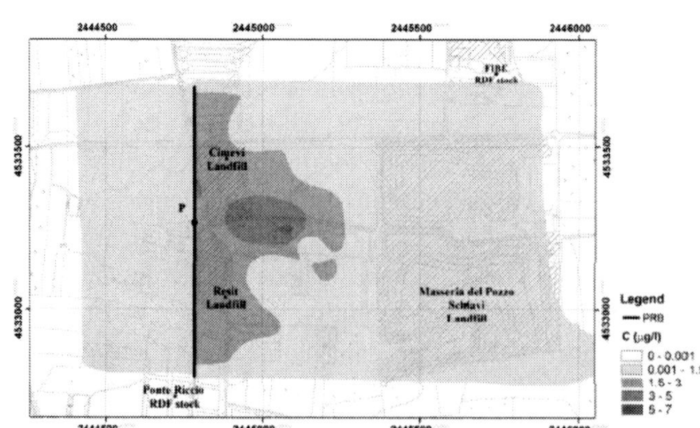

Figure 4.9. (Continued).

c)

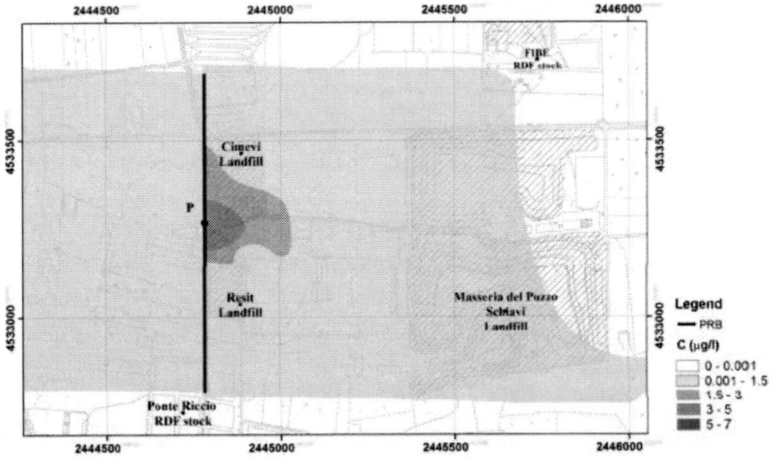

d)

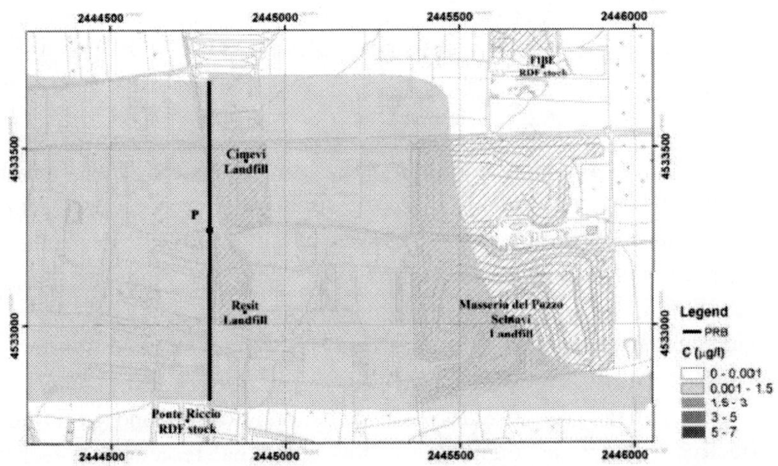

e)

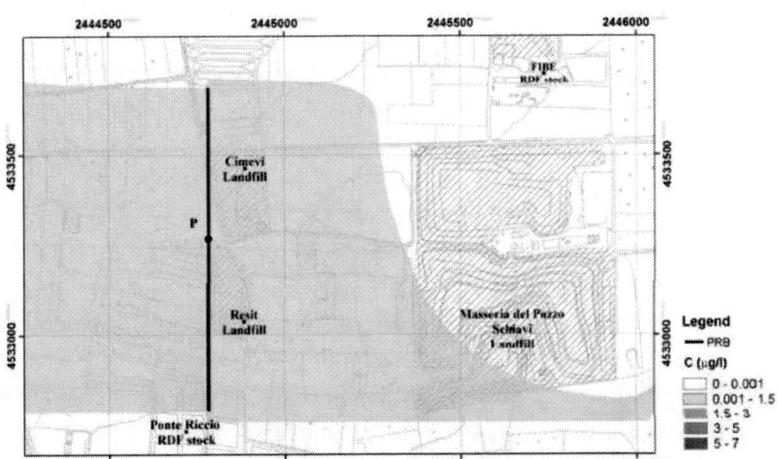

Figure 4.9. (Continued).

f)

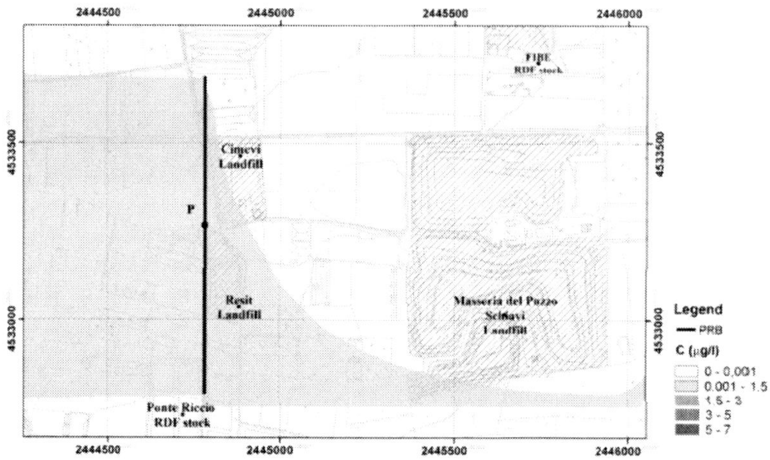

Figure 4.9. PCE iso-concentrations as a function of the working time and for PAB hickness of 3m (E=6m, L=900m and m=90°).

Conclusion

In the chapter a heuristic procedure for the optimal design of an activated carbon Permeable Adsorptive Barrier (PAB) has been described. The procedure includes various calculation steps and has the main purpose of the definition of the optimal geometrical barrier features (location, orientation and dimensions) that allow for complete pollutant plume capture with the minimum volume of adsorbing material.

The procedure has been applied as a case study to a contaminated aquifer near a solid waste landfill in the North district of Napoli (Italy) where a high level of contamination by both TCE and PCE has been recognized. Numerical simulations, proposed in the case study, confirm the effective of PABs as alternative to the traditional remediation technologies (i.e pump and treat); they also showed the evolution of the contaminated plume (both for PCE and TCE compounds) towards the barrier under the natural hydraulic gradient and the adsorptive action inside the barrier.

The results show that the procedure proposed is an effective and user-friendly tool to proper and optimize barrier design.

References

Adams C. D., Sutherland J., Kekibad J., 2002. Finale Report-Comparative Treatment Study of M. T. B. E. and Alternative Fuel Oxygenates. University of Missouri-Rolla.

Ake C. L., Wiles M. C., Huebner H. J., McDonald T. J., Cosgriff D., Richardson M. B., Donnelly K. C., Phillips T. D., 2003. Porous organoclay composite for the sorption of polycyclic aromatic hydrocarbons and pentachlorophenol from groundwater. *Chemosphere*, 51, 835–844.

American Academy of Environmental Engineers under a cooperative agreement with the US Environmental Agency, 1993. Innovative Site Remediation Technology. Vol. 3 – *Soil Washing and Soil Flashing*. E. P. A.-542-B-93-012.

American Academy of Environmental Engineers under a cooperative agreement with the US Environmental Agency, 1998. Innovative Site Remediation Technology: Design and Application. Vol. 7 – *Vacuum Extraction and Air Sparging*. Monograph Series. EPA-542-B-97-010.

Anderson M. P., 1984. Movement of contaminants in groundwater: groundwater transport-advection and dispersion. *Groundwater Contamination*. National Academy Press, Washington, D. C.

A. T. S. D. R. - Agency for Toxic Substances and Disease Registry, 1997. Toxicological Profile for Tetrachloroethylene (P. C. E.). US Public Health Service, US Department of Health and Human Services, Atlanta, G. A.

A. T. S. D. R. - Agency for Toxic Substances and Disease Registry, 2007. ATSDR CEP site count report. *www.atsdr.cdc.gov/cep/07cep.html*. US Public Health Service, US Department of Health and Human Services, Atlanta, G. A.

Bartzas G. and Komnitsas K., 2010. Solid phase studies and geochemical modelling of low-cost permeable reactive barriers. *Journal of Hazardous Materials*, 183, 301-308.

Bear J., 1979. Hydraulics of Groundwater. McGraw-Hill, New York.

Bear J., Verruyt A., 1987. Modelling Groundwater Flow and Pollution. Reidel Publishing Co., Dordrecht, Holland.

Benjamin M., 2002. Water Chemistry. McGraw Hill, New York.

Calabrò P. S., Moraci N., Suraci P., 2012. Estimate of the optimum weight ratio in Zero-Valent Iron/Pumice granular mixtures used in permeable reactive barriers for the remediation of nickel contaminated groundwater. *Journal of Hazardous Materials*, 207-208, 111-116.

California Department of Toxic Substances Control, 2006. In Situ Bioremediation of Chlorinated Hydrocarbons. An Assessment of Projects in California. O. P. P. T. D. Document No. 1218.

California Environmental Protection Agency-Department of Toxic Substances Control, 2010.Proven Technologies and Remedies Guidance-Remediation of Chlorinated Volatile Organic Compounds In Vadose Zone Soil.

Chiang W. H., Kinzelbach W., 1996. Processing Modflow: A simulation system for modeling groundwater flow and pollution. US Geological Survey.

C. L. A. I. R. E. (Contaminated Land: Applications and Real Environments), 2007. *Understanding Soil Washing*. Technical Bulletin T. B.13.

Costello C., 2003. Acid Mine Drainage: Innovative Treatment Technologies. National Network of Environmental Management Studies Fellows for US Environmental Protection Agency.

Craig J. R., Rabideau A. J., Suribhatla R., 2006. Analytical expressions for the hydraulic design of continuous permeable reactive barriers. *Advances in Water Resources*, 29 (1), 99–111.

Czurda K. A., Haus R., 2002. Reactive barriers with fly ash zeolites for in situ groundwater remediation. *Applied Clay Science*, 21, 3–20.

Dermont G., Bergeron M., Mercier G., Richaer-Laflèche M., 2008.Soil Washing for Metal Removal: A Review of Physical/Chemical Technologies and Field Applications. Québec,

Q. C., Cananda: Institut National de la Recherche Scientifique Eau Terre et Environnement (I. N. R. S.-E. T. E.).

D'Andrea P., Lai K. C., Kjeldsen P., Lo I. M., 2005. Effect of groundwater inorganics on the reductive dechlorination of T. C. E. by zero-valent iron. *Water, Air, and Soil Pollution*,162, 401-420.

Di Nardo A., Bortone I., Musmarra D., 2010a. Studio del sito contaminato di Masseria del Pozzo-Schiavi nel comune di Giugliano in Campania.Considerazioni quantitative sull'efficacia della tecnologia di bonifica pump and treat. ABBIABBè EDIZIONI, Giugliano in Campania.

Di Nardo A., Di Natale M., Erto A., Musmarra D., Bortone I., 2010b. Permeable reactive barrier for groundwater P. C. E. remediation: the case study of a solid waste landfill pollution. *Computer Aided Chemical Engineering*, 28, 1015-1020.

Di Natale F., Lancia A., Molino A., Musmarra D., 2007. Removal of chromium ions form aqueous solutions by adsorption on activated carbon and char. *Journal of Hazardous Materials*, 145 (3), 381-390.

Di Natale M., Greco R., Musmarra D., 2007. Permeable Adsorbing Barriers for groundwater remediation from hexavalent chrome pollution. *International Journal of Environmental Technology and Management*, 7 (1-2), 39-55.

Di Natale F., Di Natale M., Greco R., Lancia A., Laudante C., Musmarra D., 2008a. Groundwater protection from cadmium contamination by permeable reactive barriers. *Journal of Hazardous Materials*, 160, 428-434.

Di Natale F., Erto A., Lancia A., Musmarra D., 2008b. Experimental and modelling analysis of As(V) ions adsorption on granular activated carbon. *Water Research*, 42 (8-9), 2007-2016.

Di Natale F., Erto A., Lancia A., Musmarra D., 2009. A descriptive model for metallic ions adsorption from aqueous solutions onto activated carbons. *Journal of Hazardous Materials*, 169 (1-3), 360-369.

Di Natale F., Erto A., Lancia A., Musmarra D., 2011. Mercury adsorption on granular activated carbon in aqueous solutions containing nitrates and chlorides. *Journal of Hazarodous Materials*, 192 (3), 1842-1850.

Do D. D., 1998. Adsorption analysis: equilibria and kinetics. Imperial College Press, London.

Domenico P. A. and Schwartz F. W., 1990. Physical and Chemical Hydrogeology. John Wiley and Sons, New York.

Eljamal O., Sasaki K., Hirajima T., 2011. Numerical simulation for reactive solute transport of arsenicin permeable reactive barrier column including zero-valent iron. *Applied Mathematical Modelling*, 35, 5198-5207.

Erto A., Andreozzi R., Di Natale F., Lancia A., Musmarra D., 2009. Experimental and isotherm model analysis on T. C. E. and P. C. E. adsorption from model water solutions. *Chemical Engineering Transactions*, 17, 293-298.

Erto A., Lancia A., Musmarra D., 2010a. Natural and industrial porous material as trichloroethylene sorbents for water depuration.3rd E. C. I. International Conference on Porous Media and its Applications in Science, Engineering and Industry, 20-25 June 2010, Tuscany, Italy.

Erto A., Andreozzi R., Lancia A., Musmarra D., 2010b. Factors affecting the adsorption of trichloroethylene onto activated carbons. *Applied Surface Science*, 256 (17), 5237-5242.

Erto A., Andreozzi R., Di Natale F., Lancia A., Musmarra D., 2010c. Experimental and statistical analysis of trichloroethylene adsorption onto activated carbon. *Chemical Engineering Journal*, 156 (2), 353-359.

Erto A., Lancia A., Bortone I., Di Nardo A., Di Natale M., Musmarra D., 2011a. A procedure to design a Permeable Adsorptive Barrier (P. A. B.) for contaminated groundwater remediation. *Journal of Environmental Management*, 92, 23-30.

Erto A., Lancia A., Musmarra D., 2011b. A modelling analysis of P. C. E./T. C. E. mixture adsorption based on Ideal Adsorbed Solution Theory. *Separation and Purification Technology*, 80, 140-147.

Erto A., Lancia A., Musmarra D., 2012. A Real Adsorbed Solution Theory model for competitive multicomponent liquid adsorption onto granular activated carbon. *Microporous and Mesoporous Materials*, 154, 45-50.

E. S. T. C. P. (Environmental Security Technology Certification Program), 1999. *Technology Status Review In Situ Oxidation*. E. S. T. C. P. Project ER-0314.

E. S. T. C. P. (Environmental Security Technology Certification Program), 2003. In Situ Remediation of MTBE-Contaminated Aquifers Using Propane Biosparging. E. S. T. C. P. Project CU-0015.

E. S. T. C. P. (Environmental Security Technology Certification Program), 2010. Critical Evaluation of State of the Art In Situ Thermal Treatment Technologies for D. N. A. P. L. Source Zone Treatment. E. S. T. C. P. Project ER-0314.

E. S. T. C. P. (Environmental Security Technology Certification Program), 2011a. A Guide for Selecting Remedies for Subsurface Releases of Chlorinated Solvents. E. S. T. C. P. Project ER-200530.

E. S. T. C. P. (Environmental Security Technology Certification Program), 2011b. Application of Nucleic Acid-Based Tools for Monitoring Monitored Natural Attenuation (M. N. A.), Biostimulation, and Bioaugmentation at Chlorinated Solvent Sites. E. S. T. C. P. Project ER-0518.

European I. P. P. C. Bureau, 2003.Reference document on best available techniques in common waste water and waste gas treatment / management systems in the chemical sector - European Commission – on-line version at *http://eippcb.jrc.es/*.

Farrell J., Melitas N., Kason M., Li T., 2000. Electrochemical and column investigationof iron-mediated reductive dechlorination of trichloroethylene and percholoroethylene. *Environmental Science and Technology*. 34, 2549-2556.

F. N. A. D. E. (French Federation of Waste Management Services) and A. D. E. M. E. (French Environment and Energy Management Agency), 2007. State-of-the-Art-Practices and Implementation Recommendations for Non-Hazardous Waste Management Using Bioreactor Landfilss.

Fredlund M., 2006. S. V. F. L. U. X. and C. H. E. M. F. L. U. X.: Flow and Transport Modeling Software. A Response to Review Comments. *Ground Water*, 44 (3), 328.

Gibert O., Rotting T., Cortina J.L., de Pablo L., Ayora C., Carrera J., Bolzicco J., 2011. In-situ remediation of acid mine drainage using a permeable reactive barrier in Aznalcóllar (Sw Spain). *Journal of Hazardous Materials*, 19, 287-295.

Guan J. B, Aral M. M., 1999. Optimal remediation with well locations and pumping rates selected as continuous decision variables. *Journal of Hydrology*, 221, 20–42.

Han Y., Gallegos T. J., Demond A. H., Hayes K. F., 2011. FeS-coated sand for removal of arsenic (III) under anaerobic conditions in permeable reactive barriers. *Water Research*, 45, 593-604.

Hand D. W., Hokanson D. R., Crittenden J. C., 1999. Chapter 5. Air Stripping and Aeration. *Water Quality and Treatment*, 5th Edition, American Water Works Association, McGraw-Hill: New York, 5.1-5.68.

Hazen T. C., 2009a. Cometabolic Bioremediation. Lawrence Berkeley National Laboratory.

Hazen T. C., 2009b. In Situ: Groundwater Bioremediation. Lawrence Berkeley National Laboratory.

Hazen T. C., 2010. Biostimulation. Lawrence Berkeley National Laboratory.

He L., Huang G. H., Lu H. W., 2008a.A simulation-based fuzzy chance-constrained programming model for optimal groundwater remediation under uncertainty. *Advances in Water Resources*, 31 (12), 1622-1635.

He L., Huang G. H., Zeng G. M., Lu H. W., 2008b. An integrated simulation, inference, and optimization method for identifying groundwater remediation strategies at petroleum-contaminated aquifers in western Canada. *Water Research*, 42 (10-11), 2629-2639.

He L., Huang G. H.,. Lu H. W., 2009. A coupled simulation-optimization approach for groundwater remediation design under uncertainty: An application to a petroleum-contaminated site. *Environmental Pollution*, 157, 2485–2492.

Higgins M. R., Olson T. M., 2009. Life-cycle case study comparison of permeable reactive barrier versus pump-and-treat remediation. *Environmental Science and Technology*, 43 (24), 9432-9438.

Hsiao C. T, Chang L. C., 2002. Dynamic groundwater management with inclusion of fixed costs. *Journal of Water Resources Planning and Management*, 128 (1), 57–65.

Hosseini S. M., Ataine-Ashtiani B., and Kholghi M., 2011. Bench-Scaled Nano-Fe^0 Permeable Reactive Barrier for Nitrate Removal. *Ground Water Monitoring and Remediation*, 31 (4), 82-94. DOI: 10.1111/j1745–6592.2011.01352.x.

I. C. S. S. (International Centre for Soil and Contaminated Sites), 2006. Manual for Biological Remediation Techniques. German Federal Environmental Agency.

I. T. R. C. (Interstate Technology and Regulatory Council), 1997a. Emerging Technologies for the Remediation of Metals in Soils. Electrkinetics. MIS-4. Washington, D. C.: Interstate Technology and Regulatory Council, Metals in Soils Work Team.

I. T. R. C. (Interstate Technology and Regulatory Council), 1997b. Technical and Regulatory Guidelines for Soil Washing. MIS-1. Washington, D. C.: Interstate Technology and Regulatory Council, Metals in Soils Work Team.

I. T. R. C. (Interstate Technology and Regulatory Council), 1997c. Fixed Facilities for Soil Washing – A Regulatory Analysis. MIS-2. Washington, D. C.: Interstate Technology and Regulatory Council, Metals in Soils Work Team.

I. T. R. C. (Interstate Technology and Regulatory Council), 1997d. Emerging Technologies for the Remediation of Metals in Soils. Phytoremediation. MIS-5. Washington, D. C.: Interstate Technology and Regulatory Council, Metals in Soils Work Team.

I. T. R. C. (Interstate Technology and Regulatory Council), 2000. Dense Non-Aqueous Phase Liquids (DNAPLs): Review of Emerging Characterization and Remediation Technologies. DNAPLs-1. Washington, D. C.: Interstate Technology and Regulatory Council, DNAPLs/Chemical Oxidation Work Team.

I. T. R. C. (Interstate Technology and Regulatory Council), 2002. A Systematic Approach to In Situ Bioremediation in Groundwater Including Decision Trees on In Situ Bioremediation for Nitrates, Carbon Tetrachloride, and Perchlorate. ISB-8. Washington, D. C.: Interstate Technology and Regulatory Council, In Situ Bioremediation Team.

I. T. R. C. (Interstate Technology and Regulatory Council), 2003a. Technical and Regulatory Guidance for Surfactant/Cosolvent Flushing of D. N. A. P. L. Source Zones. DNAPLs-3. Washington, D. C.: Interstate Technology and Regulatory Council, Dense Nonaqueous Phase Liquids Team.

I. T. R. C. (Interstate Technology and Regulatory Council), 2003b. Technical and Regulatory Guidance for Constructed Treatment Wetlands. WTLND-1. Washington, D. C.: Interstate Technology and Regulatory Council, Wetlands Team.

I. T. R. C. (Interstate Technology and Regulatory Council), 2004. Strategies for Monitoring the Performance of D. N. A. P. L. Source Zone Rmedies. DNAPLs-5. Washington, D. C.: Interstate Technology and Regulatory Council, Dense Nonaqueous Phase Liquids Team.

I. T. R. C. (Interstate Technology and Regulatory Council), 2005a. Permeable Reactive Barriers: Lessons Learned/New Directions. PRB-4. Washington, D. C.: Interstate Technology and Regulatory Council, Permeable Reactive Barriers Team.

I. T. R. C. (Interstate Technology and Regulatory Council), 2005b. Technical and Regulatory Guidance for In Situ Chemical Oxidation of Contaminated Soil and Groundwater. ISCO-2. Washington, D. C.: Interstate Technology and Regulatory Council, In Situ Chemical Oxidation Team.

I. T. R. C. (Interstate Technology and Regulatory Council), 2006. Characterization, Design, Construction, and Monitoring of Bioreactor Landfills. ALT-3. Washington, D. C.: Interstate Technology and Regulatory Council, Alternative Landfill Technologies Team.

I. T. R. C. (Interstate Technology and Regulatory Council), 2007. A Decision Flowchart for the Use of Monitored Natural Attenuation and Enhanced Attenuation at Sites with Chlorinated Organic Plumes. E. A. C. O. Decision Flow chart. Washington, D. C.: Interstate Technology and Regulatory Council, Enhanced Attenuation: Chlorinated Organics Team.

I. T. R. C. (Interstate Technology and Regulatory Council), 2008a. In Situ Bioremediation of Chlorinated Ethene: D. N. A. P. L. Source Zones. BioDNAPL-3. Washington, D. C.: Interstate Technology and Regulatory Council, Bioremediation of DNAPLs Team.

I. T. R. C. (Interstate Technology and Regulatory Council), 2008b. Enhanced Attenuation: Chlorinated Organics. EACO-1. Washington, D. C.: Interstate Technology and Regulatory Council, Enhanced Attenuation: Chlorinated Organics Team.

I. T. R. C. (Interstate Technology and Regulatory Council), 2009. Phytotechnology Technical and Regulatory Guidance and Decision Trees, Revised. PHYTO-3. Washington, D. C.: Interstate Technology and Regulatory Council, Phytotechnologies Team.

I. T. R. C. (Interstate Technology and Regulatory Council), 2011a. Permeable Reactive Barrier: Technology Update. PRB-5. Washington, D. C.: Interstate Technology and Regulatory Council, P. R. B.: Technology Update Team.

I. T. R. C. (Interstate Technology and Regulatory Council), 2011b. Development of Performance Specifications for Solidification/Stabilizastion. PRB-5. Washington, D. C.: Interstate Technology and Regulatory Council, Solidification/Stabilization Team.

Jang M., Hwang J. S., Choi, I. S., Park J. K., 2005. Remediation of Arsenic-Contaminated Soils and Washing Effluents. *Chemosphere*, 60, 344-354.

Jun D., Yongsheng Z., Weihong Z., Mei H., 2009. Laboratory study on sequenced permeable reactive barrier remediation for landfill leachate-contaminated groundwater. *Journal of Hazardous Materials*, 161 (1), 224-230.

Kacimov A. R., Klammeler H., Il'yinskii N., Hatfield K., 2011. Constructal design of permeable reactive barriers: groundwater-hydraulics criteria. *Journal of Engineering Mathematics*, 71, 319-338. DOI: 10.1007/s10665-011-9457-5.

Kamath R., Rentz J., Schnoor L., and Alvarez P. J. J., 2004. Phytoremediation of Hydrocarbon-Contaminated Soils: Principles and Applications. University of Iowa.

Kamolpornwijit, W., Liang, L., West, O. R., Moline, G. R., Sullivan, B. A., 2003. Preferentialflow path development and its influence on long-term PRB performance: column study. *Journal of Contaminant Hydrology*.66, 161-178.

Komnitsas K., Bartzas G., Paspaliaris I., 2006. Modeling of reaction front progress in fly ash permeable reactive barriers. *Environmental Forensics*, 7, 219–231.

Konikow L. F., Grove D. B., 1977. Derivation of Equations Describing Solute Transport in Ground Water. *US Geol. Survey Water-Res. Inv*. 77-19.

Lambert J. M., Yang T., Thomson N. R., and Barker J. F., 2009. Pulsed Biosparging of a Residual Fuel Source Emplaced at C. F. B. Borden. *International Journal of Soil, Sediment and Water*, 2 (3), Article 6.

Leeson A., Johnson P. C., Johnson R. L., Vogel C. M. Hinchee R. E., Marley M., Pearing T., Bruce C. L., Amerson I. L., Coonfare C. T., Gillespie R. D., McWhorter D., 2002. Air Sparging Desing Paradigm. Battelle Memorial Institute, Columbus, Ohio.

Levisauskas D., Brazauskas K., Chvostova O., Baskys E., Grigiskis S., 2004. Optimization of Technological Parameters of the Contaminated Soil Washing Technology. *Environmental research, engineering and management*, 4 (30), 55-60.

Li L., Benson C. H., Lawson E. M., 2006. Modeling porosity reductions caused by mineral fouling in continuous wall permeable reactive barriers. *Journal of Contaminant Hydrology*, 83, 89-121.

Liu S., Li X., Wang H., 2011. Hydraulics Analysis for Groundwater Flow Through Permeable Reactive Barriers. *Environmental Modelingand Assessment*, 16, 591-598.

Lorbeer H., Starke S., Gozan M., Tiehm A., Werner P., 2002. Bioremediation of Chlorobenzene-contaminated groundwater on granular activated carbon barriers. *Water, Air, and Soil Pollution: Focus*, 2, 183-193.

Lu H. W., Huang G. H., Zeng G. M., Maqsood I., He L., 2007. An inexact two-stage fuzzy stochastic programming model for water resources management. *Water Resources Management*, 22 (8), 991-1016.

Mackenzie, P. D., Horney, D. P., Sivavec, T. M., 1999. Mineral precipitation and porosity losses in granular iron columns. *Journal of Hazardous Materials*, 68, 1-17.

McClain M. S., Banks M. K., and Schwab A. P., 2004. The Use of Vegetation in the Stabilization, Reclamation, and Remediation of Impacted I. N. D. O. T. Soils. Publication FHWA/IN/JTRP-2004/17. Joint Transportation Research Program, Indiana Department of Transportation and Purdue University,West Lafayette, Indiana, DOI: 10.5703/1288284313334.

McDonald M. C., Harbaugh A. W., 1988. M. O. D. F. L. O. W., A modular three-dimensional finite difference ground-water flow model. *US Geological Survey, Techniques of Water resources Investigations Book*, 6 (A1), 83-875.

McKinney D. C and Lin M. D., 1996.Pump-and-treat groundwater remediation system optimization. *Journal of Water Resources Planning and Management,* 122 (2),128–136.

Miller M. L., Bhadha J. H., O'Connor G. A., Jawitz J. W., Mitchell J., 2011. Aluminum water treatment residuals as permeable reactive barrier sorbents to reduce phosphorus losses. *Chemosphere*, 83, 978-983.

Minsker B. S, Shoemaker C. A., 1998. Dynamic optimal control of in situ bioremediation of ground water. *Journal of Water Resources Planning and Management,* 124 (3), 149–61.

Moon J. W., Moon, H. S., Kim, H., Roh, Y., 2005. Remediation of TCE-contaminated groundwater using zero valent iron and direct current: experimental results and electron competition model. *Environmental Geology*, 48 (6), 805-817.

Moretti L., 2005. In Situ Bioremediation of D. N. A. P. L. Source Zones. National Network of Environmental Management Studies Fellows for US Environmental Protection Agency.

Myers A. L., Prausnitz J. M., 1965. Thermodynamics of mixed-gas adsorption. *AIChE Journal*, 11(1), 121-127.

N. F. E. S. C. (Naval Facilities Engineering Service Centre, 1998. Application Guide for Thermal Desorption Systems. *Technical Report* TR-2090-ENV.

Ouellet-Plamondon C., Lynch R., Al-Tabbaa A., 2011. Metal Retention Experiments for the Design of Soil-Mix Technology Permeable Reactive Barriers. *Clean – Soil, Air, Water*, 39 (9), 844-852.

Paria S. and Yuet P. K., 2006. Solidification/Stabilization of Organic and Inorganic Contaminants using Portland Cement: A Literature Review. *Environmental Reviews*, 14, 217-255.

Park Y., Shin W. S., Choi S., 2011. Sorptive removal of cobalt, strontium and cesium onto manganeseand iron oxide-coated montmorillonite from groundwater. *Journal of Radioanalytical and Nuclear Chemistry*, DOI 10.1007/s10967-011-1527-7.

Pearl M., Pruijn M, Bovendeur J., 2006. The Application of Soil Washing to the Remediation of Contaminate Soils. *Land Contamination and Reclamation*, 14 (3), 713-726.

Perry R. H., Green D. W. - Perry's Chemical Engineers' Handbook 17th edition, McGraw Hill 2004.

Plagentz V., Ebert M., Dahmke A., 2006. Remediation of groundwater chlorinated and brominated hydrocarbons, benzen and chromate by sequential treatment using Z. V. I. and G. A. C. *Environmental Geology*, 49, 684-695.

P. N. N. L. (Pacific Northwest National Laboratory), 2006.Natural and Enhanced Attenuation of Chlorinated Solvents Using RT3D. PNNL-15937.

Pour A. K., Karamanev D., Margaritis A., 2005. Biodegradation of petroleum hydrocarbons in an immobilized cell airlift bioreactor. *Water Research*, 39 (15), 3704–3714.

Powell T., Smith G., Sturza J., Lynch K., Truex M., 2007. New Advancements for In Situ Treatment Using Electrical Resistance Heating. Published on line in Wiley Interscience. Remediation. DOI: 10.1002.rem.

Radke S. J., Prausnitz J. M., 1972. Thermodynamics of multi-solute adsorption from dilute liquid solutions. *AIChE Journal*, 18(4) 761-768.

Ruthven M. D., 1984. Principles of adsorption and adsorption process. John Wiley and Sons, New York.

Saichek R. R. and Krishna R. R., 2005. Electrokinetically Enhanced Remediation of Hydrophobic Organic Compounds in Solis: A Review. *Critical Reviews in Environmental Science and Technology*, 35, 115-192.

Scheidegger A. E., 1961. General theory of dispersion in porous media. *Journal of Geophysical Research*, 66 (10), 3273-3278.

S. E. R. D. P. (Strategic Environmental Research and Development Program), 2006. Improved Understanding of Fenton-like Reactions for the In Situ Remediation of Contaminated Groundwater Including Treatment of Sorbed Contaminants and Destruction of DNAPLs. S. E. R. D. P. Project ER-1288.

S. E. R. D. P. (Strategic Environmental Research and Development Program), 2007a. Improved Understanding of In Situ Chemical Oxidation Soil Reactivity. S. E. R. D. P. Project ER-1289.

S. E. R. D. P. (Strategic Environmental Research and Development Program), 2007b. Improved Understanding of In Situ Chemical Oxidation Contaminant Oxidation Kinetics. S. E. R. D. P. Project ER-1289.

S. E. R. D. P. (Strategic Environmental Research and Development Program), 2008. Using Advanced Analysis Approach to Complete Long-Term Evaluations of Natural Attenuation Processes on the Remediation of Dissolved Chlorinated Solvent Contamination. S. E. R. D. P. Project ER-1348.

S. E. R. D. P. (Strategic Environmental Research and Development Program), 2011. Enhanced Reactant-Contaminant Contact through the Use of Persulfate In Situ Chemical Oxidation (I. S. C. O.). S. E. R. D. P. Project ER-1489.

S. R. N. L. (Savannah River National Laboratory), 2006. Characterization and Monitoring of Natural Attenuation of Chlorinated Solvents in Ground Water: A Systems Approach.WSRC-STI-2006-00084, Rev.1.

S. R. N. L. (Savannah River National Laboratory), 2007. Scenarios Evaluation Tool for Chlorinated Solvent M. N. A.WSRC-STI-2006-00096, Rev. 2.

Stocking A., Eylers H., Wooden M., Herson T., Kavanaugh M.C., 2000. Chapter 2.0 Air Stripping. Treatment Technologies for Removal of Methyl Tertiary Butyl Ether (M. T. B. E.) from Drinking Water: Air Sparging, Advanced Oxidation Processes, Granular Activated Carbon, Synthetic Resin Sorbents, 2th Edition, *National Water Research Institute*, 23-105.

Stumm W., Morgan J. J., 1996. Aquatic Chemistry. John Wiley and Sons, Inc. New York.

Suzuki M., 1990. Adsorption Engineering. Elsevier, Amsterdam.

Suzuki T., Oyama Y., Moribe M., Niinae M., 2012. An electrokinetic/Fe^0 permeable reactive barrier system for the treatment of nitrate-contaminated subsurface soils.*Water Research*, 46, 772-778.

The California M. T. B. E. Research Partnership, 2006. Removal of M. T. B. E. from Drinking Water Using Air Stripping: Case Studies. National Water Research Institute.

Townsend T., Kumar D., KoJ., 2008. Bioreactor Landfill Operation: A Guide for Development, Implementation and Monitoring: version 1.0. Preparedfor the HinkleyCenter for Solid and Hazardous Waste Management, Gainesville, F. L.

Trefry M. G., Muffels C., 2007. F. E. F. L. O. W.: A Finite-Element Ground Water Flow and Transport Modeling Tool. *Ground water*, 45(5), 525-528.

Tsitonaki A., Orbicon A/S, Bjerg P. L., 2008. In Situ Chemical Oxidation – State of the Art. A. T. V. Jordog Grundvad.

U. K. Environment Agency, 2004a. Guidance on the use of Stabilisation/Solidification for the Treatment of Contaminated Soil. *Science Report*: SC980003/SR1.

U. K. Environment Agency, 2004b. Review of Scientific Literature on the Use of Stabilisation/Solidification for the Treatment of Contaminated Soil, Solid Waste and Sludges. *Science Report*: SC980003/SR2.

US Army Corps of Engineers, 2001. Engineering and Design Air Stripping. *Design Guide* No. 1110-1-3.

US D. O. E. (Department of Energy), 1995. In Situ Air Stripping Using Horizontal Wells. *Innovative Technology Summary Report.*

US D. O. E. (Department of Energy), 1998. Transportable Vitrification System. *Innovative Technology Summary Report.*

US D. O. E. (Department of Energy), 2000. Remediation of DNAPLs in Low Permeability Soils. D. O. E./E. M.-0550.

US D. O. E. (Department of Energy), 2001. Ultralow Concentration Mercury Treatment Using Chemical Reduction and Air Stripping. WSRC-MS-2001-00388.

US E. P. A. (Environmental Protection Agency), 1986. Handbook for Stabilization/Solidification of Hazardous Wastes. E. P. A./540/2-86/001.

US E. P. A. (Environmental Protection Agency), 1988. Updated Health Assessment Document for Tetrachloroethylene. E. P. A.600/8-82/005B.

US E. P. A. (Environmental Protection Agency), 1989a. Stabilization/Solidification of C. E. R. C. L. A. and R. C. R. A. Wastes. E. P. A. /625/6-89/022.

US E. P. A. (Environmental Protection Agency), 1989b. Performance of Air Stripping and G. A. C. for S. O. C. and V. O. C. Removal from Groundwater. E. P. A. /600/S2-88/053.

US E. P. A. (Environmental Protection Agency), 1990. Basics of Pump-and-Treat Ground-Water Remediation Technology. E. P. A. /600/8-90/003.

US E. P. A. (Environmental Protection Agency), 1991. Guide for Conduction Treatability Studies under C. E. R. C. L. A.: Soil Washing. E. P. A. /542/2-91/010B.

US E. P. A. (Environmental Protection Agency), 1993. Engineering Bulletin: Solidification/Stabilization of Organic and Inorganics. E. P. A. /540/S-92/015.

US E. P. A. (Environmental Protection Agency), 1995a. *Manual Bioventing Principles and Practices*. Volume 1: Bioventing Principles. E. P. A. /540/R-95/534a.

US E. P. A. (Environmental Protection Agency), 1995b. *Manual Bioventing Principles and Practices.* Volume I1: Bioventing Design. E. P .A. /625/XXX/001.

US E. P. A. (Environmental Protection Agency), 1996a. Pump-and-Treat Ground-Water Remediation: A Guide for Decision Makers and Practitioners. E. P. A. /625/R-95/005.

US E. P. A. (Environmental Protection Agency), 1996b. Presumptive Response Strategy and Ex-Situ Treatment Technologies for Contaminated Ground Water at C. E. R. C. L. A. sites. E. P. A. /540/R-96/023.

US E. P. A. (Environmental Protection Agency), 1997a. Permeable Reactive Subsurface Barriers for the Interception and Remediation of Chlorinated Hydrocarbon and Chromium (IV) Plumes in Ground Water. E. P. A. /600/F-97/008.

US E. P. A. (Environmental Protection Agency), 1997b. Analysis of Selected Enhancements for Soil Vapor Extraction. E. P. A./542/R-97/007.

US E. P. A. (Environmental Protection Agency), 1997c. Ground Water Issue. How Heat Can Enhance In-Situ Soil and Aquifer Remediation: Important Chemical Properties and Guidance on Choosing the Appropriate Technique. E. P. A. /540/S-97/502.

US E. P. A. (Environmental Protection Agency), 1997d. Ground Water Issue. Design Guidelines for Conventional Pump-and-Treat Systems. E. P .A. /540/S-97/504.

US E. P. A. (Environmental Protection Agency), 1997e. Best Management Practices (BMPs) of Conventional Pump-and-Treat Systems. E. P. A ./530/R-97/007.

US E. P. A. (Environmental Protection Agency), 1997f. Engineering Forum Issue Paper: Thermal Desorption Implementation Issues. E. P. A. /540/F-97/031.

US E. P. A. (Environmental Protection Agency), 1998. Permeable Reactive Barrier Technologies for Contaminant Remediation. E. P. A./600/R-98/125.

US E. P. A. (Environmental Protection Agency), 1999. Monitored Natural Attenuation of Chlorinated Solvents. E. P. A. /600/F-98/022.

US E. P. A. (Environmental Protection Agency), 2000a. Solidification/Stabilization Use at Superfund Sites. E. P. A./542/R-00/010.

US E. P. A. (Environmental Protection Agency), 2000b. Engineered Approach to In Situ Bioremediation of Chlorinated Solvents: Fundamentals and Field Applications. E. P. A. /542/R-00/008.

US E. P. A. (Environmental Protection Agency), 2000c. Introduction to Phytoremediation. E. P. A. /600/R-99/107.

US E. P. A. (Environmental Protection Agency), 2001a. Use of Bioremediation at Superfund Sites. E. P. A. /542/R-01/019.

US E. P. A. (Environmental Protection Agency), 2001b. Brownfields Technology Primer: Selecting and Using Phytoremediation for Site Cleanup. E. P. A. /542/R-01/006.

US E. P. A. (Environmental Protection Agency), 2002a. Groundwater Remedies Selected at Superfund Site. E. P. A./542/R-01/022.

US E. P. A. (Environmental Protection Agency), 2002b. Elements for Effective Management of Operating Pump and Treat Systems. E. P. A. /542/R-02/009.

US E. P. A. (Environmental Protection Agency), 2002c. Arsenic Treatment Technologies for Soil, Waste, and Water. E. P. A. /542/R-02/004.

US E. P. A. (Environmental Protection Agency), 2002d. Ground Water Issue. Calculation and Use of First-Order Rate Constants for Monitored Natural Attenuation. E. P. A. /540/S-02/500.

US E. P. A. (Environmental Protection Agency), 2004a. In Situ Thermal Treatment of Chlorinated Solvents. Fundamentals and Field Applications. E. P. A. /542/R-04/010.

US E. P. A. (Environmental Protection Agency), 2004b. Minergy Corporation Glass Furnace Technology Evaluation. E. P. A. /540/R-03/500.

US E. P. A. (Environmental Protection Agency), 2004c. How to Evaluate Alternative Cleanup Technologies for Underground Storage Tank Sites. E. P. A./510/R-04/002.

US E. P. A. (Environmental Protection Agency), 2004d. Performance Monitoring of M. N. A. Remedies for VOCs in Ground Water. E. P. A. /600/R-04/027.

US E. P. A. (Environmental Protection Agency), 2005a. Cost-Effective Design of Pump and Treat Systems. E. P. A. /542/R-05/008.

US E. P. A. (Environmental Protection Agency), 2005b. Evaluation of Phytoremediation for Management of Chlorinated Solvents in Soil and Groundwater. E. P. A. /542/R-05/001.

US E. P. A. (Environmental Protection Agency), 2005c. Measurement if Fugitive Emissions at a Bioreactor Landfill. E. P. A. /542/R-05/001.

US E. P. A. (Environmental Protection Agency), 2006a. Off-Gas Treatment Technologies for Soil Vapor Extraction Systems: State of the Practice. E. P. A./542/R-05/028.

US E. P. A. (Environmental Protection Agency), 2006b. In Situ Treatment Technologies for Contaminated Soils. E. P. A. /542/F-06/013.

US E. P. A. (Environmental Protection Agency), 2006c. Engineering Issue: In Situ Chemical Oxidation. E. P. A. /600/R-06/072.

US E. P. A. (Environmental Protection Agency), 2007a. Treatment Technologies for Mercury in Soil, Waste, and Water. E. P. A./542/R-07/003.

US E. P. A. (Environmental Protection Agency), 2007b. A Cost Comparison Framework for Use in Optimizing Ground Water Pump and Treat Systems. E. P. A. /542/R-07/005.

US E. P. A. (Environmental Protection Agency), 2007c. Optimization Strategies for Long-Term Ground Water Remedies (with Particular Emphasis on Pump and Treat Systems). E. P. A. /542/R-07/005.

US E. P. A. (Environmental Protection Agency), 2007d. Mine Waste Technology Program. Acid/Heavy Metal Tolerant Plants. E. P. A. /600/R-07/114.

US E. P. A. (Environmental Protection Agency), 2007e. Bioreactor Performance. E. P. A. /530/R-07/007.

US E. P. A. (Environmental Protection Agency), 2008a. A Systematic Approach for Evaluation of Capture Zones at Pump and Treat Systems. E. P. A. /600/R-08/003.

US E. P. A. (Environmental Protection Agency), 2008b. Green Remediation: Incorporating Sustainable Environment Practices into Remediation of Contaminated Sites. E. P. A. /542/R-08/002.

US E. P. A. (Environmental Protection Agency), 2009a. D. N. A. P. L. Remediation: Selected Projects Where Regulatory Closure Goals Have Been Achieved. E. P. A. /542/R-09/008.

US E. P. A. (Environmental Protection Agency), 2009b. Green Remediation Best Management Practices: Pump and Treat Technologies. E. P .A ./542/F-09/005.

US E. P. A. (Environmental Protection Agency), 2010a. Green Remediation Best Management Practices: Bioremediation. E. P. A. /542/F-10/006.

US E. P. A. (Environmental Protection Agency), 2010b. Phytotechnologies for Site Cleanup. E. P. A. /542/F-10/009.

US E. P. A. (Environmental Protection Agency), 2011a. Optimization Evaluation Lee Chemical Superfund Site – City Of Liberty, Clay Country, Missouri. E. P. A. /542/R-11/013.

US E. P. A. (Environmental Protection Agency), 2011b. An Approach for Evaluating the Progress of Natural Attenuation in Groundwater. E. P. A. /600/R-11/204.

US N. R. L. (Naval Research Laboratory), 1999. Thermal Plasma Waste Remediation Technology: Historical Perspective and Current Trends. N. R .L. /MR/6170-99-8335.

Vogan J. L., Focht R. M., Clark D. K., Graham S. L., 1999. Performance evaluation of a permeable reactive barrier for remediation of dissolved chlorinated solvents in groundwater. *Journal of Hazardous Materials*, 68, 97–108.

Wang W., Ahlfeld D. P., 1994. Optimal groundwater remediation with well location as a decision variable: model development. *Water Resources Research*, 30 (5), 1605–1618.

Wisconsin Department of Natural Resource, 2003. Guidance for Design, Installation and Operation of Soil Venting Systems. PUB-RR-185.

Zhang M., He F., Zhao D., Hao X., 2011. Degradation of soil-sorbed trichloroethylene by stabilized zero valent iron nanoparticles: Effects of sorption, surfactants,and natural organic matter. *Water Research*, 45, 2401-2414.

Zheng C., 1990. MT3D, a modular three-dimensional transport model. S. S. Papadopulosand Associates, Inc., Rockville, Maryland.

In: Landfills
Editors: G. B. C. Cabral and B. A. E. Botelho
ISBN: 978-1-62257-121-5

Chapter 4

TECHNOLOGICAL ADVANCEMENTS FOR WASTE MANAGEMENT IN VALLEY FILLS/LANDFILLS

Innocent Pumure[1*], John J. Renton[2] and Ronald B. Smart[3]
[1]College of Health, Science and Technology, School of Environmental, Physical and Applied Sciences, University of Central Missouri, Warrensburg, MO, US
[2] Department of Geography and Geology, West Virginia University, Morgantown, WV, US
[3]C. Eugene Bennett Department of Chemistry, West Virginia University, Morgantown, WV, US

ABSTRACT

This chapter introduces readers to the processes involved in the creation of valley fills and how valley fills may affect the geochemistry of surface and underground waters.

An ideal placement of mining waste in valley fills/landfills often requires a greater depth of prior knowledge on the chemical and physical properties of mining waste. Such knowledge allows for the development of appropriate methodologies and then procedures that can be designed to avoid landfill collapse and/or leaching of specific chemical substances.

Solid mining waste or overburden containing substantial amounts of bioavailable and bio-accessible chemical species can be treated in-situ during placement in landfills to ameliorate subsequent release of certain targeted chemical species. Assessment of leaching characteristics using various laboratory analytical procedures is presented.

The leaching processes are discussed with reference to myriad applications of a variety of instrumental methods of analysis. Emerging technologies on proactive and reactive control measures that include the sequestration and immobilization of dissolved minerals ions that encompass selenate, selenite, arsenate, and arsenite are also presented.

---

* Tel 1-660 543 4143; Fax 1-660 543 4843.

1. Introduction

Overview of MTR/VF

Mountaintop removal/valley fill (MTR/VF) coal mining is a process that involves the removal of mountain tops to expose beds of coal. MTR/VF begins with the removal of the available flora and fauna that is found in areas identified to contain seams of coal beneath the surface. Organic materials are also removed from nearby valleys designated for the creation of valley fills. After clearing trees, soil and waste rock are unearthed in search of coal beds. This rock material generated by the mining phase is then placed in adjacent valleys and streams creating valley fills. Deposition of the rock material into the stream beds often covers and suffocates benthic micro-organisms that are needed for the sustenance of the natural aquatic environments. Frequently, valleys are engineered in such way that they would secure the mining waste rock and allow water to drain through the valley fill into associated main waterways. MTR/VF is considered an attractive mining method by coal operators because it allows for an efficient exposure of coal beds while minimizing operational costs. After valleys are filled, the resulting horizontal surfaces may be used for a variety of purposes that include building houses, shopping malls, and recreational facilities. The crushed rock waste contains a variety of clay minerals that may expand and then compact during wet/dry seasonal variations. The hydro compression of the mining waste may inadvertently create problems in some areas with filled valleys that are being used for housing and recreational activities[1]. Valley fills have been found to leach a variety of metals that may disturb the ecological balance in natural aquatic systems.[2-6] Toxicological assays on leachates from mine spoils have shown some genotoxicity associated with their chemical composition. This chapter will focus more on the environmental chemistry and bio-geochemical transformations that may occur within the mine overburden placed in valley fills.

Economical Aspects of MTR/VF

According to the National Mining Association (NMA)[7], Mountaintop mining operations produce desirable, high-wage jobs that are the economic engine of local communities, states and the region in which they operate. The average mining wage is more than $66,000/year, excluding overtime – 57 percent higher than the average for industrial jobs. There are more than 14,000 surface coal miners in the Appalachian region, and for every mining job an additional 3.5 downstream jobs are created through mining services, sales, and other related business. Accordingly, mountaintop mining provides almost 60,000 jobs in Appalachia. It is believed that eliminating MTR/VF will have a devastating economic impact on the communities where these mines are located. It is therefore imperative for a cost benefit analysis to be performed to assess the environmental impacts associated with MTR/VF and the economical benefits of this mining technology[7]. MTM/VF mining procedures yearn for continuous improvement in the design of efficient overburden collection and containment in valley fills. This must be followed by an effective monitoring process that allows for immediate problem solving in the event of a valley fill failure or the leaching of effluent containing dissolved and suspended substances beyond the levels stipulated by Section 404(a) of the Clean Water Act.

2. Containment of Solid Mining Waste in Valley/land fills

The design of an effective engineered mining waste containment fill that allows for continuous monitoring of compaction and leaching before and after reclamation is desirable. An example of a process leading to the construction of an engineered valley fill is shown in Figures 1-4. The creation of an under drain allows for water to flow through and continuous monitoring of the leachates. However, in some situations there is intermittent flow of water that creates seasonal streams that only exist during rainy seasons. The resulting flattened surfaces may then be used for a variety of construction endeavors.

Figure 1. The design of an under drain. [8] From, The Truth About Surface Mining, http://www.truthaboutsurfacemining.com/Pages/default.aspx, accessed 3/9/2012.

Figure 2. Placing rock waste into a valley fill. [8] From, The Truth About Surface Mining, http://www.truthaboutsurfacemining.com/Pages/default.aspx, accessed 3/9/2012.

Figure 3. Grading the surface. [8] From, The Truth About Surface Mining, http://www.truthaboutsurfacemining.com/Pages/default.aspx, accessed 3/9/2012.

Figure 4. Revegetation of flattened surfaces. [8] From, The Truth About Surface Mining, http://www.truthaboutsurfacemining.com/Pages/default.aspx, accessed 3/9/2012.

3. Biotic and Abiotic Chemical Reactions Occurring in Mine Wastes

When the overburden remains unexcavated from the underground and not exposed to any atmospheric conditions, natural processes will take place to maintain the ecological balance/equilibrium between chemical species in the solid phase and the ones dissolved or suspended in the water column. However, exposing the mining waste materials to ambient environmental conditions often results in redox and dissolution chemical reactions occurring at the solid-solution interface that can affect stream water quality. These reactions can occur through physical, chemical, or biogeochemical transformations occurring on particles surfaces. Hartman et. al.[9] found reduced densities of micro invertebrates in streams affected by MTR/VF mining which they proposed was due to the elevated conductivity caused by increased concentrations of ions (Na, K, Mn, Mg, Ca, Ni, and Fe) and fine sediment compared to reference streams.

The increase in fine sediment can suffocate benthic micro-organisms and could also cause a high concentration of suspended particles. Small particles have a low settling velocity and might take several days before they accumulate at the bottom of streams, which creates reduced visibility for aquatic organisms such as fish. The dissolution or particle disintegration of the mine overburden often results in leaching of both organic and inorganic materials from the solid phase into the water column. The chemical reactions may also be facilitated by resident micro-organisms.

Once released from valley fills, organic and inorganic substances can enter many different pathways that are dependent on the physical and chemical properties of the water and sediment.

The combined concentrations of (i) soluble amounts released immediately into the liquid phase and (ii) amounts retained by the solid phase that can be easily transferred into the liquid phase is considered to be bioaccessible.[10] The release rates from the solids depend on their relative bioavailabilities. Hence as illustrated in Figure 5, phases 2 - 4 represent the labile fractions that are bioavailable compared to the inert material in phase 1. The inert material in phase 1 may become bioavailable at some time if the solid material continues to react with other environmental constituents. The metalloids of primary concern are Arsenic and Selenium.

EPA Water Quality Regulations

The USEPA National Interim Primary Drinking Water Regulations maximum contaminant level (MCL) for selenium is 50 ng/mL. A concentration higher than the MCL has the potential to cause the following effects from a lifetime exposure: hair and fingernail loss; damage to kidney and liver tissue, and the nervous and circulatory systems.[11]The USEPA criterion continuous concentration (CCC) for selenium, which is an estimate of the highest concentration of selenium in surface water to which an aquatic community can be exposed indefinitely without resulting in an unacceptable effect, has been set as 5 ng/mL.[12] In a 2001 USEPA study undertaken in a MTR/VF mining area in southern West Virginia, measured selenium concentrations were higher than 5 ng/mL in 66 out of the 213 water samples collected.

All 66 sampling locations were located downstream from filled sites.[13] No detectable arsenic concentrations were found in all 213 water samples collected. [13] Individuals who drink water containing arsenic in excess of EPA's MCL of 10 ng/mL[14] over many years could experience skin damage or problems with their circulatory system, and may have an increased risk of cancer. The EPA CCC for arsenic is 150ng/mL. Interestingly, the CCC for Se is 30 times less than that of arsenic and on the other hand the MCL for selenium is five times more than that of arsenic.

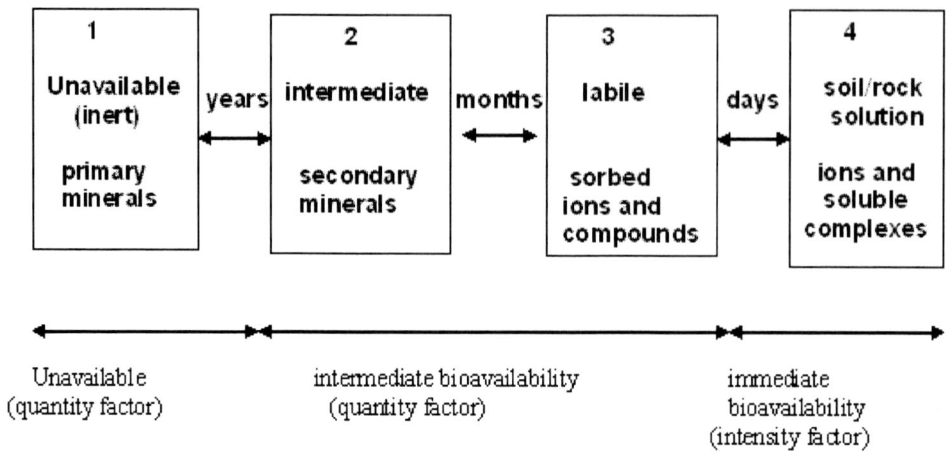

Figure 5. Bioaccessibility of solid phase chemical substances at the solid-solution interface. [10].

The fate of the released inorganic and organic chemical species depend on the prevailing environmental conditions that include pH, electrochemical properties, presence or absence of complexing agents such as natural organic matter (humic and fulvic acids), and other reactive organic and inorganic materials. [15] Such environmental conditions can result in changes of oxidation states and the sequestration or remobilization of selenium and arsenic, thereby controlling their bioavailabilities and bioaccessibilities. Chemical reactions occurring on the surfaces of the rock particles include acid rock drainage (ARD)/acid mine drainage (AMD) which occurs through the oxidation of iron containing minerals such as pyrite and pyrrhotite. ARD originates from the sulfide mineral oxidation in the presence of water and oxygen, and to some extent bacteria, to yield primarily sulfuric acid. The acid generated mobilizes some metals resulting in their increased bioavailabilities, creating a hazardous environment for aquatic animals and plants. Oxidation and hydrolysis of metal sulfides in landfills results in the formation of several soluble hydrous iron sulfates, formation of acidity and the subsequent leaching of toxic heavy metals.[16] [17] Low pH, high acidity, high salinity, turbidity, elevated concentrations of Fe and sulfate and a variety of other metals and metalloids are usually used as indicators for the presence of AMD. [16, 18] The pH values can be so low such that interfering ions may cause negative pH readings to be recorded, an acid potentiometric error. The lower the pH, the more severe the potential effects of mine drainage on aquatic life and the environment.

Pyrite minerals have been extensively studied in many mining environments but not as much attention has been placed on pyrrhotite. Here is an autocatalytic mechanism that has been proposed by Pumure and Muzenda[19] for the oxidation of pyrrhotite that was modeled from the steps involved in pyrite oxidation.

$$FeS + 2\,O_2 \rightleftarrows Fe^{2+} + SO_4^{2-} \quad (1)$$

$$Fe^{2+} + 0.25\,O_2 + H^{+} \rightleftarrows Fe^{3+} + 0.5H_2O \quad (2)$$

$$Fe^{3+} + 3H_2O \rightleftharpoons Fe(OH)_3 + 3H^+ \quad (3)$$

$$FeS + 8Fe^{3+} + 4H_2O \rightleftharpoons 9Fe^{2+} + SO_4^{2-} + 8H^+ \quad (4)$$

Reactions 1-3 occur in tandem and the synthesized ferric ions are used up in reaction 4 to produce ferrous ions that are available for use in reactions 2 and 3. In reaction 4, although one cycle of pyrrhotite oxidation produces 8 moles of acidity compared to 16 moles of acidity(Hydronium ions) produced from a similar pathway from one mole of pyrite oxidation, the kinetics for pyrrhotite oxidation are much faster compared to that of pyrites. Recent studies performed in a laboratory confirmed that pyrrhotite is much more reactive compared to pyrite.[20, 21] On a mass to mass basis, pyrrhotite was found to produce more sulfate ions in solutions compared to the same amount of pyrite. Hence special attention should also be placed on the pyrrohotite composition in mining waste. Nonetheless, the relationship between pyrite oxidation and ARD/AMD is the one that has been extensively studied. The acid produced from ARD/AMD can mobilize chemical species that are sequestered in the in the solid mining waste. The acid aided leaching of both organic and inorganic substances from the solid phase into the solution phase may alter the quality of both surface and underground waters.

Acidophiles such as *Thiobacillus thio-oxidans* and *Thiobacillus ferro-oxidans* and can catalyze the oxidation of sulfiide minerals resulting in the formation of acidity. Micro-organism based oxidation of sulfide minerals can occur at the same time together with chemical oxidation enhancing the production of sulfuric acid.[22] Several other archea and bacillus microorganisms have been identified to participate in the bio-geochemical transformation and cycling of several minerals including their participation in AMD/ARD. The microbiological based production of acidity occurs at a faster rate than the chemical transformations. Mineral sulfide oxidation requires the alternating of dry and wet spells for the oxidation to occur. Upon neutralization by carbonate or silicate chemical species in water and when pH is raised to around 4.5, the ferric ions are converted into iron oxyhydroxides commonly known as yellow boy. Mine effluent containing yellow boy is indicative of the presence of ARD/AMD. Once AMD/ARD has started to occur in mine spoils, it is almost very difficult to contain or stop.

It would appear that the leaching of metals is only associated with acidic environments, recently there have been studies that are pointing towards base catalyzed dissolution of rocks. It has been reported that bicarbonate (HCO_3^-) ions may accelerate the leaching of metalloids such as selenium and arsenic from clay minerals and other waste compounds. [23-26] Since mine wastes contain a variety of clay minerals (claystones, mudstones, siltstones and shales), base catalyzed dissolution of clay minerals may result in the release of arsenic and selenium oxo-anions from the solid phase into the water column. Substantial amounts of arsenic and selenium have been found in alkaline natural waters that are in contact with mining waste rock. Both acid catalyzed and base catalyzed reactions may selectively leach specific types of metals and metalloids from the solid phase into the solution phase. The acid based reactions would preferentially release positively charged cations that do not form precipitates in acidic environments. On the other hand, negatively charged oxo-anions can be dislodged from the particle surfaces through ion exchange with negatively charged bicarbonate ions through ion exchange. Therefore, appropriate in-situ or ex-situ protocols have to be put in place to deal with specific group of metals associated with specific mine wastes. Hence prior to designing a treatment system, a complete characterisation of effluent mine drainage is needed to determine an appropriate system and how to design that system to meet treatment goals.[27]

The formation of polymeric iron oxyhydroxide precipitate can be produced through the following mechanism

$$Fe(H_2O)_6^{3+} \rightleftarrows Fe(H_2O)_5OH^{2+} + H^+, \textit{ mononuclear} \quad (5)$$

$$2Fe(H_2O)_5OH^{2+} \rightleftarrows [(H_2O)_5Fe\text{-}O\text{-}Fe(H_2O)_5]^{4+} + H_2O, \textit{ dinuclear} \quad (6)$$

$$2[Fe(H_2O)_5\text{-}O\text{-}(H_2O)_5Fe]^{4+} \rightleftarrows$$
$$[(H_2O)_5Fe\text{-}O\text{-}(H_2O)_4Fe\text{-}O\text{-}Fe(H_2O)_4\text{-}O\text{-}Fe(H_2O)_5]^{8+} + H_2O \textit{ tetranuclear} \quad (7)$$

Hydrolysis of ferric solutions result initially to mono- and binuclear species that further interact to produce species of higher nuclearity. As these polynuclear species age they eventually form crystalline compounds or amorphous precipitates (amorphous iron (III) hydroxide hydrate).[28, 29] As many Iron nuclei coagulate together, the resulting polynuclear precipitate will become denser and fall to the bottom of the water reservoirs. As the heavy precipitates fall to the bottom, they often "gather" or co-precipitate with other chemical substances resulting in the removal of iron and other chemical species from the water column into the sediments. Such self purification mechanisms can also occur with hydrous aluminium substances contained in river and mine effluent systems. It is desirable to have conditions in valley fill effluent that favour precipitation of polymeric hydrous iron to enhance the co-precipitation of other contaminants thereby reducing metal load in effluent streams.

Several forms of primary and secondary minerals have been identified on a variety of mining waste that had been contained in a fill. X-ray methods have been applied to obtain semi-quantitative analysis of primary and secondary minerals on surfaces of particles obtained from fills. Both X-ray diffraction (XRD) and X-ray fluorescence (XRF) techniques have found several applications in studying the mechanistic physical and chemical processes occurring on particle surfaces. A variety of X- ray methods have been used in the characterization of many mineral phases from mining waste however, some of their applicabilities are somewhat limited because due to matrix effects. Other atomic spectrometric techniques such as Inductively Coupled Plasma –Mass Spectrometry (ICP-MS) and Graphite Furnace Atomic Absorption Spectrometry (GFAAS) can also be used to complement results from XRF techniques. For example, ng/g levels for arsenic and selenium levels were found to be below the limit of detection of an XRF method (using a Philips 1420 XRF spectrometer) however, GFAAS and hydride generation atomic absorption spectrometry (HGAAS) techniques were successfully applied to analyze for such low concentrations in aqueous extracts from solid samples. In some instances, data from X-ray methods is combined with analytical data from other methods in order to obtain a better understanding of the minor and major elemental and mineral compositions of materials being studied.[30] Raman spectroscopy has also been used in the identification of several natural and synthetic mineral phases. In laboratory environments, Raman spectroscopy has also been applied in the identification of rutile and anatase mineral forms of titanium dioxide in synthetic materials. In some applications micro Raman spectroscopy was deemed to provided complementary information when compared to X-ray methods on the characterization of several iron bearing natural and synthetic mineral phases commonly found in ARD/AMD environments.[30] Although Infrared spectroscopy (IR) is one of the most mature analytical techniques, its applications in the study of rocks and minerals are being taken over by Raman Spectroscopy. One disadvantage being its inability to effectively distinguish between the Anatase and Rutile structures of titanium dioxide. However, attenuated total reflectance spectroscopy and diffuse reflectance infrared spectroscopic techniques are commonly used in the analysis of surface Al-OH and Si-OH vibrational modes that are very important in the characterization of phyllosilicate clays and their dissolution products.[31] For excellent mineral phase analysis, an XRD instrument is used.[32] Although Infrared and Raman spectroscopic techniques can be used to qualitatively identify the presence of different mineral phases on the surfaces of mining waste, complementary information can be obtained with the use of an XRD instrument. XRD instruments are the most widely used instruments that have long since been applied in the characterization of primary and secondary mineral phases occurring on the surfaces of mining waste.[32]

Combinations of Scanning electron microscope (SEM) and Energy Dispersive Spectrometer (EDS) /Wavelength Dispersive Spectrometer (WDS) have been applied in the analysis of different phases occurring on mining waste. More precisely, the environmental SEM (ESEM) is often employed to provide images of different distinguishable phases occurring on mineral surfaces while the EDS/WDS which essentially are X–ray fluorescence detectors, are used to provide elemental compositions of the identified mineral phases. Only major elemental inclusions may be detected while the minor inclusions, which might be of primary interest, may not be detected using X-ray fluorescence methods. Recent developments of XRF techniques have led to better instruments that can be used to analyze for substances in the ppb to ppt levels.

Laser Ablation ICP- MS is another powerful tool that has found immense application in the direct analysis of solids.[33] Laser ablation was used to analyze surfaces of pyritic mining waste collected from areas affected with ARD/AMD. The particles showed enrichment of the surfaces with high concentrations of some non-pyritic elements.[33] The elements that showed an increased surface concentration include arsenic, cadmium, copper and zinc. The Laser ablation method has the advantage in that there is minimal sample pre-treatment before analysis.

Water leachates from engineered landfills can be collected from wells that are dug at predetermined monitoring locations. Parameters such as temperature, pH, conductivity, turbidity, total dissolved solids (as mg/L NaCl) can be analyzed on-site using portable instruments. Dissolved anions such as chloride, sulphate, sulphide, selenate, arsenate, selenite, arsenite, phosphate can be analyzed for using a high throughput Ion Chromatography. Such a technique may also be used in the analysis of a wide range cationic metallic species. In some instances portable ion selective probes may also be used to analyze for some of the anions and cations. One limitation, however, is the transportation of several ion selective electrodes to each monitoring site. Elemental species in the drainage may also be analyzed using a variety of atomic spectroscopic techniques that include atomic absorption, atomic emission, and mass spectrometry. The focus of the next sections are going to be based on the release of selenium and arsenic from mining overburden and the associated practical technologies that have been developed to reduce the concentrations of selenium in the mine drainages.

4. Estimation of Bioaccessibility, Bioavailability and Toxicology of Mining Waste

Chemical Speciation of Arsenic and Selenium

Bioavailability and toxicity of arsenic and selenium depend not only on their total concentration but also on their physicochemical nature. Inorganic forms of both arsenic and selenium are generally considered to be more potent toxins compared to organic forms.

Selenium Speciation

Selenium exists in –2, -1, 0, +4 and +6 oxidation states. Most organic forms are in the –1 and -2 oxidation states and are mainly found in biological systems and also in reducing swampy environments.[34] In biological systems selenosugars, selenoproteins, and selenoamino acids have been reported.[34] Biotransformation through alkylation of selenium is a detoxification process that changes the more poisonous inorganic selenium to alkylselenium species , which are less toxic and are easily excreted through the urine and sweat in animals and also as volatile gases in plants.[35] Volatile compounds such as dimethylselenide, dimethyldiselenidc [35, 36] and hydrogen selenide are produced by some plants and microorganisms[37] through the detoxification process. Humans excrete trimethylselenonium ion [35] in urine.

Depending on the prevailing pH and redox potentials in natural waters, selenium is mainly found as selenite (SeO_3^{2-}) or selenate (SeO_4^{2-}). Selenite is less mobile in water compared to selenate and is mainly found in acidic waters with reducing redox potentials. Selenates are found in mainly high pH environments under oxic conditions. Selenites are generally considered to be more toxic compared to selenates and are more easily bioaccumulated.[38]

Arsenic Speciation

In natural waters arsenic is found mainly as arsenite (AsO_3^{-3}) or arsenate (AsO_4^{-3}). Arsenite is more mobile because it is neutral (H_3AsO_3) at pH 5 to 8 which is very common in water and sediments. [39] Arsenite is mainly found in reducing environments whereas arsenate is mainly found under oxidizing conditions. In general arsenite is more toxic and reactive compared to arsenate. [40] Both arsenic and selenium are used as terminal electron acceptors during anaerobic respiration. [41]

Selenium and Arsenic Toxicity to Fish and other Wildlife

In the 1980s fish and wildlife died at Kesterson National Wildlife Refuge in California due to selenium poisoning as a result of direct and indirect consumption of polluted agricultural drainage emanating from nearby farms.[42, 43] Selenium concentrations in bird livers were found to be in the range associated with adverse reproductive effects. Elevated selenium concentrations in fish and bird eggs are associated with higher teratogenic and mortality rates.

All arsenic species have profound toxic and teratogenic effects on living organisms. The toxicity of arsenic in different types of fish has been widely studied and 1.02 μg/ml of arsenic was reported to kill 50 % of Tilapia fish.[44]

5. Processes Affecting the Natural Release of Chemical Species from Solid Mining Waste

Common mathematical models that are used to describe the kinetics for the release of inorganic and organic substances from solids into solution are Parabolic diffusion, Elovich, Power function and First and Second order kinetic models. [45, 46] These models are often used to approximate the overall chemical kinetics of intricate reactions occurring in complex matrices that are often unknown. Both parabolic diffusion and first order kinetics are used to describe a desorption process that is controlled by surface film and intraparticle diffusion. [47]Although the first order and parabolic diffusion models are both very important in describing the surface film diffusion or intraparticle diffusion of materials from rocks into solution [47], the first order kinetic model can be most useful in comparing the rates of accelerated rock disintegration and dissolution with the concomitant release of arsenic and selenium from solid mining waste.

For a general sorption/desorption reaction;

$$A_{(solid)} \underset{k_2}{\overset{k_1}{\rightleftharpoons}} B_{(aqueous)} \qquad (8)$$

Upon the application of steady-state approximation, the differential first order kinetic equation for a reversible sorption-desorption chemical process can be written as follows,

$$ln\,(1- B_t/ B_\infty) = -\,(k_1 + k_2)t. \qquad (9)$$

Where B_t is the concentration of extracted/released species at time t. k_1 and k_2 are the first order reaction constants for the forward (release) and backward (sorption) reactions and B_∞ is the concentration of released species at infinite time.

With respect to the increasing concentration of the desorbed species B in the mixture, k_1 has a negative slope and k_2 has a positive slope. Equation 8 clearly shows that there are two competing rate constants that bring about change in the sorption/desorption kinetics. For desorption to occur, the condition $k_2 >> k_1$ is imperative otherwise sorption will become dominant. Equation 9 can be simplified to

$$ln\,(1- B_t/B_\infty) = -\,k\,(observed).\,t \qquad (10)$$

Where k (observed) = $k_1 + k_2$.

Approximation of B_∞

$\mathbf{B_t}$ is approximated by plotting $\mathbf{B_t}$ vs. $\mathbf{B_{t+x}}$ (Kezdy-Swinebourne method [48]) where **x** is a constant temporal incremental factor used to pair up early data points with the later ones. The point at which this line intersects a 45° line on the Cartesian plane is used to estimate the $\mathbf{B_\infty}$ value. An appropriate value of **x** has to be chosen. It is recommended that **x** should 2-3 times the half-life of the total (approximated) concentration (Appendix 1B). A plot of **ln(1-Bt/B$_\infty$) vs. t** yields a slope equal to $\mathbf{k_{observed}}$.

A well characterized kinetic equation and method can be used to estimate the amounts of chemical species originating from a specific mining region. Such studies would that allow for suitable deposition of overburden in appropriate environments that allow for control and containment of any leached chemical substances projected to be released over a certain period of time.

Ultrasound Based Leaching

Ultrasound extraction often lead to the release metals from the solid into the liquid phase on a much shorter time scale, thereby decreasing analysis times.[49-52] Ultrasonic extraction enhances the dissolution process by producing acoustic cavitations that result in bubble formation and their subsequent collapse, which generates high temperature and pressure gradients, resulting in increased chemical reactivities.[53]

The use of ultrasonic energy has proven to be an easy, convenient, and a fast way of desorbing inorganic [49, 51, 54] and organic pollutants [55] from sediments, soil and biological samples.[50,56] Ultrasound extraction is also an aseptic method and it has been suggested that high frequency ultrasound (20 kHz) in low volumes (200 mL) of bacterial suspensions resulted in a continuous reduction of bacterial cell numbers.[57-58] Hence microbial sample alteration or modification after ultrasonic extraction is considered to be minimal if samples are placed in previously sterilized containers. Temperature-controlled ultrasound extractions provide important kinetic data that can be useful in describing the chemistry at the solid-solution interface. Focused ultrasonic probes enable efficient adjustment and monitoring of temperature in a water-jacketed extraction vessels. The solid–solution chemistry can then be used to infer the bioavailabilities of metals released into associated aquatic systems. Most traditional batch extraction methods utilize lengthy mechanical shaking or soxhlet extractions that may take hours or days for a single extraction to be performed.

Ultrasonic First Order Kinetics

Since natural rock dissolution is a lengthy process [47], laboratory based ultrasound extraction can be used to mimic the natural weathering of pulverized rock waste on a much shorter time scale.

The diffusivity rate or velocity (**v**) of the extracted species is proportional to the driving force acting on the species as they move out of the particles into the solution. The ultimate velocity (**v**) in sonochemical extractions is considered to be proportional to three main factors namely; the inherent mobility, **u** (which depends on solution pH, surface area and lithology of rocks), sonication power **B** and the chemical potential, μ (which is dependent on the Gibbs free energy function for a chemical reaction, ΔG_r between the adsorbed and the desorbed species).

$$v = \mu .\ F. \qquad (11)$$

Where v is the velocity in $cm\ s^{-1}$, μ is the mobility of the extracted species in $cm\ s^{-1}/N\ mol^{-1}$ and F the extraction force in newtons (N).

Heterogeneous or homogenous extractions are governed by the ability of a chemical to move from one phase into another down a concentration gradient until equilibration or when the chemical potentials (μ) of the species in the interacting phases are equal. [59,60] Therefore, the extraction force F can be defined as

$$F = d\mu/dx \qquad (12)$$

Where $d\mu/dx$ is a chemical potential gradient across a film of desorbing species.

In sonochemical extractions, the extracting force F is magnified by the ultrasound power B as follows

$$F = B.\ d\mu/dx \qquad (13)$$

Hence combining 3, 4 and 5

$$v = u.\ B.\ d\mu/dx. \qquad (14)$$

Therefore, the velocity of substances leaching from solid particles in ultrasound assisted extractions is proportional to the following parameters: 1) pH, 2) the mobility of the selenium and arsenic compounds from the solid to liquid phase, 3) the chemical potential of these species, 4) lithology of rocks extracted, 5) the surface area of particles, and 6) the level of ultrasonication.

7. Environmental Consequence of Released Selenium and Arsenic

Several authors have explored the adsorption and desorption techniques using a variety of pure adsorbents such as clays or goethite in the presence or absence of humic and fulvic substances.[61-65] Desorption hysteresis studies provide insights into the remobilization of previously sorbed arsenic and selenium. The release of sorbed ions depends largely on their physicochemical nature and, to some extent on the solution properties of the extracting media. Su and Suarez [66] indicated that selenate adsorption to iron oxide was reversible while selenite was not. The binding of selenate was shown to depend on ionic strength and that of selenite was not. Such information is especially important in determining the bioavailable fraction of selenium that originates from solid waste under natural environmental conditions.

Manning and Goldberg [67] studied the stability of the binding of arsenate and arsenite at the clay mineral-water interface and concluded that arsenite is more stable between pH values 4 and 9 and the arsenite is oxidized to arsenate above pH 9. However, there was evidence of heterogeneous oxidation of arsenite on the clay surface that resulted in strongly bound arsenate. Most clay minerals and activated alumina have been shown to have more affinity towards arsenate than arsenite. Thus, even though arsenite species could be stable in water at natural water pH values that range from 6.5-8.5, their concentrations were quickly diminished due to heterogeneous oxidation. They also concluded that modelling of arsenite oxidation on many common mineral matrices may not always be valid due to the uncontrollable catalytic oxidation processes that occur on the mineral surfaces. MnO_2 impurities in clay minerals for example, have been shown to be responsible for the observed catalytic behaviour in lake sediments.[68]

River sediment has been identified as an accumulator of both organic and inorganic pollutants and is an important indicator of environmental pollution.[69, 70] It is often the remobilization of the sorbed selenium and arsenic from the sediments that can lead to environmental problems in river waters, and the widespread incidences of sediment contamination. [71-73] A large number of sediment quality guidelines (SQGs) have been developed to aid regulators in dealing with contaminated sediments and Burton has reviewed their worldwide use.[74]

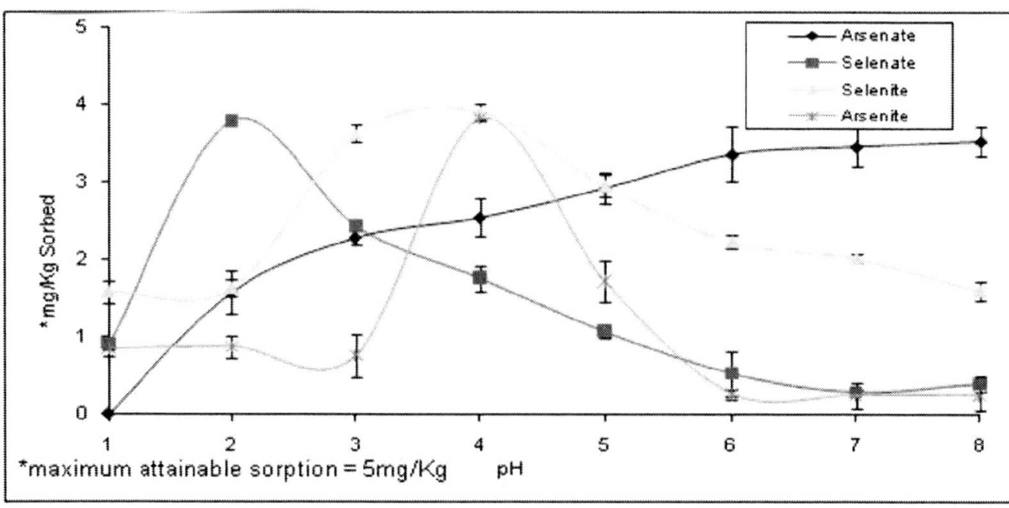

Figure 6. Sorption of selenite, selenate, arsenite and arsenate on kaolin as a function of pH. *Conditions:* 1 g treated kaolin, Ionic strength = 2 mM as NaCl, with 50ng/mL arsenite, arsenate, selenate or selenite standards [75].

Batch sorption experiments of arsenate, arsenite, selenate and selenite on Kaolin, a clay mineral, show that different chemical species have pH depended sorption capabilities. Kaolin is the major component of sediments that are found on many floors of water bodies such as streams, rivers, lakes and oceans. Such sorption characteristics could mirror the actual chemical dynamics occurring on river beds. River beds are the natural sinks for many inorganic pollutants that include arsenic and selenium. The amounts sorbed by kaolin as a function of pH for arsenate, arsenite, selenate and selenite is shown in Figure 6. Maximum sorption for all species was around 80% but occurred at different pH values. The sorption for selenate increased sharply from pH 1 to a maximum of 76 % at pH 2 and then decreased steadily to less than 5 % between pH values ranging from 6 to 8. The selenite sorption increased from about 35% at pH 1 to 80% between pH 3 and 4 and the decreased gradually to around 40% between pH 6 and 8. Arsenite sorption started from 20 % between pH values of 1-3 and then increased sharply to a maximum 78% at pH 4 followed by an abrupt decrease between pH 6 and 8 to values less than 5%. A gradual increase in arsenate sorption from 0.4% at pH 1 to about 75% at pH 8 is shown in Figure 6.

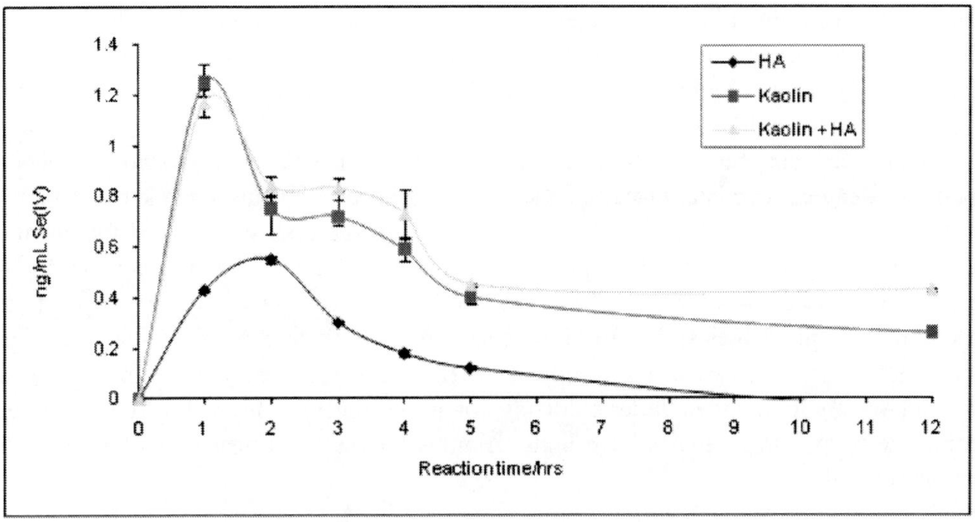

Figure 7. Reduction of 50 ng/mL of Se (VI) on Kaolin and Humic acids at pH 4. *Conditions:* 1 g treated kaolin, Ionic strength = 2mM as NaCl, with 50ng/mL, selenate. Se (IV) concentration was monitored as a function of time [75].

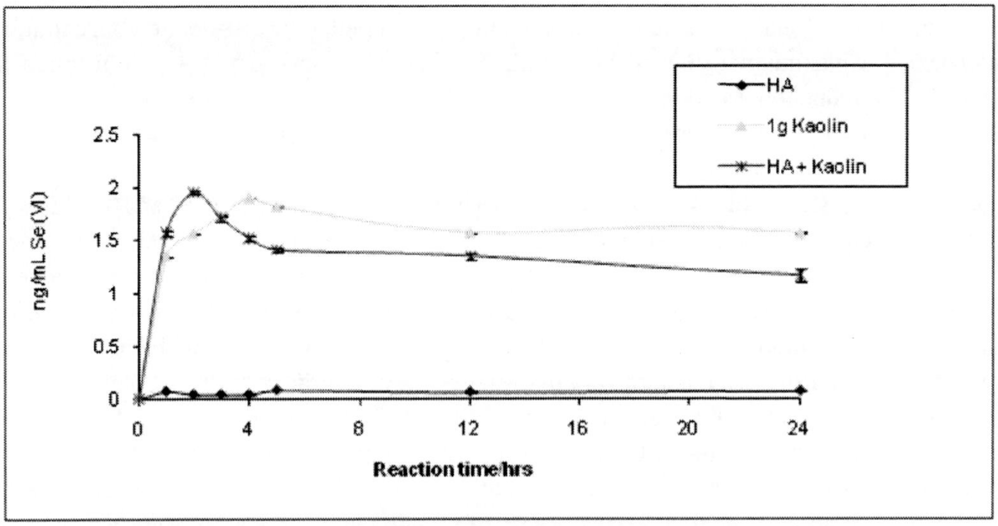

Figure 8. Reduction of 50 ng/mL of Se (VI) on Kaolin and Humic acids at pH 7. Conditions: 1 g treated kaolin, Ionic strength = 2mM as NaCl, with 50ng/mL, selenate. Se (IV) concentration was monitored as a function of time [75].

Information obtained from the kaolin sorption studies is important in determining the extent to which arsenic and selenium can sorb onto solid surfaces. Considering that kaolin is ubiquitous in stream sediments, there is a greater chance of finding more Se(VI) and As(III) in river waters at pH > 5 compared to Se(IV) and As(V) because of their low affinities shown in Figure 6. It follows that selenates are more mobile and predominant in river systems compared to selenites; however, selenites are more toxic[76]. Recently it was reported that arsenite could be more prevalent in natural waters than previously thought because it exists as the neutral H_3AsO_3.[77] The neutrality of the arsenite species could be responsible for the low affinity with the kaolin at high pH values. The results suggests that As (III) is more mobile in reducing aquatic systems compared to As (V).[78] In oxidizing environments arsenic species can be oxidized to As (V) which has a higher affinity for the sediments and this could explain why arsenic concentrations in all 213 stream water samples collected by the EPA was below the limit of detection. Such sorption characteristics can also be used to study procedures suitable for cleaning up contaminated industrial and domestic effluents using kaolin and other adsorbents. Silica gel was found to be a poor choice for the removing any of the chemical species studied because of poor sorption efficiency. Kaolin is a good candidate for the explorations because it has a higher affinity for all the selenium and arsenic species studied at different optimal pH values and at 2mM ionic strength. However in oxidizing natural aquatic systems use of kaolin may not be beneficial because most selenium will be in the selenate form which is highly mobile. Under oxidizing conditions most arsenic will be converted into the arsenate form, and arsenates have high affinities for the solid phase at environmentally important pH values.

8. Kinetic Reduction of 50 ng/mL Se (VI) at pH 4 and pH 7

Kinetic experiments for the reduction of selenate with humic acid (HA) and kaolin were carried out at pH 4 and 7. An initial increase in the accumulation of selenite concentration that reached a maximum value of 2.0 ng/mL between 1-2 hours at pH 4 as illustrated in Figure 7. The reduction reaction is faster at pH 4 than at pH 7 as shown in Figures 7 and 8 respectively. It took approximately 4 hours at pH 7 and 2 hours at pH 4 to reach the respective maximum concentrations. The follow up disappearance of the transient Se (IV) peak was also faster at pH 4 than at pH 7. A combination of humic acid and kaolin was superior in reducing selenate to selenite species compared to kaolin alone at both pH values. This could be as a result of competitive sorption between humic acid and the produced selenite species onto the surfaces of kaolin particles. Such competitive sorption can also be responsible for the mobilization and reduced affinity of selenium species with kaolin. Hence it could be the reason why natural organic matter plays a crucial role in controlling the concentrations of water soluble selenium and arsenic [79] in the environment. In reactions with kaolin alone the sorption of selenite was not hindered resulting in reduced solution selenite concentrations. Reactions involving both kaolin and HA produced higher aqueous selenite concentrations probably because some of sorption sites on kaolin were occupied by HA and therefore not available for selenite adsorption/sorption. Selenate reduction is dependent on the pH, sediment (kaolin) and presence of HA. It has been shown that HA alone cannot reduce selenate at high pH probably due to electrostatic repulsions. However, there is significant reduction at low pH. The presence of TiO_2 impurities on kaolin provides a catalytic pathway for the reduction of selenate. The presence of a Se (IV) transient peak could suggest that the selenate reduction passes through a selenite intermediate. This immediate disappearance of selenite from solution suggests that selenite species are intermediates that are used up or complexed in a follow up reaction. The transient appearance of the Se (IV) can be used to confirm the assertions made by Johnson (2004) [80] who suggested that Se(VI) is reduced to Se(-II) in a stepwise fashion through Se(IV) and Se(0) intermediates.

Hence kaolin, a major component of stream sediments plays a critical role in the sorption, redox conversions and the remobilization of selenium species.

It has been reported that compounds with polycarboxylic acids functional groups such as ascorbic acid can reduce Se (IV) to Se (0).[81] It appears kaolin enhances the reduction probably due to the presence of catalytic TiO_2 impurities.[82] This could be the reason why more selenite was produced in kaolin-based experiments compared to HA alone. HA alone at a concentration of 4.4 mg/L is not effective in reducing selenate ions at pH 7, as shown in Figure 8 but at pH 4 as shown in Figure 7.

9. Analytical Extraction of Selenium and Arsenic from Solid Samples

Bioavailable concentrations in sediments are usually obtained using standard and modified Tessier's extraction methods[83] and total available by both open vessel and closed vessel acid digestion procedures. Traditionally, selenium and arsenic in solid samples have been extracted with water, NaOH, phosphate buffers, oxidizing agents such as $K_2S_2O_8$, using open and closed vessel methods that use concentrated acids digestions.[84,54]. Open vessel digestions may result in the loss of volatile analyte species that may be formed during digestion process. Acid digestions of selenium may form volatile species that include $SeOCl_2$ which may be easily lost during the heating process and thus closed vessel methods are more preferred compared to open vessel methods. Some arsenic species may also be lost when open vessel acid digestion methods are used.

Pressure controlled or temperature controlled closed vessel microwave digestions that use combinations of strong acids often result in complete or near complete dissolution of the solid samples while preventing loss of analytical species of interest during the digestion process. Acid combinations range from aqua regia (3 parts HCl: 1 part HNO_3) to a variety of triple acid attack procedures consisting of HF, HCl and HNO_3. HF is included to help break up the silicate matrix found in most natural solids through the formation of volatile SiF_4. The destruction of siliceous materials often results in the release of encapsulated chemical species into solution. Acid digestions result in the liberation into solution of both loosely and strongly bound metals together with residual metals that are closely associated with silicate matrices. Such extraction methods, although useful in estimating the entire amounts of available metals in solids, cannot give the essential information on the readily bioavailable concentrations that are immediately leached into aquatic systems soon after valleys are filled. It is imperative to use analytical methods that extract the loosely bound metals, which are readily mobile because their concentrations can quickly build up under conditions that favour high residence times. Acid leaching, Standard Tessier, and the Standards, Measurements and Testing program (SMT) formerly known as Community Bureau of Reference, Commission of the European communities (BCR) methods are very useful in estimating the amount of metals available in different phases.[85, 86]The flow diagrams for the BCR, Standard Tessier and modified Tessier extraction methods are shown in Figures 9, 10 and 11, respectively.

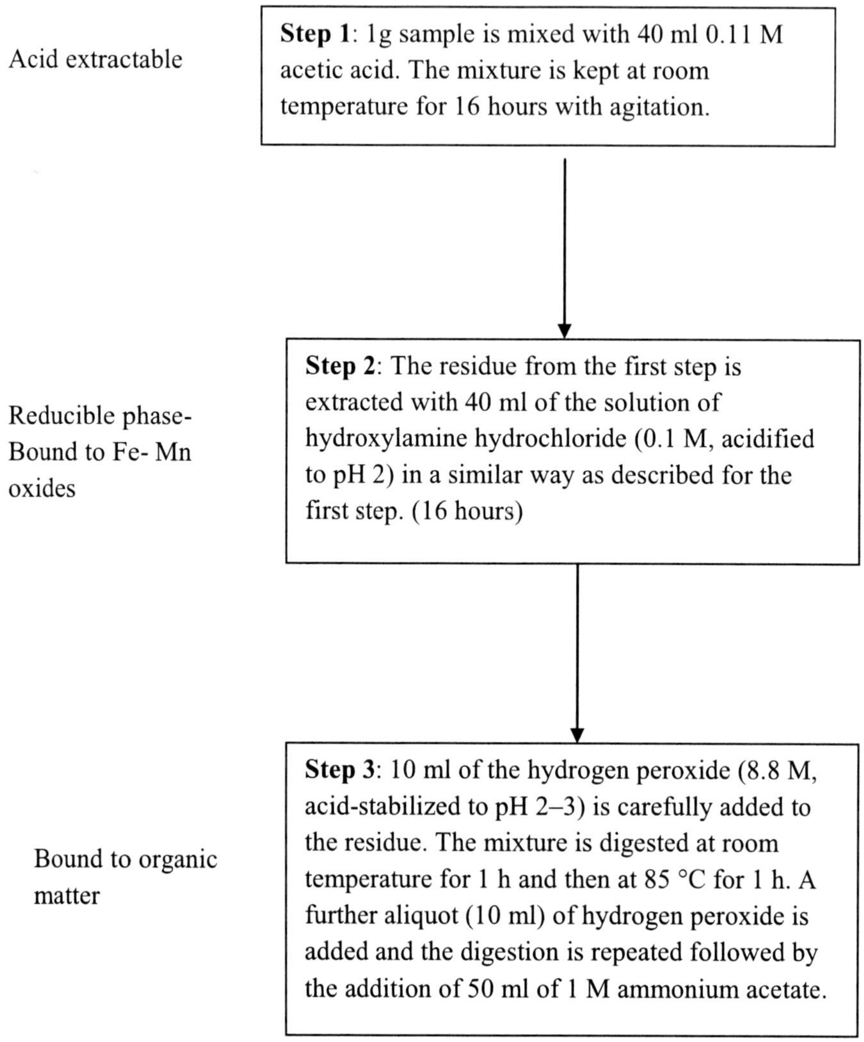

Figure 9. SMT (BCR) extraction procedure [85].

The initial stage of the BCR method uses a weak acid (acetic acid) and in Standard Tessier extraction methods a highly concentrated chloride salt (NaCl, KCl or $MgCl_2$) is used to estimate both water soluble and exchangeable metals in solid samples. In the Acid leaching method, 0.01– 4M HCl, [87] is used to represent an HCl extractable fraction, which is often related to total concentration of selenium that can be released over an infinite period of time. Caution must be exercised when using acids in acidic-based digestions, appropriate use of hoods (especially for perchloric acid digestions) and personal protective equipment must be properly used to ascertain safety. Although these HCl extractions have been widely used for positively charged metallic species, their applicability to the extractions of selenium and arsenic oxoanions is debatable because these anions are closely associated with both HA and FA.[88] HA is soluble in alkaline media and insoluble in acidic media whereas FA is soluble in both acidic and alkaline media. Hence base extractions using strong bases such as NaOH and KOH often result in the extraction of metallic species in the organic fractions that are closely associated with NOM.

It is well-known that the toxicity and the bioavailability of any chemical species depend on both the total concentration and, often more importantly, on the chemical form. The use of acidic media, oxidizing, and reducing agents will often change the oxidation state of metals. Thus these methods are only appropriate for the quantitation of species stable in the chosen extraction medium. This also explains why there is a general consensus that deionized water should be used as an extracting solvent either hot or cold because it has the ability to preserve the natural oxidation state of metals being leached.[82, 89-91]

Attempts have been made to use alkaline phosphate buffers and sodium hydroxide solutions to leach out and to preserve the oxidation states of selenium species.[91]Phosphate buffer extractions have been found to remove both exchangeable and strongly bound selenium is soils with insignificant modifications of the oxidation states. The loosely bound portion that comprises the water-soluble and exchangeable fractions are the most bioavailable. The high chloride concentrations used in BCR and HCl-based leaching protocols, although very essential in displacing the oxy-anions from the solid surfaces by ion exchange, may also reduce Se(VI) to Se (IV) [92]during the course of extraction thus altering the natural distributions of species in the samples important in assessing the immediate release and bioavailability of metals in the aquatic environment.

In other applications, as mandated by the USEPA, a TCLP (toxicity characteristic leaching procedure) is used to approximate both organic and inorganic materials that can be easily leached from solid mining waste. The TCLP method uses a pH 4.93 $\pm$ 0.05 acetic acid-acetate buffer. However, a special extraction vessel is required to perform TCLP tests. In some studies, [93] the TCLP method was found to extract arsenic concentrations that were ten times less that what was obtained using other methods. In general, all static batch extraction methods although they provide information on substances that are immediately bioaccessible, they do not offer any details about substances that may not be leached at the time of extraction but may become bioavailable when the waste materials are exposed to changing environmental conditions. Kinetic tests would therefore become important in characterizing the potential mobilities of different chemical substances as they move from the solid phase into the aqueous phase.

Examples of a standard Tessier [94] and modified Tessier sequential extraction [95] methods are shown in Figures 10 and 11 respectively. The advantage of the modified Tessier method being that it can be used to speciate the loosely bound and the strongly bound forms of oxy-anions of arsenic and selenium.

Ultrasonic Extractions

Ultrasonic extractions are much faster and more convenient compared to the sequential extraction procedures. Even though there is insignificant changes in Se(IV) and Se(VI) distributions under the application of ultrasound energy, As (III) and As(V) distributions change. Thus ultrasound energy can be used to convert As (III) into As(V).[96] The As(V) readily binds to solid surfaces and hence such a technology can be used to remove both As(III) and As(V) from aqueous solutions provided a suitable cleaning up reagent such as kaolin or better is used. Ultrasonic extraction extractions may be used with any extracting media that include acids, bases or water. Water based extractions can be used to provide conditions that approximate accelerated natural weathering conditions.

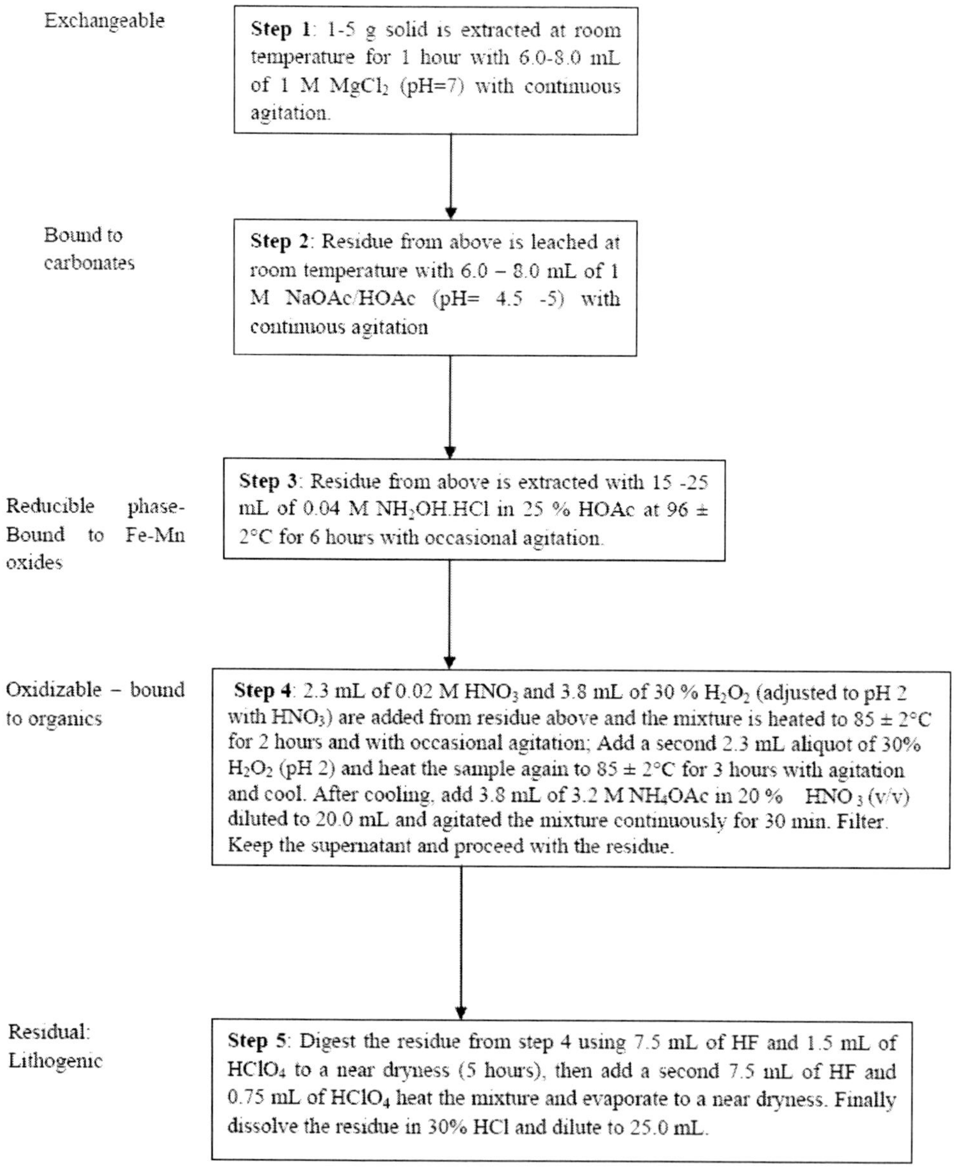

Figure 10. Standard Tessier sequential extraction procedure [94].

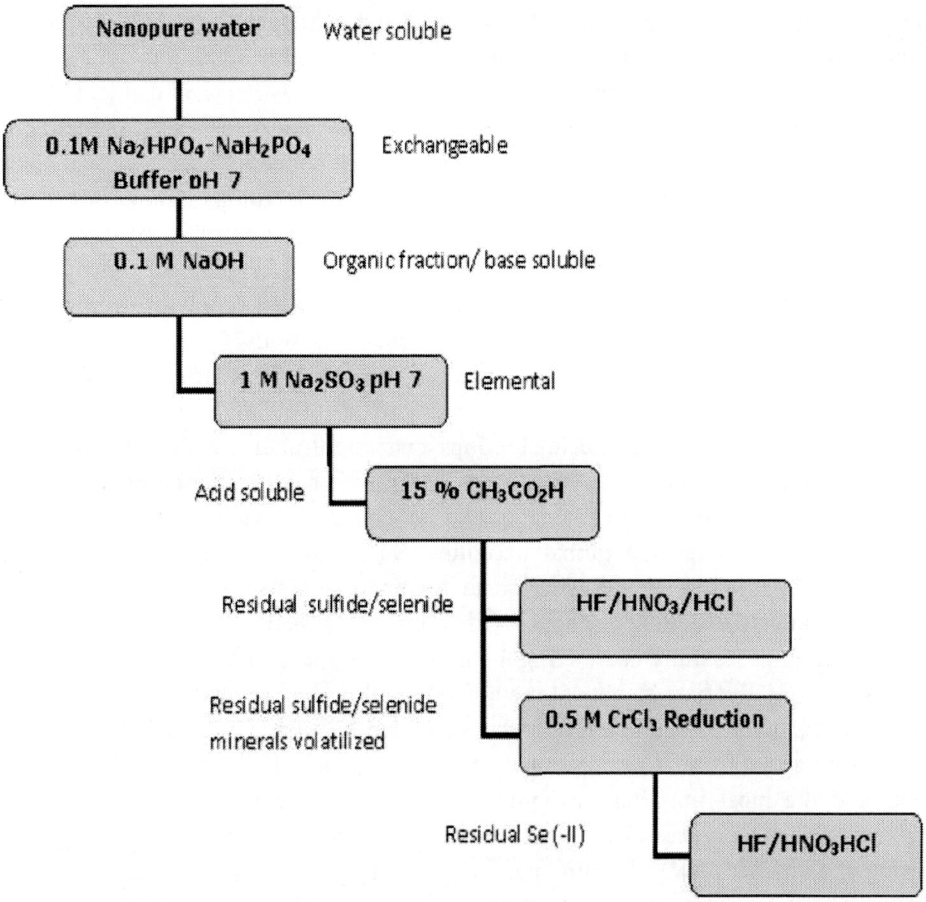

Figure 11. A flow diagram for the Modified Tessier sequential extraction method for selenium. [95]

10. Toxicology of Liquid and Solid Mining Waste

Water flowing out of a landfills or valley fills need to be frequently monitored for compliance with standing governmental regulations such as section 404 (a) of the US Clean Water Act. The water quality is usually monitored by determining the individual concentrations of dissolved and suspended organic and inorganic chemical substances. Such a monitoring technique, although useful, does not provide information with regards to the global toxicities associated with either the effluent waters or the solid materials. One chemical present in water may interact with other chemical species through antagonistic, additive, or synergistic effects resulting in either lowering or enhancement of its toxicity. Thus, the overall toxicity associated with one individual material might not mirror its effects when combined with other materials in solution. Hence, a more global approach in analyzing water quality indices is imperative in order to fully understand the overall toxicology of mine effluent waters. One approach is to study the biodiversity of flora and fauna inhabiting the affected areas and then comparing the types of benthic macro and microorganisms thriving in affected areas to environments that are considered to be natural. Certain biota are would thrive in polluted waters while other sensitive organisms might find it challenging to grow in the affected areas. Another approach is to use the Ames salmonella test, [97-100] a rapid *in vitro* screening procedure used to determine the genotoxicity of solid and liquid materials. The technique involve the use bacterial strains of *Salmonella typhimurium* that lack the ability to repair DNA and unable synthesize histidine, an amino acid necessary for their growth. Thus, the bacterial cultures cannot grow in media that does not contain any external source of histidine. However, if the cultures are grown in media containing mining waste that is potentially genotoxic, then the bacterial DNA undergoes mutation. The resulting mutation enables the bacteria to now start producing the amino acid histidine *de novo*. Thus the bacteria will now be able to grow without the need for an external supply of histidine. Such a genetic conversion that occurs *in vitro* can be extended to study the potency of the mine wastes to cause genetic damage in living organisms. The tests offer a global toxicity associated with the effluent waters which takes into account any occurring metal-metal, antagonistic, additive, or synergistic interactions that might alter the chemical toxicities of individual chemical substances. Ames tests only offer data pertaining to potential genotoxicity associated with mine drainage, and hence there is need to develop other water quality indices that provide the global toxicities. Such global approaches can enable the development of appropriate or targeted waste treatment protocols that may be applied in specific geographical or geological areas. Other bioassays such as the Comet assays[101] have been applied successfully in studying the mutagenicity of acidic drainages.

Site specific multivariate statistical methods that allow for simultaneous monitoring of a variety of water quality parameters can be instituted to assess the global water quality emanating from a mine site. The importance of chosen variables depends on the lithology of the rocks and the most important variable that determine the water quality in a specific geographical region. Advances have already been made in using Hotelling based statistical monitoring of a number of water quality parameters simultaneously.

Several US states have developed water quality indices that consist of a range numbers from 1-100 that are used to determine drinking and wastewater quality. All the selected parameters are analyzed for and results obtained are transformed into a single number that falls within a scale that ranges 1-100 scale. Corrective action to be taken will then depend on value of the score obtained.

11. Proactive and Reactive Measures (In-situ and ex-situ) of Mining Waste Treatment

Proactive treatment of mine waste requires a thorough chemical compositional analysis of the wastes involved. Different regions have different geological composition of the associated waste. Hence mining wastes from different mining environments might have different priorities of waste containment. Analysis of core samples obtained from MTM/VF from the Appalachian region in the USA resulted in the quantitation of both minor and major mineral inclusions using X-ray spectroscopic techniques. XRD was used to analyze for mineral phases and X-ray fluorescence was used to analyze for elemental inclusions. Unfortunately, arsenic and selenium could not be detected using the X–ray methods; however ultrasound extractions were performed followed by the analysis of extracts for using GFAAS. The results were analyzed using principal component analysis (PCA). The application of a two step PCA to the chemical data (Figure 13), suggested that ultrasound leachable selenium concentrations were associated with 14Å d spacing phyllosilicate clays (chlorite, montmorillonite and vermiculite all 2:1 layered clays) while ultrasound leachable arsenic concentrations were closely related to the concentration of Illite, another 2:1 phyllosilicate clay.

Negative correlations between leachable arsenic and selenium with kaolinite a 1:1 layered clay, were also observed. The observed negative correlations were used to rule out the presence of selenium or arsenic in 1:1 kaolinite. Hence mining waste from MTM/VF sites containing substantial amounts of illite and 14 Å d spacing clays may require to be placed in priority landfills or valley fills.[24]

The correlation between illite and arsenic suggested that there are substantial amounts of arsenic that could be located in either the interlayer or the octahedral lattice of illite (Figure 14). The Illite and 14 Å d spacing clays are major constituents of claystones, shales, and siltstones. The information obtained indicates that the interlayer in 14 Å d spacing and illite clays contains significant concentrations of selenium and arsenic, respectively. In general, "shales" (shales, claystones, and siltstones) were found to contain more leachable selenium compared to sandstones. The entry of these elements into the interlayer could have occurred during intercalation and lithification stages of sedimentary rock formation.

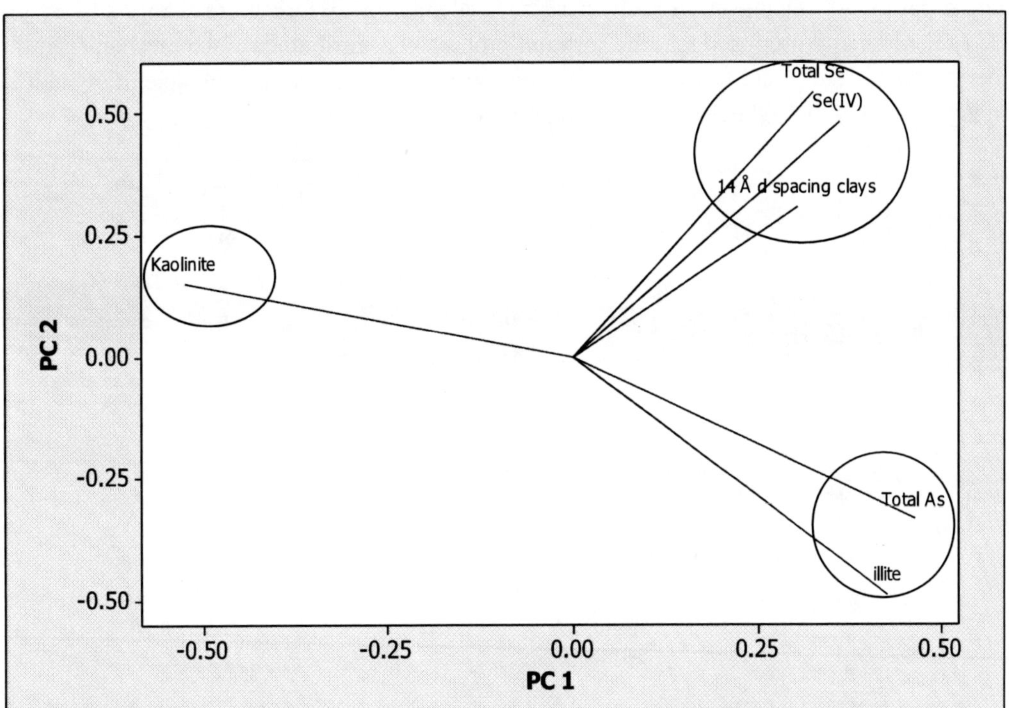

Figure 13. Two step principal component analysis of major and minor mineral and elemental inclusion in phylosilicate clays [24] .

12. Prioritization of Waste Disposal

Ziemkiewicz et. al.[102] developed an in-situ treatment of mining waste that immobilizes leached selenium on goethite (iron hydroxide) surfaces. A layer of goethite is deliberately placed first in a landfill before the introduction of selenium laden mining waste. The process resulted in 70% decrease in the concentration of dissolved selenium in the associated mine drainage. Both selenium [103]and arsenic species can get absorbed through inner and outer sphere coordination onto the goethite chemical surface. [104] Thus mining waste containing substantial amounts of illite and 14 Å d spacing clays would have to be placed in landfills or valley fills containing more goethite because 14 Å d spacing clays contain more selenium and illite has more arsenic. Using a combination of goethite and nanoparticles of titanium dioxide in the *in-situ* treatment could result in a greater percentage of immobilized selenium. In laboratory environments, both rutile and anatase titanium dioxide nanoparticles have been used to sorb more than 90% of arsenic and selenium [105-106]. TiO_2 has high affinitities for Se(IV), Se(VI), As(III), and As(V) that span from pH 3 to 9 [106-107]. Such a wide pH range covers the 6.5 – 8.5 pH values that are found in most natural aquatic environments. The binding of selenium and arsenic oxoanion species to TiO_2 nanoparticles has been shown to occur through inner sphere and outer sphere surface complexation via pH dependent surface hydroxyl groups [108]. In other applications zero valent Iron was used to sorb selenate in batch experiments. Zhang et. al.[109] used zerovalent iron to sorb more than 90 % of the added selenium and proposed that the sorption process was attributed to two mechanisms; the direct sorption of selenate to iron particles and the reduction of selenate to selenite which in turn is then sorbed onto the zerovalent iron particles.

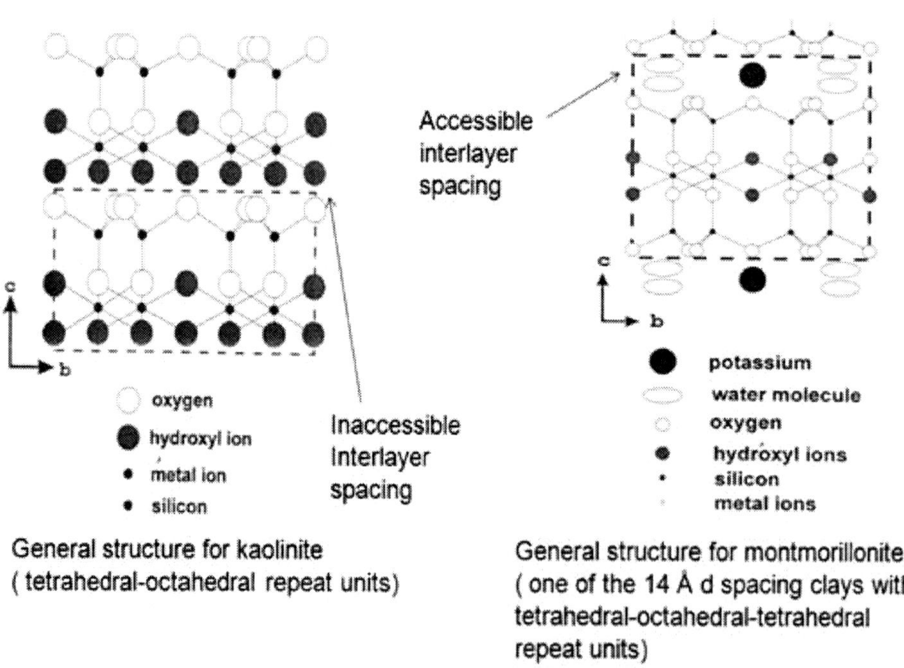

Figure 14. General chemical structures of kaolinite and montmorillonite [24].

Coal fly ash[110-112] has been deliberately added to mining waste *in situ* in order to improve the neutralization potential. Such a proactive treatment allows for immediate neutralization of anticipated ARD/AMD that may occur. Most positive charged metal ions are precipitated or co-precipitated together with Iron and aluminium hydroxides leading to their immobilization onto solid phases. Although coal fly ash is very effective in neutralizing the acidic drainage, it also has inherent significant quantities of arsenic, selenium and sulphate anions that are soluble in alkaline waters. So even though the pH is raised there will sometimes be considerable amounts of negatively charged oxoanionic species that may leached into the water column. Appropriate site specific monitoring need to be enforced to ensure adherence to government discharge regulations such as section 404 of the US Clean Water Act. Recently there have been assertions that alkaline media containing bicarbonate may leach arsenic and selenium from phyllosilicate clays.[23,52] Coal fly ash has also been used to reactively control the pH of ARD/AMD. The chemical reactions occurring in situ are similar to the chemical reaction that may occur *ex-situ*. The difference being that ARD/AMD has to happen first before the introduction coal fly ash in reactive measures whereas in proactive control, the coal fly ash is added well in advance during the placement of mining waste landfills/valley fills. Zeolites have also found prominent use in the treatment of acidic drainages. Natural Zeolites[113-114] offer have high ion exchange capacities that enable the sorption of heavy metals from mine drainages. Appropriate zeolites can be chosen that enable the entrapment of ionic species within their interstitial pores.

Other neutralizing substances that include $CaCO_3$, CaO, $Ca(OH)_2$ and ammonium hydroxide may be deliberately added to mining waste to increase the neutralization potential. These chemicals can be used in the proactive and reactive treatment of acidic drainages. Environmentally friendly bactericides or detergents in the form of biodegdradable linear alkyl sulfonates have been used to help destroy *Thiobacillus thiooxidans* and *Thiobacillus Ferrooxidans*, which promote ARD/AMD production. Phosphate ions may also be added to precipitate both ferric and ferrous ions. Effluent from the ARD/AMD may be redirected to flow through engineered wetlands containing sulphate-reducing bacteria to decrease sulphate concentrations. The wetlands may also contain hyperaccumulator plants, which selectively remove specific metals from the water bodies. [27] Topsoiling and revegetation is also an effective method that is currently being used in remediating mining areas affected by ARD/AMD. The method results in the reduction of oxygen infiltration into the waste thereby reducing the chances for the oxidation of sulfidic minerals and thus alleviating the formation of ARD/AMD. Revegetation using hyperaccumulator plants can help target the removal of specific metals or metalloids from the water and soil columns.

CONCLUSION

As long as there is a sustained demand for coal and minerals use in technological advancements, there will always be a need for the appraisal of current mining protocols and provisions for the development of environmentally friendly waste treatment procedures that are engineered to protect the livelihoods of the future and present generations. Such concerted efforts will usher new and sustainable avenues for the placement and containment of waste in valley fills or landfills. It is therefore imperative to expedite the dissemination of the practical emerging technologies to the global mining personnel and to make the technological information easily accessible. Such a process will help create a more environment conscious world that always engineer current and available solutions into their mining, waste disposal and reclamation before the commencement of mining activities. Proactive control measures done in-situ will help ameliorate environmental problems that might happen years after the beginning or cessation of mining activities. A combination of engineering, chemical, biological, and toxicological assays of both liquid and solid mining waste is always essential to help manage the environmental effects of contaminants and pollutants discharged from mine overburden.

REFERENCES

[1] Moberly, C.; Workman, S. R.; Warner, R. C., Calibration and Evaluation of a Hydrologic Model for Loose-Dump Mine Spoil. *International Journal of Surface Mining, Reclamation and Environment* 2001, 15, (3), 147.
[2] Hartman, K. J.; Kaller, M. D.; Howell, J. W.; Sweka, J. A., How much do valley fills influence headwater streams? *Hydrobiologia* 2005, 532, 91-102.
[3] Dreher, G. B.; Finkelman, R. B., Selenium mobilization in a surface coal mine, Powder River Basin, Wyoming, U.S.A. *Environmental Geology* 1992, 19, (3), 155.
[4] Coleman, L.; Bragg, L. J.; Finkelman, R. B., Distribution and mode of occurrence of selenium in US coals. *Environmental Geochemistry and Health* 1993, 15, (4), 215.
[5] Lussier, C.; Veiga, V.; Baldwin, S., The geochemistry of selenium associated with coal waste in the Elk River Valley, Canada. *Environmental Geology* 2003, 44, (8), 905.
[6] Pond, G., Biodiversity loss in Appalachian headwater streams (Kentucky, USA): Plecoptera and Trichoptera communities. *Hydrobiologia* 2011, 679, (1), 97.
[7] Association), N. N. M., Mountaintop Mining factbook. In 2009; Vol. 2012.
[8] Resources, A. N. *The truth about surface mining*; Alpha Natural resources: 3/9/2012, 2012.
[9] Hartman, K.; Kaller, M.; Howell, J.; Sweka, J., How much do valley fills influence headwater streams? *Hydrobiologia* 2005, 532, (1), 91.
[10] Rieuwerts, J. S.; Thornton, I.; Farago, M. E.; Ashmore, M. R., Factors influencing metal bioavailability in soils: preliminary investigations for the development of a critical loads approach for metals. *Chemical Speciation and Bioavailability* 1998, 10, (2), 61-75.
[11] USEPA, Ground Water & Drinking Water. *http://www.epa.gov/ safewater/ contaminants/dw_contamfs/selenium.html* 2006.
[12] USEPA, Current National Recommended Water Quality Criteria. *http://www.epa.gov /waterscience/criteria/wqctable/index.html#R* 2008.
[13] Bryant, G.; McPhilliamy, S.; Childers, H. *A Survey of the water Quality of Streams in the Primary Region of Mountaingtop /Valley Fill Coal Mining*; WVGES: 2002, 2002.
[14] Sylvester, P.; Westerhoff, P.; Mooller, T.; Badruzzaman, M.; Boyd, O., A hybrid sorbent utilizing nanoparticles of hydrous iron oxide for arsenic removal from drinking water. *Environmental Engineering Science* 2007, 24, (1), 104-112.

[15] Myneni, S. C. B.; Tokunaga, T. K.; Brown, G. E., Abiotic selenium redox transformations in the presence of Fe(II,III) oxides. *Science* 1997, 278, (5340), 1106-1109.
[16] Gray, N. F., Acid mine drainage composition and the implications for its impact on lotic systems. *Water Research* 1998, 32, (7), 2122.
[17] Protection, P. D. o. E., Science of Acid Mine Drainage and Passive Treatment. In 2012.
[18] White, W. W.; Jeffers, T. H., Chemical Predictive Modeling of Acid Mine Drainage from Metallic Sulfide-Bearing Waste Rock. In *Environmental Geochemistry of Sulfide Oxidation*, American Chemical Society: 1993; Vol. 550, pp 608-630.
[19] Muzenda, C.; Pumure, I., Effects of Nickel mining activities on water quality. *Reclamation with a purpose, proceedings of a joint conference of American society of mining and reclamation 19 th annual national and international conference of acid rock drainage, Lexington Kentucky* 2002.
[20] Nicholson Ronald, V.; Scharer Jeno, M., Laboratory Studies of Pyrrhotite Oxidation Kinetics. In *Environmental Geochemistry of Sulfide Oxidation*, American Chemical Society: 1993; Vol. 550, pp 14-30.
[21] Chinchon-Paya, S.; Aguado, A.; Chinchon, S., A comparative investigation of the degradation of pyrite and pyrrhotite under simulated laboratory conditions. *Engineering Geology* 127, (0), 75.
[22] Evangelou, V. P., Pyrite oxidation inhibition in coal waste by PO4 and H2O2pH-buffered pretreatment. *International Journal of Surface Mining, Reclamation and Environment* 1996, 10, (3), 135.
[23] Ryan, P. C.; Kim, J.; Wall, A. J.; Moen, J. C.; Corenthal, L. G.; Chow, D. R.; Sullivan, C. M.; Bright, K. S., Ultramafic-derived arsenic in a fractured bedrock aquifer. *Applied Geochemistry* 26, (4), 444.
[24] Pumure, I.; Renton, J. J.; Smart, R. B., The interstitial location of selenium and arsenic in rocks associated with coal mining using ultrasound extractions and principal component analysis (PCA). *J Hazard Mater* 198, 151-8.
[25] Khanra, S.; Mallick, D.; Dutta, S. N.; Chaudhuri, S. K., Studies on the Phase Mineralogy and Leaching Characteristics of Coal Fly Ash. *Water, Air, & Soil Pollution* 1998, 107, (1), 251.
[26] Palumbo-Roe, B.; Klinck, B.; Cave, M.; Prosun Bhattacharya, A. B. M. J. B. R. Z. a. R. H. L., Arsenic speciation and mobility in mine wastes from a copper-arsenic mine in Devon, UK: a SEM, XAS, sequential chemical extraction study. In *Trace Metals and other Contaminants in the Environment*, Elsevier: 2007; Vol. Volume 9, p 441.
[27] Zipper, C.; Jage, C., Passive treatment of acid-mine drainage with vertical-flow systems. *Virginia Cooperative Extension, Powell River Project* 2001, 460, (133), 6.
[28] Cornell, R. M.; Giovanoli, R.; Schneider, W., Review of the hydrolysis of iron(III) and the crystallization of amorphous iron(III) hydroxide hydrate. *Journal of Chemical Technology & Biotechnology* 1989, 46, (2), 115.
[29] D. Kirk, N., Hydrogeochemical processes governing the origin, transport and fate of major and trace elements from mine wastes and mineralized rock to surface waters. *Applied Geochemistry* 26, (11), 1777.
[30] Das, S.; Hendry, M. J., Application of Raman spectroscopy to identify iron minerals commonly found in mine wastes. *Chemical Geology* 290, (3-4), 101.

[31] Madejova, J., FTIR techniques in clay mineral studies. *Vibrational Spectroscopy* 2003, 31, (1), 1.
[32] Srodon, J.; Drits, V. A.; McCarty, D. K.; Hsieh, J. C. C.; Eberl, D. D., Quantitative X-ray diffraction analysis of clay-bearing rocks from random preparations. *Clays and Clay Minerals* 2001, 49, (6), 514-528.
[33] Ohlander, B. r.; Muller, B.; Axelsson, M.; Alakangas, L., An attempt to use LA-ICP-SMS to quantify enrichment of trace elements on pyrite surfaces in oxidizing mine tailings. *Journal of Geochemical Exploration* 2007, 92, (1), 1.
[34] Zawislanski, P. T.; Benson, S. M.; Terberg, R.; Borglin, S. E., Selenium speciation, solubility, and mobility in land-disposed dredged sediments. *Environmental Science & Technology* 2003, 37, (11), 2415-2420.
[35] Pyrzynska, K., Speciation of selenium compounds. *Analytical Sciences* 1998, 14, (3), 479-483.
[36] Ranjard, L.; Prigent-Combaret, C.; Nazaret, S.; Cournoyer, B., Methylation of inorganic and organic selenium by the bacterial thiopurine methyltransferase. *Journal of Bacteriology* 2002, 184, (11), 3146-3149.
[37] Nelson, D. C.; Casey, W. H.; Sison, J. D.; Mack, E. E.; Ahmad, A.; Pollack, J. S., Selenium uptake by sulfur-accumulating bacteria. *Geochimica Et Cosmochimica Acta* 1996, 60, (18), 3531- 3539.
[38] Maier KJ and Knight, A. W., Comparative acute toxicity and bioconcentration of selenium by the midge Chironomus decorus exposed to selenate, selenite, and seleno-DL-methionine. *Archives of environmental contamination and toxicology* 1993, 25, (3), 365.
[39] Krupka, K. M.; Serne, R. J.; Kaplan, D. I., Geochemical Data Package for the 2005 Hanford Integrated Disposal Facility Performance Assessment, PNNL-13037 rev2. *Pacific Northwest National Laboratory, Richland. WA.http://www.pnl.gov/main/ publications/ external/technical_reports* 2004.
[40] Hsia, T. H.; Lo, S. L.; Lin, C. F.; Lee, D. Y., Characterization of Arsenate Adsorption on Hydrous Iron-Oxide Using Chemical and Physical Methods. *Colloids and Surfaces a-Physicochemical and Engineering Aspects* 1994, 85, (1), 1-7.
[41] Stolz, J. E.; Basu, P.; Santini, J. M.; Oremland, R. S., Arsenic and selenium in microbial metabolism. *Annual Review of Microbiology* 2006, 60, 107-130.
[42] Lemly, A. D., Teratogenic Effects of Selenium in Natural-Populations of Fresh-Water Fish. *Ecotoxicology and Environmental Safety* 1993, 26, (2), 181-204.
[43] Ohlendorf, H. M.; Hoffman, D. J.; Saiki, M. K.; Aldrich, T. W., Embryonic Mortality and Abnormalities of Aquatic Birds - Apparent Impacts of Selenium from Irrigation Drainwater. *Science of the Total Environment* 1986, 52, (1-2), 49-63.
[44] Liao, C. M.; Chiang, K. C.; Tsai, J. W., Bioenergetics-based matrix population modeling enhances life-cycle toxicity assessment of tilapia Oreochromis mossambicus exposed to arsenic. *Environ Toxicol* 2006, 21, (2), 154-65.
[45] Kithome, M.; Paul, J. W.; Lavkulich, L. M.; Bomke, A. A., Kinetics of ammonium adsorption and desorption by the natural zeolite clinoptilolite. *Soil Science Society of America Journal* 1998, 62, (3), 622-629.
[46] Goh, K. H.; Lim, T. T., Geochemistry of inorganic arsenic and selenium in a tropical soil: effect of reaction time, pH, and competitive anions on arsenic and selenium adsorption. *Chemosphere* 2004, 55, (6), 849-859.

[47] Sparks, D. L., *Soil Physical Chemistry*. Second ed.; CRC Press: 1998.
[48] Espenson, J. H., *Chemical kinetics and reaction mechanisms*. 1 st ed.; Mcgraw-Hill: 1981; p 24-25.
[49] Luque-Garcia, J. L.; de Castro, L., Ultrasound: a powerful tool for leaching. *Trac-Trends in Analytical Chemistry* 2003, 22, (1), 41-47.
[50] Mierzwa, J.; Adeloju, S. B.; Dhindsa, H. S., Ultrasound accelerated solid-liquid extraction for the determination of selenium in biological samples by electrothermal atomization atomic absorption spectrometry. *Analytical Sciences* 1997, 13, (2), 189-193.
[51] Perez-Cid, B.; Lavilla, I.; Bendicho, C., Use of ultrasonic energy for shortening the sequential extraction of metals from river sediments. *International Journal of Environmental Analytical Chemistry* 1999, 73, (2), 79-92.
[52] Pumure, I.; Renton, J. J.; Smart, R. B., Ultrasonic extraction of arsenic and selenium from rocks associated with mountaintop removal/valley fills coal mining: Estimation of bioaccessible concentrations. *Chemosphere* 78, (11), 1295-300.
[53] Brunori, C.; Ipolyi, I.; Macaluso, L.; Morabito, R., Evaluation of an ultrasonic digestion procedure for total metal determination in sediment reference materials. *Analytica Chimica Acta* 2004, 510, (1), 101-107.
[54] Perez-Cid, B.; Bola, C., Comparison between total determination and extractable heavy metals from river sediments using conventional and microwave accelerated leaching tests. *International Journal of Environmental Analytical Chemistry* 2001, 81, (2), 101-115.
[55] Tse, K. K. C.; Lo, S. L., Desorption kinetics of PCP-contaminated soil: effect of temperature. *Water Research* 2002, 36, (1), 284-290.
[56] Arruda, S. C. C.; Rodriguez, A. P. M.; Arruda, M. A. Z., Ultrasound-assisted extraction of Ca, K and Mg from in vitro citrus culture. *Journal of the Brazilian Chemical Society* 2003, 14, (3), 470-474.
[57] Mason, T. J.; Lorimer, J. P.; Bates, D. M., Quantifying Sonochemistry - Casting Some Light on a Black Art. *Ultrasonics* 1992, 30, (1), 40-42.
[58] Ince, N. H.; Belen, R., Aqueous Phase Disinfection with Power Ultrasound:Process Kinetics and Effect of Solid Catalysts. *Environmental Science & Technology* 2001, 35, (9), 1885.
[59] Atkins, P., *Physical Chemistry*. 4th ed.; W.H. Freeman Company, New York: 1990; p 210.
[60] Atkins, P.; De Paula, J., *Physical Chemistry*. 7 th ed.; W.H. Freeman: 2002; p 836.
[61] Lin, Z.; Puls, R. W., Adsorption, desorption and oxidation of arsenic affected by clay minerals and aging process. *Environmental Geology* 2000, 39, (7), 753-759.
[62] Mohan, D.; Pittman, C. U., Arsenic removal from water/wastewater using adsorbents - A critical review. *Journal of Hazardous Materials* 2007, 142, (1-2), 1-53.
[63] Gimenez, J.; Martinez, M.; de Pablo, J.; Rovira, M.; Duro, L., Arsenic sorption onto natural hematite, magnetite, and goethite. *Journal of Hazardous Materials* 2007, 141, (3), 575-580.
[64] Trivedi, P.; Axe, L., Modeling Cd and Zn sorption to hydrous metal oxides. *Environmental Science & Technology* 2000, 34, (11), 2215-2223.
[65] Thanabalasingam, P.; Pickering, W. F., Arsenic Sorption by Humic Acids. *Environmental Pollution Series B-Chemical and Physical* 1986, 12, (3), 233-246.

[66] Suarez, D. L.; Goldberg, S.; Su, C.; Manning, B. A., Evaluation of oxyanion adsorption mechanisms on oxides using FTIR spectroscopy and electrophoretic mobility. *Abstracts of Papers of the American Chemical Society* 1997, 213, 59-Geoc.

[67] Manning, B. A.; Goldberg, S., Adsorption and stability of arsenic(III) at the clay mineral-water interface. *Environmental Science & Technology* 1997, 31, (7), 2005-2011.

[68] Scott, M. J.; Morgan, J. J., Reactions at Oxide Surfaces.1. Oxidation of as(Iii) by Synthetic Birnessite. *Environmental Science & Technology* 1995, 29, (8), 1898-1905.

[69] Wrobel, K.; Wrobel, K.; Caruso, J. A., Pretreatment procedures for characterization of arsenic and selenium species in complex samples utilizing coupled techniques with mass spectrometric detection. *Analytical and Bioanalytical Chemistry* 2005, 381, (2), 317-331.

[70] Gao, S.; Tanji, K. K.; Dahlgren, R. A.; Ryu, J.; Herbel, M. J.; Higashi, R. M., Chemical status of selenium in evaporation basins for disposal of agricultural drainage. *Chemosphere* 2007, 69, (4), 585-594.

[71] Watzin, M. C.; McIntosh, A. W.; Brown, E. A.; Lacey, R.; Lester, D. C.; Newbrough, K. L.; Williams, A. R., Assessing sediment quality in heterogeneous environments: A case study of a small urban harbor in Lake Champlain, Vermont, USA. *Environmental Toxicology and Chemistry* 1997, 16, (10), 2125-2135.

[72] Burton, A., ASSESSING THE TOXICITY OF FRESHWATER SEDIMENTS. *Environmental Toxicology and Chemistry* 1991, 10, (12), 1585–1627.

[73] USEPA, Incidence and severity of sediment concentration in surface waters of the U.S. *EPA 823-R-97-006* 1997, 1-3.

[74] Burton, A., Sediment quality criteria in use around the world. *Limnology* 2002, 3, 65-75.

[75] Pumure, I. Investigation of selenium and arsenic in coal-mining associated rocks and sediments using ultrasonic and sequential extractions techniques. Dissertation, West Virginia University, Morgantown, 2008.

[76] Fernandez, M. G. C.; Palacios, M. A.; Camara, C., Flow-Injection and Continuous-Flow Systems for the Determination of Se(Iv) and Se(Vi) by Hydride Generation Atomic- Absorption Spectrometry with Online Prereduction of Se(Vi) to Se(Iv). *Analytica Chimica Acta* 1993, 283, (1), 386-392.

[77] Arai, Y.; Sparks, D. L.; Davis, J. A., Arsenate adsorption mechanisms at the allophane - water interface. *Environmental Science & Technology* 2005, 39, (8), 2537-2544.

[78] Cullen, W. R.; Reimer, K. J., Arsenic Speciation in the Environment. *Chemical Reviews* 1989, 89, (4), 713-764.

[79] Redman, A. D.; Macalady, D. L.; Ahmann, D., Natural organic matter affects arsenic speciation and sorption onto hematite. *Environ Sci Technol* 2002, 36, (13), 2889-96.

[80] Johnson, T. M., A review of mass-dependent fractionation of selenium isotopes and implications for other heavy stable isotopes. *Chemical Geology* 2004, 204, (3-4), 201-214.

[81] Shaker, A. M., Kinetics of the reduction of Se(IV) to Se-sol. *Journal of Colloid and Interface Science* 1996, 180, (1), 225-231.

[82] Tan, T. T. Y.; Beydoun, D.; Amal, R., Photocatalytic reduction of Se(VI) in aqueous solutions in UV/TiO2 system: Kinetic modeling and reaction mechanism. *Journal of Physical Chemistry B* 2003, 107, (18), 4296-4303.

[83] Tessier A, C. P., Sequential extraction procedure for the speciation of particulate trace metals. *Analytical Chemistry* 1979, 51, 844-851.

[84] Quevauviller, P., BCR, turning over a new leaf. *Trac-Trends in Analytical Chemistry* 2003, 22, (4), 187-190.

[85] Mossop, K. F.; Davidson, C. M., Comparison of original and modified BCR sequential extraction procedures for the fractionation of copper, iron, lead, manganese and zinc in soils and sediments. *Analytica Chimica Acta* 2003, 478, (1), 111-118.

[86] Van Herreweghe, S.; Swennen, R.; Vandecasteele, C.; Cappuyns, V. r., Solid phase speciation of arsenic by sequential extraction in standard reference materials and industrially contaminated soil samples. *Environmental Pollution* 2003, 122, (3), 323.

[87] Ochsenkuhn-Petropoulou, M.; Tsopelas, F., Speciation analysis of selenium using voltammetric techniques. *Analytica Chimica Acta* 2002, 467, (1-2), 167.

[88] Selby, F.; Potin Gautier, M.; Lespes, G.; Astruc, M., Selenium speciation in soils after alkaline extraction. *Science of The Total Environment* 1997, 207, (2-3), 81.

[89] Bujdoš, M.; Muľová, A.; Kubová, J.; Medveď, J., Selenium fractionation and speciation in rocks, soils, waters and plants in polluted surface mine environment. *Environmental Geology* 2005, 47, (3), 353.

[90] Zhang, Y.; Frankenberger, W. T., Determination of Selenium Fractionation and Speciation in Wetland Sediments by Parallel extraction. *International Journal of Environmental Analytical Chemistry* 2003, 83, (4), 315.

[91] Sharmasarkar, S.; Vance, G. F., Soil and Plant Selenium at a Reclaimed Uranium Mine. *J. Environ. Qual.* 2002, 31, (5), 1516-1521.

[92] Lu, C.-Y.; Yan, X.-P., Capillary electrophoresis on-line coupled with hydride generation-atomic fluorescence spectrometry for speciation analysis of selenium. *ELECTROPHORESIS* 2005, 26, (1), 155.

[93] Amlan Ghosh, Muhammed Mukiibi and, Wendell Ela. TCLP Underestimates Leaching of Arsenic from Solid Residuals under Landfill Conditions *Environmental Science & Technology* 2004, 38, (17), 4677-4682

[94] Pizarro, J.; Rubio, M. A.; Castillo, X., Study of chemical speciation in sediments: An approach t o vertical metals distribution in Rapel reservoir (Chile). *Journal of the Chilean Chemical Society* 2003, 48, (3), 45-50.

[95] Kulp, T. R.; Pratt, L. M., Speciation and weathering of selenium in Upper Cretaceous chalk and shale from South Dakota and Wyoming, USA. *Geochimica Et Cosmochimica Acta* 2004, 68, (18), 3687-3701.

[96] Pumure, I.; Renton, J.; Smart, R., Accelerated aqueous leaching of selenium and arsenic from coal associated rock samples with selenium speciation using ultrasound extraction. *Environmental Geology* 2009, 56, (5), 985.

[97] Klaunig, E. J.; Kamendulis, L. M., *Casarett and Doull's Toxicology, The Basic Science of Poisons*. 7 th ed.; Mcgraw Hill: 2008.

[98] White, P. A.; Claxton, L. D., Mutagens in contaminated soil: a review. *Mutat Res* 2004, 567, (2-3), 227-345.

[99] Miadokova, E.; Duhova, V.; Vlckova, V.; Sladkova, L.; Sucha, V.; Vlcek, D., Genetic risk assessment of acid waste water containing heavy metals. *Gen Physiol Biophys* 1999, 18 Spec No, 92-8.

[100] Mortelmans, K.; Zeiger, E., The Ames Salmonella/microsome mutagenicity assay. *Mutation Research/Fundamental and Molecular Mechanisms of Mutagenesis* 2000, 455, (1-2), 29.

[101] Leffa, D.; Damiani, A.; da Silva, J.; Zocche, J.; dos Santos, C.; Boufleur, L.; Dias, J.; de Andrade, V., Evaluation of the Genotoxic Potential of the Mineral Coal Tailings through the Helix Aspersa; (Müller, 1774). *Archives of Environmental Contamination and Toxicology* 59, (4), 614.

[102] Ziemkiewicz, P.; O'Neal, M.; Lovett, R., Selenium Leaching Kinetics and In situ Control. *Mine Water and the Environment* 2010, 30, (2), 141.

[103] Sharrad, M. O. M.; Liu, H.; Fan, M., Evaluation of FeOOH performance on selenium reduction. *Separation and Purification Technology* 84, 29.

[104] Pena, M.; Meng, X.; Korfiatis, G. P.; Jing, C., Adsorption Mechanism of Arsenic on Nanocrystalline Titanium Dioxide. *Environmental Science & Technology* 2006, 40, (4), 1257.

[105] Peak, D.; Sparks, D. L., Mechanisms of Selenate Adsorption on Iron Oxides and Hydroxides. *Environmental Science & Technology* 2002, 36, (7), 1460.

[106] Li, S.; Deng, N., Separation and preconcentration of Se(IV)/Se(VI) species by selective adsorption onto nanometer-sized titanium dioxide and determination by graphite furnace atomic absorption spectrometry. *Analytical and Bioanalytical Chemistry* 2002, 374, (7), 1341.

[107] Dutta, P. K.; Ray, A. K.; Sharma, V. K.; Millero, F. J., Adsorption of arsenate and arsenite on titanium dioxide suspensions. *Journal of Colloid and Interface Science* 2004, 278, (2), 270.

[108] Zhang, Y.; Amrhein, C.; Chang, A.; Frankenberger Jr, W. T., Effect of zero-valent iron and a redox mediator on removal of selenium in agricultural drainage water. *Science of The Total Environment* 2008, 407, (1), 89.

[109] Perez-Lopez, R.; Nieto, J. M.; de Almodovar, G. R., Utilization of fly ash to improve the quality of the acid mine drainage generated by oxidation of a sulphide-rich mining waste: Column experiments. *Chemosphere* 2007, 67, (8), 1637.

[110] Rios, C. A.; Williams, C. D.; Roberts, C. L., Removal of heavy metals from acid mine drainage (AMD) using coal fly ash, natural clinker and synthetic zeolites. *Journal of Hazardous Materials* 2008, 156, (1-3), 23.

[111] Gitari, W. M.; Petrik, L. F.; Etchebers, O.; Key, D. L.; Okujeni, C., Utilization of fly ash for treatment of coal mines wastewater: Solubility controls on major inorganic contaminants. *Fuel* 2008, 87, (12), 2450.

[112] Motsi, T.; Rowson, N. A.; Simmons, M. J. H., Adsorption of heavy metals from acid mine drainage by natural zeolite. *International Journal of Mineral Processing* 2009, 92, (1-2), 42.

[113] Motsi, T.; Rowson, N. A.; Simmons, M. J. H., Kinetic studies of the removal of heavy metals from acid mine drainage by natural zeolite. *International Journal of Mineral Processing* 2011, 101, (1-4), 42.

In: Landfills
Editors: G. B. C. Cabral and B. A. E. Botelho
ISBN: 978-1-62257-121-5

Chapter 5

Reduction of Environmental Impact of Municipal Landfill Leachate during Oxidative Treatment: The Importance of Phytotoxicity Assessment

Andreja Žgajnar Gotvajn, Gabriela Kalčíková and Jana Zagorc-Končan
University of Ljubljana, Faculty of Chemistry and Chemical Technology, Ljubljana, Slovenia

Abstract

In the last decades sustainable waste management has become a high priority due to increased municipal waste production. In most countries landfilling of the waste still plays the most important role in waste management. As a result of stabilization processes, landfill leachate and biogas are formed. Heavily polluted landfill leachate must be collected and treated properly. Due to the complex matrix of the leachate there are no general recommendations of how to treat landfill leachate and usually a combination of physical, chemical and biological methods is used. Very promising are advanced oxidation processes (AOPs) with high efficiency for organics removal because of the formation of extremely reactive and non-selective hydroxyl radicals. Fenton's oxidation is one of these AOPs with high efficiency and low capital costs. Its main disadvantage is possible formation of different by-products with hazardous environmental impact.

This is the reason that physico-chemical methods alone are not enough for reliable monitoring of treatment efficiency. They must be accompanied by different biotests to assure actual removal and decomposition of harmful pollutants prior to release of the leachate into the environment.

Toxicity assessment should be performed using species belonging to different levels of aquatic ecosystem: decomposers (bacteria), producers (like duckweed *Lemna minor*) and consumers (like brine shrimp *Artemia salina* or water flea *Daphnia magna)* as well as species belonging to terrestrial ecosystem (like producers; white mustard *Sinapis alba* and white onion *Allium cepa)*. Toxicity of landfill leachates to plants, especially to higher plants, has not received much study.

In this chapter an overview of different treatment methods and methods for ecotoxicity assessment for landfill leachates has been presented. In a case study municipal landfill leachate was treated by Fenton oxidation and its physico-chemical parameters, biodegradability and toxicity were determined prior to, and after treatment. Fenton oxidation was not successful for treatment of the investigated leachate. Maximal treatment efficiency was 58% in terms of COD removal at higher molar ration (1:10) and phytotoxicity decreased for just 30%. In some cases, formation of phytotoxic pollutants after Fenton oxidation was identified. Phytotoxicity tests were confirmed to be a necessary supplement for reliable evaluation of hazardous impact of landfill leachate. They are also important to confirm the efficiency of the selected treatment methods. Commonly applied toxicity tests with producers and consumers in the ecosystem do not assure enough data on the impact of hazardous landfill leachates to avoid long-term pollution problems.

Keywords: Fenton oxidation, biological treatment, landfill leachate, municipal landfill, oxidative treatment, phytotoxicity

INTRODUCTION

Landfill Leachate

Landfilling is still the primary method currently used for municipal solid waste disposal in many countries [1]. This method generally offers lower costs of operation and maintenance as compared to other viable thermal processes. Besides households and other urbane activities, industry is directly associated with the production of large amounts of solid wastes. Several methods and working strategies have been developed for the integrated management of these wastes. They started by pollution prevention, waste minimization (zero waste management), materials recovery and reuse as well as recycling. In most of the countries, composting, recycling or incineration of waste are still secondary options for waste handling and thus landfilling still plays the most important role in waste management. Moreover, continuing economic growth, technological developments and changing consumption and production patterns in many countries all over the world in the past decades have been accompanied by a rapid increase in municipal and industrial waste generation [2]. For example in the European Union, 524 kg of municipal waste was generated per person in the year 2008. Most of the waste is generated in Denmark (802 kg per person) and the least in the Czech Republic (306 kg per person) and Poland (316 kg). 40% of municipal waste produced has been disposed in landfills (20% incinerated, 23% recycled, 17% composted) [3] and thus landfilling is still the most widespread method for disposal of European solid wastes. In Slovenia, the situation is comparable. The amount of municipal wastes collected dropped 5.4% in the last year resulting in 623,223 tonnes in 2010. 194,078 tonnes of wastes were disposed at industrial landfills (23.5%) and 8,000 tonnes at landfills for hazardous wastes (less than 1%). At the same time, 54,923 tonnes of yard trimmings and 71,459 tonnes of packaging materials were collected separately [4]. This means, that in the year 2010, 272 kg of municipal wastes were deposited per capita, while the amount of all deposited wastes per capita was 304 kg. However, production of municipal waste in Slovenia in 2010 was 449 kg per capita [5]. 62% of them were deposited, 1% incinerated, 34% recycled and 2% composted. Available data rank Slovenia together with Denmark as the third among EU member states regarding the level of recycling of municipal wastes. The most efficient are Germany (48%), Belgium and Sweden (both 36%) [5].

There is not only scarcity of available landfill sites and long-term deterioration of the landscape as well as its limited usability, but there is also a large amount of leachate and biogas generated at a landfill sites, which is a major problem in managing the disposal of municipal solid waste [2]. Biogas formed during the ageing of wastes at least has some economic value, because it should be entrapped and utilised for energy purposes, while landfill leachate itself represents only an economic burden in any waste management strategy. The total amount of CH_4 recovery for a (final or temporary) covered landfill site could be up to 80-85% of the total methane production. In some cases, this rate can reach even more than 90%. For an HDPE covered landfill, the total methane emissions of the surface of the cell could be lower than 1%, significantly reducing the environmental impact of biog [6].

Landfill leachate is liquid, generated by infiltration and percolation of rainfall, groundwater, run-off or floodwater into and through the waste layer of an existing or abandoned landfill site [7]. It is also a consequence of waste moisture and degradation processes. Different constituents are released into the solution mainly as a result of chemical and/or biological decomposition [8] due to a combination of physical, chemical and microbial processes in the waste [7]. Because of the high concentrations of various pollutants, municipal landfill leachates are considered one of the wastewaters with the highest environmental impact. Careful site management can reduce the quantity and pollution of leachate produced, but it still can not completely eliminate it. Even within the single landfill site variability in the landfill leachate quality is frequently evident due to changing atmospheric conditions affecting degradation processes. During waste degradation, landfills go through four main phases: i) initial aerobic phase, ii) anaerobic acidic phase, iii) initial methanogenic phase and iv) intensive and stable methanogenic phase. Those phases define composition of the leachate [9]. Significant components of leachate are heavy metals and degradable organics at the beginning of the landfill operation, while persistent organic pollutants usually appear later as a result of biotic and abiotic processes in the system [10]. Among these substances are several compounds classified as potentially hazardous: bioaccumulative, toxic, and genotoxic. They could also have an endocrine disruptive effect [11]. They could be presented in very low ($\mu g \cdot L^{-1}$) concentrations and hardly detectable.

Municipal landfill leachates generally contain various types of pollutants. Different organic substances are present mainly as dissolved organic matter with a large portion of man-made xenobiotic organic substances. Inorganic compounds such as ammonium, sodium, chlorides and bicarbonates form the second group of pollutants. The concentration of metals is strongly dependent upon processes in the landfill. Dissolved organic matter could be biodegradable or resistant/persistent to biodegradation. A great concern is focused on xenobiotic substances, which are usually present in very low concentrations, but among them a lot of hazardous pollutants such as aromatic hydrocarbons, phenols, pesticides and plasticizers can be found [12]. Concentrations of metals in landfill leachate are usually quite low, especially in older, more mature landfill leachates, formed after the initial phase of waste stabilization. The data reported by several researchers showed that most heavy metal concentrations in landfill leachate are at or below US drinking standards [13,14]. The reason for the low concentrations of heavy metals in older methanogenic leachates is not a lack of heavy metals present in the waste, but it is a consequence of physico-chemical conditions in the body of the landfill. Sorption and precipitation are determined as significant mechanisms for metal immobilization and the subsequent low concentration of metals in the leachate.

Wastes containing soils and certain types of organic matter have a significant sorptive capacity especially at the neutral to high pH values prevailing in methanogenic leachate. In addition, the solubility of many metals, sulphides and carbonates at higher pHs is low. Sulphides and carbonates are capable of forming precipitates with Cd, Ni, Zn, Cu and Pb. While carbonates are abundant in landfill leachate, the solubility of metal carbonates is generally higher than the solubility of metal sulphides. Cr is an exception because it does not form an insoluble sulphide precipitate but it forms precipitates with hydroxide. Determination of the sulphur content of the landfill waste has shown that the waste does not contain sufficient sulphur to bind all the heavy metals (only 5%). Occasionally, phosphates and hydroxides will also precipitate metals. Hydroxide precipitates are formed at pHs at/above neutral, which is typically in the case of older methanogenic leachates [14]. Heavy metal balances for landfills have shown that less than 0.02% of heavy metals received at landfills are leached from the landfill after 30 years.

The presence and concentration of pollutants vary at the site significantly and it depends upon the composition of solid wastes, physico-chemical conditions at the landfill, local rainfall regime, geology and especially upon landfill age [8, 15]. In most of the European Union countries the problem of selection of effective leachate treatment strategy has existed for some time, but a universal solution has not yet been offered [16]. Main priorities of sustainable waste management are to reduce the amount of solid waste, to reuse and recycle them or to recover them as a valuable energy source. The least favorable option is their disposition at the landfill [17]. Due to this approach, wastes disposed on the landfill in the near future will contain only non-recyclable and non-biodegradable fractions. They are expected to be very stable, resistant to microbial degradation and free of residual organic biodegradable matter. This could lead to generation of leachates with similar composition as problematic stabilized mature leachates (old ones) from traditional landfills followed by useless biological treatment systems. For environmental protection and human health it is essential to assess the effect of such sustainable waste management on the composition of future landfill leachate and to apply and/or to develop suitable treatment methods for their effective treatment.

Some authors had used mathematical models to predict changes in concentrations of selected pollutants in the leachates of studied landfill versus time on the basis of the 5-years of monitoring data [9]. Predictions clearly showed that the concentration of chloride in the leachate decreases slowly from 1000 – 1200 $mg \cdot L^{-1}$ to below 200 $mg \cdot L^{-1}$ in 20 years. The same is also true for total dissolved solids, while the concentration of ammonium N decreased from 1000 mg/l to almost 0 mg/l within a 10 year period [9]. In the same time period, the concentration of iron and chlorobenzene is expected to increase from 10 to 17 – 23 $\mu g \cdot L^{-1}$, as well as the concentration of bicarbonates: from average 4000 $mg \cdot L^{-1}$ of $CaCO_3$ to average 5000 $mg \cdot L^{-1}$. This study clearly confirms that stability of the leachate could be confirmed by analyses of landfill leachate. Landfill is considered stabile when parameters in the leachate have negligible values or they are bellow required limits [9]. Processes of stabilization are proven to take decades to finish and thus landfill monitoring and management is considered a lengthy task, so any decision regarding treatment procedures could have major environmental and economic consequences.

LANDFILL LEACHATE TREATMENT

In recent years, there have been different treatment approaches for landfill leachate treatments [18]. The complexity of landfill leachates and the fact that one single leachate changes its characteristics with time complicate the selection and performance of the treatment process and thus there are no general recommendations for successful landfill leachate treatment. Proper treatment of particular municipal landfill leachate is still a challenge for local authorities [2] facing legislative requirements as well as a variety of demands from their local micro-environment. Conventional landfill leachate treatments can be classified into three major groups: i) recycling and combined treatment with domestic sewage, ii) biological treatment employing aerobic and anaerobic biodegradation processes and iii) chemical and physical methods, such as chemical oxidation, adsorption, coagulation/flocculation, membrane techniques and air stripping [18]. The process currently used often required combined techniques which are designed as multistage units optimized for treatment of contaminants, which varies in concentration over years [16]. As a result, regulatory demands at particular areas must be met. One of the important legislative parameters is also toxicity to different organisms in the environment. If this final effluent is unacceptably toxic, a comprehensive program to reduce the toxicity must be started. Such a program [19] typically includes a sequence of steps: firstly measurement of the toxicity and then fine-tuning the operation of the existing treatment facility. Following toxicity identification and evaluation (TIE) procedures in the plant, three kinds of activities are simultaneously used: implementation of "best management practices", evaluation of individual-source treatment technologies and upgrading the existing wastewater treatment system.

Biological Treatment

Biological treatment – aerobic biodegradation process is commonly used for wastewater management due to its reliability, simplicity, environmental friendliness and high cost-effectiveness. Biodegradation under aerobic conditions is carried out by microorganisms, which are able to completely destroy pollutants in landfill leachate and not just to transform or concentrate them into some other medium. Aerobic treatment can also achieve the removal of nitrogen through ammonium nitrogen nitrification followed by denitrification [18]. Biological treatment, often as an activated sludge system, is commonly used for fresh landfill leachate (less than 1-2 years old) containing high concentrations of biodegradable matter (BOD_5/COD ratio > 0.6) [1, 12]. However, with increasing landfill age the biodegradability of leachate decreases and more biorefractory compounds occur. They are products of microbial degradation of waste and they remain in landfills because microorganisms are unable to degrade them further, so they are typical for mature landfill leachate (more than 10 years old) [20]. Mature landfill leachate is poorly biodegradable and usually contains a high amount of compounds (such as different micro-pollutants, ammonium or chlorides) which act as inhibitors of biological nitrification and thus increased landfill age usually leads to replacement of biological treatment with other physico-chemical methods [12].

The application of anaerobic digestion technology is growing wordwide because of its economic and environmental advantages. However, the use of anaerobic biological processes to treat landfill leachate is not very effective, because after the initial aerobic (acetogenic) phase, landfills actually become anaerobic digesters by themselves. Once this has occurred, the leachate produced has already been subjected to anaerobic digestion, so there is a little additional treatment efficiency obtained by anaerobic treatment of such leachates [21]. At the same time, an anaerobic system would produce an effluent still containing a very high concentration of ammonium nitrogen. A concentration so high that a further aerobic stage would still be necessary to nitrify the anaerobic effluent to be suitable for a watercourse discharge, whereas no such second stage is needed for the aerobic process [22]. At the same time, anaerobic digestion is not very effective for wastewaters - leachates with less than 6,000 $mg \cdot L^{-1}$ of COD.

Some authors confirmed that leachates with 12,000 $mg \cdot L^{-1}$ of COD could be treated the most cost effectively (to be self-efficient in energy) and reliably. A too low load of organics in the leachate for effective performance of anaerobic treatment, and costs for additional removal of ammonium are important disadvantages. However, well managed landfills do not produce adequate quantities of leachate at these high COD concentrations for the minimum of 10 years needed for the pay-back of the additional capital cost of anaerobic systems. Such high CODs in the 6,000 to 12,000 $mg \cdot L^{-1}$ last no more than 2 or 3 years in each landfill cell [22], due to short-lived acetogenic phase [23]. Therefore, anaerobic treatment of landfill leachates is more efficiently provided through recirculation of acetogenic leachate within the landfill body itself, where optimal conditions and temperature are likely to be present. At the same time, formed biogas could be collected by an existing gas collection system [23].

Combined biological treatment of landfill leachates with municipal sewage was a favorable option a few years ago due to low operational costs and easy maintenance of the system, but due to presence of persistent and toxic compounds especially in the mature landfill leachate, this option was recently omitted [24]. However, presence of nitrogen (originated from landfill) and phosphorous (from municipal wastewater) in such a mixture could be an important advantage of such approach regarding biological treatment processes [18, 25].

Landfill leachate is thus in many cases treated in the common municipal sewage treatment plant. In the year 2000, 62% of landfill leachates in France were treated biologically, while in 2002, this number was reduced to 46% as a consequence of implementation of membrane and different physico-chemical treatment techniques [18]. Those methods were necessary to overcome the problems related to high ammonia concentrations and its toxicity to microorganisms of activated sludge as well as their ability to remove or degrade persistent (micro)pollutants present in the fresh and especially aged (mature) landfill leachates.

Leachate recycling as a possible treatment method is effective in terms of increasing moisture content in the body of the landfill. It has been explored by many researchers from early 1970. As a result, larger amounts of present biodegradable matter are going to degrade and gas production is expected to increase. The stabilization period to reach limit values for emission to environmental compartments is shortened. The transport of water is also important for removal of inhibitors, present within the wastes or formed during degradation processes [26]. In some cases activated sludge and nutrients were also added into the leachate [27] to optimize physico-chemical conditions for biodegradation of pollutants. The general objective of such approach is production of stabile (inert) waste within a reasonable time scale to reduce the environmental risk to an acceptable level in case of any liner failure.

Chemical Oxidation – Advanced Oxidation Processes (AOPs)

Chemical oxidation is one of the widely studied methods for the treatment of effluents containing refractory compounds such as mature landfill leachate. Growing interest has been recently focused on advanced oxidation processes (AOPs) [18]. AOPs are attractive methods for removal of color, for reduction of organic load and/or to improve the biodegradability of recalcitrant contaminants in mature leachate.

They are also required for treatment of leachates containing toxic soluble organics to enable further biological treatment [16]. Some of the processes are based on direct reaction of oxidant (ozone) with contaminants while the majority is based on production of a powerful oxidizing agent, mainly the hydroxyl radical (·OH), which is the second strongest oxidant [28]. It is also a non-specific one and it is therefore able to rapidly oxidize a variety of persistent organic molecules [16]. Hydroxyl radicals react with organics and break them down gradually into smaller fragments with higher biodegradation potential. Sometimes organics are even completely degraded mainly into CO_2 and water. Processes could involve energy of light (UV photochemical methods: i) ozone/UV, ii) H_2O_2/UV, iii) ozone/H_2O_2/UV, iv) Fe^{2+}/H_2O_2/UV (photo-Fenton process) and v) photocatalysis (TiO_2/UV). There is also a variety of non-photochemical methods applicable: i) Fe^{2+}/H_2O_2 (Fenton process), ii) ozonation, iii) ozone/ H_2O_2, iv) ozone/catalyst and many others [16, 29].

Fenton Process

Most of AOPs use a combination of strong oxidants like ozone, oxygen or hydrogen peroxide with catalysts like transition metals, irons, semiconductor powders, radiation or ultrasound. One common feature of such systems is the high demand on electrical energy for devices such as ozonizers, UV lamps, and ultrasounds and this results in higher treatment costs [30]. The only exception is the Fenton process, which is one of the AOPs with high efficiency and low cost [31].

In the Fenton process hydroxyl radicals are generated as a result of a chain reaction between ferrous ion and hydrogen peroxide (Reaction (1)) [30]. The use of Fenton reagent is favorable due to the decomposition of H_2O_2 into harmless substances after the process accompanied by a catalytical amount of Fe^{2+} used in the reaction [29]. The equipment used for the performance of the process is also relatively simple, because there are no high temperatures and pressures involved.

The formed radicals (Reaction (1)) are able to oxidize a wide range of chemicals in aquatic medium as well as to remove non-oxidized pollutants by coagulation, which proceeds during Fenton process [31]. However, it was found that efficiency is strongly pH dependent [29]. Low pHs between 3.0 and 5.0 have been the most effective in terms of oxidation of organic matter [32]. At low pH < 3.0 removal of organics decreased due to the lower reaction rate (Reaction (1)), while the higher ones (pH = 5.0 – 7.0) favor coagulation [33].

$$Fe^{2+} + H_2O_2 \rightarrow Fe^{3+} + OH^- + OH^{\bullet}$$
$$k = 76.5 \ mol^{-1}s^{-1} \quad (1)$$

Formed Fe(III) can react with H_2O_2 in the so-called Fenton-like reactions (Reactions (2) – (4)) regenerating Fe^{2+} and thus supporting the Fenton process:

$$Fe^{3+} + H_2O_2 \rightarrow Fe^{2+} + HO_2\bullet + H^+$$
$$k = 2 \times 10^{-3} \ mol^{-1}s^{-1} \quad (2)$$

$$Fe^{3+} + HO_2\bullet \rightarrow Fe^{2+} + O_2 + H^+ \quad (3)$$

$$Fe^{3+} + R\bullet \rightarrow Fe^{2+} + R^+ \quad (4)$$

Several studies have demonstrated that the best oxidation efficiency is achieved when neither H_2O_2 nor Fe^{2+} is overdosed, so that the maximal amount of OH• radicals is available for the oxidation of organics. If there is too much peroxide, some undesired reactions (Reactions (5) and (6)) could take place, resulting in destruction of OH• radicals [29]:

$$Fe^{2+} + HO\bullet \rightarrow Fe^{3+} + HO^-$$
$$k = 3 \times 10^8 \ 1 \ mol^{-1} \ s^{-1} \quad (5)$$

$$H_2O_2 + HO\bullet \rightarrow HO_2\bullet + H_2O$$
$$k = 2.7 \times 10^7 \ 1 \ mol^{-1} \ s^{-1} \quad (6)$$

Because landfill leachates are ultimately complex mixtures it is difficult to establish an adequate stechiometric dose of reagents [34]. Not only H_2O_2/Fe^{2+} molar or mass ratio, but also the dosages of both reagents must be optimized. COD and color removal efficiencies increase gradually with the increase of reagent dosages [32]. However, the further increased dosage results only in a small increase of treatment efficiency, which could be explained as a result of formation of short chain organic acids that are difficult to oxidize completely into water and CO_2 [35]. At the same time, the formation of waste sludge also increased with increased COD removal, so this parameter is also an important consideration when optimization of the process is being taken into account. However, mass or molar ratio of Fe^{2+} to H_2O_2 is the most important factor in determining the particular wastewater to minimize scavenging effects [36]. Some authors reported molar ratios of Fe^{2+}: H_2O_2 from 1 : 1.5 to 1 : 12 [32, 34] as appropriate. These differences could be explained by variations in landfill leachate characteristics.

Wet Air Oxidation

Another possible oxidative pretreatment method for landfill leachates is wet air oxidation (WAO). It is considered one of the promising and simple techniques for partial oxidation of organic pollutants into more biologically amenable intermediates [37]. It has been shown to be a rapid and effective method in many cases. It is the liquid phase thermo-chemical process that takes place at elevated pressures (2 – 20 MPa) of air or pure oxygen and at high temperatures (125 – 320°C) [38]. Under these conditions, complex organic compounds are mostly gradually oxidized into carbon dioxide, hydrogen, nitrogen and water. As a result, an increase in biodegradability of the effluent by transformation of hydrophobic and toxic substances to biodegradable (by)products occurs [37, 38]. Unfortunately in many cases, lower molecular weight carboxylic acids (acetic acid e.g.) are also formed as in the case of Fenton oxidation and their resistance to further oxidation could be evident [39].

In spite of many advantages (e.g. wide application range, good efficiency, high reaction rate), WAO is among the more expensive methods especially due to high maintenance requirements [40]. Costs could be significantly reduced by implementation of an appropriate catalyst to promote the process under milder operating conditions and shorter retention times [39]. As catalysts, different metals are usually applied, while in some cases oxidation promoters such as hydrogen peroxide could also be implemented [41]. Homogenous catalysts are generally more active than heterogenic ones, but the problem of recovering possible toxic metal salts could affect the economics of the treatment [39].

Ozonation

Molecular ozone is a strong oxidizer having high reactivity towards a large number of organic pollutants. It has been used for the oxidation of pollutants in wastewaters and has also been suggested as an alternative for the treatment of landfill leachates. It can be used at the beginning, the intermediate phase or at the end of the treatment procedure [42]. If ozonation is used as a pretreatment process in a complex treatment scheme, it is aimed mainly at the decomposition of large organic molecules in order to improve feasibility for further treatment (biodegradation, adsorption, etc.). Ozone is very effective for removal of color and odor [42].

Ozone reacts with organic matter in landfill leachates via two pathways: directly or via different derived oxygen radicals which can act as secondary oxidants [43]. As a result, different products could be formed according to reaction conditions. Mechanism of ozonation is strongly dependent upon pH. Ozonation at high pH (O_3/OH^-) or in combination with H_2O_2 results in formation of hydroxyl radicals with oxidation potential higher than that of ozone [28, 44]. However, for complete mineralization of the organic pollutants present in the leachates, high doses of ozone would be required which would result in increased treatment cost [44].

Photocatalytic Oxidation

Photoactivated chemical reactions are characterized by a radical mechanism initiated by photons in the presence of a catalyst or oxidant [45]. Hydroxide radicals can be generated when the photocatalyst, such as TiO_2, is illuminated and these ultra-reactive species lead to successful mineralization of pollutants to CO_2, H_2O and other minerals [46]. Heterogenic catalysis using TiO_2 is one of the most studied processes among AOPs, due to TiO_2 chemical stability, price, non-toxicity and of course high efficiency [45]. The main obstacle in using this process in industrial scale is removal of the catalyst after treatment from effluent. As a viable option immobilization of TiO_2 on films, membranes or glass has been studied [47]. In such a case, mass transport is the limiting phase, because the contact between catalyst and UV radiation is limited.

However, the photocatalytic oxidation with UV/TiO_2 has seldom been investigated for landfill leachate so far and there is not much information about conditions and design of photocatalytic experiments [16]. The main work has been focused on COD removal but not on other compounds also prevailing in the leachates (ammonia, trace metals, etc.) [47].

Assessment of Ecotoxicity

Due to their complex composition, landfill leachates are known as pollutants with important environmental impact [48] and their release into the environment represents a significant risk for ecosystems [49]. Evaluation of the hazardous impact of landfill leachate was for a long time based mostly on evaluation of the hazardous impact of individual chemicals identified through chemical analyses, which can determine a range of potentially toxic compounds in such a complex matrix [10]. In recent years, it has become generally accepted that results from physico-chemical analyses by themselves do not allow adequate evaluation of the toxic effects of leachates due to important limitations. Chemical analyses do not reveal complex interaction as additive effects of present compounds and/or, synergism or antagonism; they do not reflect the possible transformations of hazardous substances as well as their bioavailability, affected by many parameters [50]. As a result, increasing attention has been put into incorporation of bioassay – toxicity tests for comprehensive evaluation of landfill leachate hazard. In bioassays, different organisms can be employed for detection of toxic compounds in the sample. Rapidity of response combining with sensitivity and reproducibility make bioassays a pragmatic approach for assessing the intrinsic toxicity as a complement to chemical analyses [51]. Nevertheless, in many studies the toxic impact of landfill leachates is assessed only by one type of organism [52]. The environmental impact of leachates is limited to only one acute toxicity test using the water flea *Daphnia magna* by the Slovenian legislation [53]. For reliable assessment of the effect of landfill leachate on the environment a battery of biotests, including organisms from different trophic levels of the food chain, have to be used. Application of decomposers, producers and consumers from aquatic and/or terrestrial ecosystems seems to be particularly appropriate, because they provide a desirable approach due to the different sensitivities of various species to a number of potential toxicants and they allow us to evaluate the impact of leachates on various spheres of the ecosystem that they may get into [13]. However, it is unfeasible and often unethical to test many species (especially vertebrates) and thus a minimal number of tests suited to the hazard assessment should be selected. Recently, many toxicity tests, mostly acute ones, using organisms from different trophic levels have been investigated for assessment of landfill leachate toxicity (Table 1). *Daphnia magna* and other fresh water crustaceans have been employed for assessment of landfill leachate toxicity due to their high sensitivity and widespread application in monitoring programs for quality assessment of wastewaters that discharge into a receiving stream [15]. *Artemia salina* has also gained recent popularity as an organism used for ecotoxicological evaluation of wastewaters as well as various chemicals [54]. It is a crustacean that lives in salt lakes and therefore it is tolerant to the high concentration of chloride and ammonium nitrogen ions often found in landfill leachate [55]. Toxicity of landfill leachate, as determined by *Artemia,* seems to be fairly connected to the toxic impact of present organics and not only to inappropriate environment in the test of e.g. osmotic stress for freshwater organisms due to high salt concentration. Marine bacteria *Vibrio fischeri* is also very often employed as a microorganism for assessment of toxicity due to its many advantages of simplicity of an acute test, short duration of the test and the small volume of the sample needed [10]. On the other hand, mixed bacterial culture – activated sludge from an aerobic biological treatment plant has not been used so far for ecotoxicological evaluation of landfill leachate.

Table 1. Organisms applied for assessment of toxicity of landfill leachates

Organisms		Ecosystem	Reference
Decomposers	Bacteria *Vibrio fischeri*	Marine	[10, 13, 31, 44, 52, 56, 57, 58]
	Protozoa *Spirostomum ambiguum*	Saline, fresh water	[52]
	Mixed bacterial culture Activated sludge	Fresh water	[31, 57]
Producers	Micro-algae *Scenedesmus subspicatus*	Fresh water	[52]
	Duckweed *Lemna minor*	Fresh water	[52, 56, 57, 59, 60,]
	Garden cress *Lepidium sativum*	Terrestrial	[56]
	White mustard *Sinapis alba*	Terrestrial	[57]
	Turnip rape *Brassica rapa*	Terrestrial	[56]
Consumers	Water flea *Daphnia magna*	Fresh water	[13, 49, 52, 57]
	Water flea *Ceriodaphnia dubia*	Fresh water	[52, 61]
	Water flea *Daphnia similis*	Fresh water	[44]
	Thamnocefalus platyurus	Fresh water	[13, 52]
	Rotifer *Brachyonus calyciflorus*	Saline, Brackish, fresh water	[13]
	Brine shrimp *Artemia salina*	Saline, Brakish	[34, 44, 55, 57, 62]
	Gammarus pulex	Fresh water	[63]
	Asellus aquaticus	Fresh water	[63]
	Fish *Brachydanio rerio*	Fresh water	[44]

However, the application of a toxicity test using activated sludge which contains microorganisms involved in biodegradation processes in the environment and in the biological treatment plants should be used for predicting the effect of landfill leachates released into the environment or its impact on performance of biological treatment plants. For a few years, the free floating aquatic plant *Lemna minor* has been prioritized for testing of phytotoxicity of different wastewater due to its many advantages (described in following chapter).

Most of applied toxicity tests are for evaluation of toxicity to the aquatic ecosystem. However, leakage of landfill leachate or release of untreated leachate into the ecosystem can significantly affect the terrestrial ecosystem. The use of higher plants have seldom been employed in comparison to other organisms, even though they are routinely used in chemical testing and they could provide a lot of useful information on the toxicity of leachates to producers in the ecosystem [50, 56].

Toxicity tests have also been used for the assessment of treatment efficiency of different methods used for landfill leachate treatment. However, only several authors used the battery of biotests with various organisms from different trophic levels for evaluation of treatment efficiency [44, 57], more widely one [34, 64] or two [31, 63] biotests were applied and most often ecotoxicological assessment of treated leachate was not included in the study [1, 28, 33, 46, 65, 66]. During AOPs, formation of toxic by-products was already noticed by several authors [31, 44] and in spite of good COD or DOC, removal rates of toxicity have increased. Higher toxicity after coagulation treatment when aluminum salts are used as a coagulant has been observed as well [44, 64]. For these reasons, monitoring of toxicity should be always included in the assessment of treatment efficiency of various methods to achieve a comprehensive evaluation of the environmental impact of treated landfill leachate.

Evaluation of Phytotoxicity

As indicated in the previous chapter, many authors have intensively studied the environmental impact of landfilling focused on landfill leachate generation. However, toxicity to plants, especially to higher plants, has not yet been studiedextensivelly.

Plants, as photosynthetic organisms, produce reactive oxygen species (ROS) during photosynthesis and other metabolic processes. ROS involve free radicals as superoxide radicals, hydroxyl radical, perhydroxy radicals, alkoxy radicals and non-radical molecular forms as hydrogen peroxide and singlet oxygen. Under physiological conditions, ROS are readily removed by antioxidant systems. However, various stress conditions lead to overproductions of ROS, which cannot be removed by antioxidant mechanisms. They are highly reactive and toxic and cause damage to proteins, lipids, carbohydrates and DNA which ultimately results in oxidative stress [67]. Production of an undesirable amount of oxygen reactive species in plants has been proved after exposure to landfill leachate, which results in growth inhibition, chlorophyll inhibition and oxidative stress [68].

Phytotoxic effects of leachate have been usually evaluated by an algae toxicity test [52, 63] or using duckweeds [52, 60]. However, landfill leachates can be significantly dark and turbid and the light necessary for the algae growth can be absorbed by colored substances and particles found in the leachate [69].

Absence of light can inhibit algal growth, not as the result of a toxic action and thus such results can not be reliable. At the same time, results obtained in toxicity tests with algae cannot be used for prediction of toxicity of the investigated sample to higher plants due to the fact that same significant phytotoxic pollutants to higher plants were found to be stimulatory to algae [70]. Toxicity of effluents to duckweed *Lemna minor* (Picture 1) has been intensively studied, especially due to its attractiveness for colored, turbid, oxygen-demanding samples, its high tolerance to wide pH ranges [69] and its high sensitivity to different landfill leachates [56, 60].

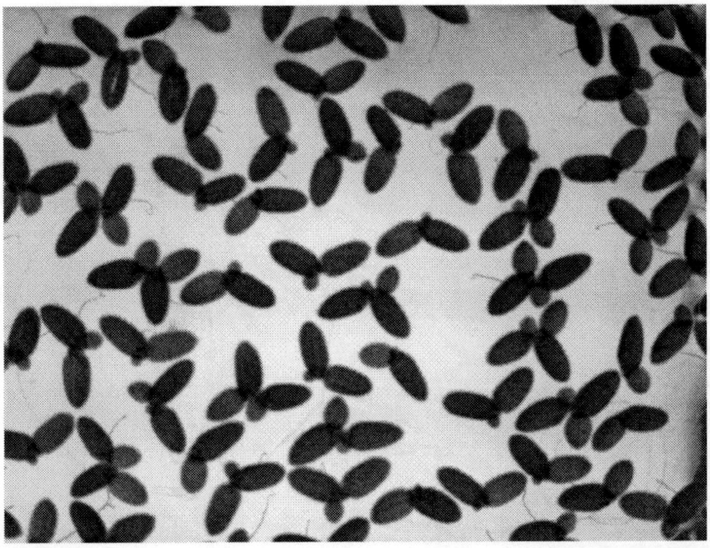

Picture 1. *Lemna minor*.

Toxicity of landfill leachate to terrestrial producers has been evaluated as well [56, 57, 71]. Phytotoxic effect of leachates to seeds of higher plants can be evaluated by different approaches, usually by:

A) Standardized method (ISO 11269-1, 1993) focused on a root growth inhibition of a pre-germinated plant in soil [72],
B) Root growth inhibition or root elongation tests using a hydroponic system - growth in the test tube [56],
C) Root growth inhibition tests focused on the inhibition of root growth in the first three days of germination in a Petri dish [73].

However, when using pre-germinated seeds (A – standardized method), important processes in the plant such as nutrient transport, cell division, as well as rapid metabolism have already been accomplished, because they are occurring in the most intensive phase of a germination period [74] and thus the standardized test does not cover the most sensitive phase of plant life [75]. The hydroponic growth system (B) was mostly used in root growth inhibition test using the *Allium cepa* onion bulb [76] (Picture 2). Other plants can also grow hydroponically and thus plant seeds can be used for long duration root elongation tests [56]. In spite of covering the most important periods of the life cycle of plant, root elongation tests using seeds is not often applied, probably due to its high demand on quantity of tested samples in comparison to root growth inhibition tests in the Petri dish (C), which has many advantages. The root growth inhibition test focuses on the root growth in the first three days of germination in Petri dish and has been used in short-term phytotoxicity tests to provide valuable information about root growth inhibition, plant enzymes activation, cell expansion and other parameters [74]. The test is fast, simple, cost-effective and sensitive with minimal operational cost. All the advantages make the test ideally suitable as a rapid stand-by toxicity test and cost-effective tool to evaluate the ecotoxicological risk of different pollutants to higher plants [50, 70, 74].

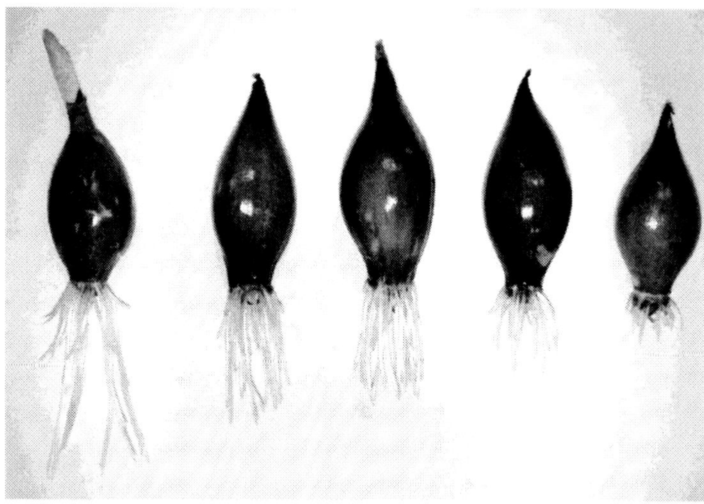

Picture 2. The effect of landfill leachate on root growth of onion *Allium cepa*.

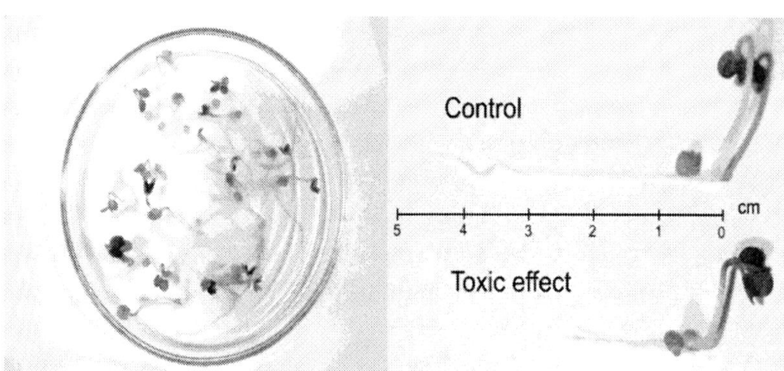

Picture 3. The effect of landfill leachate on root growth of Sinapis alba after three days of incubation in Petri dish in comparison to control sample (without leachate).

Evaluation of phytotoxicity tests using seeds depend on the chosen test, however germination [77], root biomass determination [70], biomass production [51], seedlings emergence [51, 71], photosynthetic pigment determination [68, 78], measurement of enzyme and protein activity [68] have been selected as endpoints in several tests. But the measurement of the length of roots is determined most often [56, 57, 71, 78, 79] (Picture 3).

Any plant seeds can be used for phytotoxicity tests. Many crop and non-crop species from the monocotyledon or dicotyledon groups have been recommended by the U.S. Environmental Protection Agency [80], or by the Organization for Economic Cooperation and Development [81]. Barley (*Hordeum vulgare*) is a prioritized species in test methods for the measurement of inhibition of root growth in soil [72], while white mustard (*Sinapis alba*) is a plant imposed by the Guidelines of the Department of Waste Management [73] for evaluation of ecotoxicity of waste using root growth inhibition tests focused on the first three days of germination. Garden cress is another plant (*Lepidium sativum*) that has been used for assessment of phytotoxicity of landfill leachate [56], heavy metals [79], effluent, surface waters and sediments [50] or as a detection system for cyanotoxins in water [82] and it has thus become one of most often applied plant species. Results of many studies demonstrated that toxicity tests using higher plants are a valuable and indispensable part of ecotoxicology.

CASE STUDY

The objective of the case study was to evaluate phytotoxicity of leachate from one of the regional municipal landfills and to assess the potential hazard of municipal landfill leachate for aquatic and terrestrial plants. The study also focused on evaluation of phytotoxicity prior to, and after one of the advanced oxidation treatment processes – the Fenton oxidation.

Description of the Site

The investigated landfill leachate used in the study was generated in a municipal landfill opened in 1977 and reconstructed in 2004. The whole landfill area covers 1.50 ha. 175,000 tonnes of wastes were deposited there by 2011. The landfill is expected to be closed in 2013. Leachates are collected in an open retention basin of 750 m^3. There is no treatment plant and the raw leachate is released into a nearby stream. The maximal yearly amount of landfill leachate is 8,000 m^3 and the maximal daily amount is 56 m^3. The biogas produced is also not utilized for energy purposes due to its negligible amounts [83]. The quality of the leachate has been monitored since 2010. During this period leachate was sampled several times (October 2010, November 2010, March 2011 and November 2011), at the effluent from the retention basin. Samples used in this study were obtained in March 2011 and in November 2011 and they represent an average sample. The sample collected in March 2011 was treated by Fenton oxidation while the sample obtained in November 2011 was used for assessment of phytotoxicity. All samples were entrapped as random samples and transported in high-density polyethylene containers to the laboratory. Some parameters were determined immediately while some parts of the samples were frozen for other experiments at -28 ± 2 °C.

Physico-Chemical Analyses

The quality of landfill leachate was evaluated by different physico-chemical parameters as pH, BOD_5 (Biochemical Oxygen Demand) [84], COD (Chemical Oxygen Demand) [85],

DOC (Dissolved Organic Carbon) [86], concentration of ammonium nitrogen [87], concentrations of chlorides and alkalinity [88]. Samples were not filtrated prior to analysis.

Phytotoxicity and Toxicity Tests

For assessment of phytotoxicity of raw leachates two methods using higher plants from aquatic and terrestrial ecosystem were chosen. First, phytotoxicity of raw landfill leachates to aquatic plants was evaluated by using duckweed *Lemna minor* in a growth inhibition test [89]. The leachate was diluted by the Steinberg medium [89] to obtain different test concentrations. For the test, 10 healthy, young leaves of *Lemna minor* were placed onto the surface of the samples and illuminated in a 16h light/8h dark cycle. The number of leaves was counted at the end of the test and EC_{50} values were calculated.

The second phytotoxicity test uses terrestrial plants and it is based on the measurement of inhibition of root growth in the first three days of the germination period [73]. Several crop plants from monocotyledon or dicotyledon groups were chosen. White mustard *Sinapis alba*, garden cress *Lepidium sativum*, cucumber *Cucumis sativum*, wheat *Tritium aestivum*, cabbage *Brassica oleracea*, millet *Panicum miliaceum* and radish *Raphnaus sativus* were chosen due to their widespread application in agriculture and/or they have been found to be appropriate species for phytotoxicity assessment. Twenty seeds of each plant were germinated in each Petri dish with the filtration paper on the bottom. The leachate was diluted by growth medium [73] to different concentrations. The lengths of the roots were measured after three days of incubation in the dark at a constant temperature 21°C (± 0.2 °C) and IC_{50} values were calculated.

To compare the acquired data to the commonly applied toxicity test, the toxicity of leachates to the luminescent bacteria *Vibrio fischeri* was also determined. The test, using freeze-dried luminescent bacteria *Vibrio fischeri* (DR. LANGE LUMIStox, 2001) [90], is based on measuring the intensity of bioluminescence affected by the toxicity of the sample. The pH of the samples was corrected to 7.0 ± 0.2 and subsequently the samples were diluted by 2% NaCl solution to obtain different test concentrations. The bioluminescence was measured prior to, and after incubation (30 min) by a LUMIStox luminometer. All tests were carried out in duplicates and definitive tests were performed twice.

For evaluation of the treatment efficiency of the Fenton oxidation for removal of phytotoxic pollutants, one phytotoxicity test using white mustard *Sinapis alba* was chosen. A toxicity test using luminescent bacteria *Vibrio fischeri* was applied as well, because it is a widely used test and very often applied for assessment of efficiency of different treatment methods [15, 37, 91]. A root growth inhibition test with *Sinapis alba* used for evaluation of phytotoxicity prior to, and after treatment was chosen due to the high sensitivity of the plant to the investigated leachate, using a small volume of samples (in comparison to test using duckweed *Lemna minor*) and usage of the species in the Guideline of the Department of Waste Management [73]. For assessment of treatment efficiency of Fenton oxidation, landfill leachate was diluted to 30 % v/v by growth medium for phytotoxicity tests and by 2% NaCl solution to 2.5 % v/v for the luminescent inhibition tests.

Fenton Oxidation

Fenton oxidation was performed according to Žgajnar Gotvajn et al. (2011) [31] with some modifications. Landfill leachate was not diluted prior to experiments and not filtered, and pH was adjusted to 3.0 (± 0.2) using sulfuric acid (Merck, Germany). 200 mL of the sample was placed in a 2 L beaker, which was submerged in a temperature-controlled batch to attain constant temperature (25 ± 5 °C) during the whole experiment. $FeSO_4$ (Merck, Germany) was dissolved in the landfill leachate to obtain a concentration 0.3 M. The Fenton reaction was started by adding an appropriate amount (0.9 M, 1.5 M, 2.4 M and 3 M) of H_2O_2 (30 % v/v) to obtain different molar ratios of Fenton reagents $M(Fe^{2+})$:$M(H_2O_2)$ (1:3, 1:5, 1:8 and 1:10). The mixture was constantly stirred during all experiments at 400 rpm. Fenton reactions proceeded for 30 minutes. Prior to all analyses, the Fenton reaction was stopped by addition of a few pellets of NaOH and boiled for a few minutes to eliminate the excess H_2O_2. It was filtered through black ribbon paper to remove and determine the concentration of the formed iron sludge after drying at 105° C.

The concentration of organic carbon content in the waste iron sludge was also determined (Shimazu TOC 5000A Analyzer, SSM) after filtration samples were neutralized to pH 7.0 (± 0.5) and frozen for toxicological evaluation. To assess the impact of changing pH and boiling as the first two stages of the Fenton procedure, one portion of the landfill leachate was also acidified to pH 3.0 (± 0.2), stirred at 400 rpm for 30 minutes and subsequently, a few pellets of NaOH were added and the sample was boiled, cooled down, filtered through black ribbon paper and neutralized.

At the end of each experiment the remaining ferric sludge was dried in a dryer at 105 ± 1 °C for 2 days until a constant mass was obtained. Its mass was determined to calculate its concentration in g L^{-1}. The amount of organic fraction (in mass %) was determined by oxidation of dry waste sludge at 900°C in SSM Shimadzu TOC 5000A.

EXPERIMENTS AND RESULTS

Leachates were sampled twice in the landfill where municipal waste has been disposed for more than 30 years. The first leachate was sampled in March 2011, while the second one in November 2011. Both leachates were physico-chemically characterized (Table 2). They contained high concentrations of chlorides, ammonium nitrogen and high alkalinity, but they represent average samples when compared to minimal and maximal values of parameters in Table 2.

However, significant differences between leachates can be observed (Table 2). Leachate sampled in March 2011 was more polluted by organics, which can be seen from the higher COD and BOD_5 contents. Leachates sampled in November 2011 were less polluted by organics, but contained more inorganic compounds, the concentration of ammonium nitrogen was almost three times higher, and the concentration of chlorides and alkalinity was twice as high than in the sample from March 2011.

Samples were significantly different, most probably due to the different time of sampling. Leachate sampled in March 2011 was more polluted by organics due to the intensive degradation processes developed within the body of the landfill during spring season when increased temperatures and humidity, due to the rainy season start, intensify decay of wastes [92]. The BOD_5/COD ratio of investigated leachates was higher (0.33, Table 2) than expected for the leachate from old landfill probably due to the burial of a higher amount of biodegradable organic wastes at the time of deposition, when no separation of organic biodegradable wastes was practiced.

However, both raw leachates demonstrated similar qualities (high pH, high concentrations of ammonium nitrogen and chlorides), which classify them among mature leachates with a significant content of biorefractory compounds. According to the Slovenian effluent limit [53], leachate from a non-hazardous landfill could be released into surface waters in a case of maximal COD below 300 mg L^{-1}, BOD_5 must not exceed 30 mg L^{-1}, and ammonium nitrogen must be less than 50 mg L^{-1}. Leachates from the landfill are not allowed to be released into the environment and they have to be properly treated.

Table 2. Physico-chemical parameters of landfill leachates sampled in March 2011 and November 2011 at regional municipal landfill

Parameter	March 2011	November 2011	Minimal and maximal values measured since 2010 (N=4)
pH	8.1 ± 0.1	8.4 ± 0.1	6.8 – 8.4
COD (mg L^{-1})	3263 ± 150	2574 ± 44	784 - 3411
BOD_5 (mg L^{-1})	1089 ± 105	227 ± 18	209 - 1195
BOD_5/COD	0.33	0.09	0.08 – 0.35
DOC (mg L^{-1})	1075 ± 42	888 ± 53	835 - 1518
N-NH_4^+ (mg L^{-1})	519 ± 7	1408 ± 237	513 - 1650
Cl^- (mg L^{-1})	1100 ± 100	1902 ± 27	750 - 1930
Alkalinity $CaCO_3$ (mg L^{-1})	8862 ± 111	15250 ± 750	8750 - 16000

Toxicity to bacteria *Vibrio fischeri* was determined for both leachates. Leachate sampled in March 2011 was more toxic to luminescent bacteria (30minEC50 = 1.4% v/v) than the leachate sampled in November 2011 (30minEC50 = 3.6% v/v). One of the average samples was also used for determination of toxicity to *Daphnia magna* as requested by authorities [53]. *Daphnia* acute immobilization test [93] was used for evaluation of toxicity of leachates sampled in the same landfill in previous study [94]. 24hEC_{50} was 12.9% v/v, while allowed value for release of the leachate in the surface water has to reach more than 25 % v/v. Leachate was confirmed to be too toxic for direct environmental disposal.

The second sample (November 2011) was also used for phytotoxicity testing and results are presented in Figure 1. The landfill leachate was less toxic to plants than to bacteria *Vibrio* fischeri. The most sensitive species to the investigated leachate was aquatic plant duckweed *Lemna minor*. Similar sensitivities showed terrestrial species white mustard *Sinpais alba*, garden cress *Lepidium sativum* and cabbage *Brassica oleracea*. Higher tolerance was observed in wheat *Tritium aestivum*, millet *Panicum miliaceum* and radish *Raphanus sativus*. The most tolerant terrestrial plant seems to be cucumber *Cucumis sativum*. As it can be seen from Figure 1, sensitivity of different organisms from the same ecosystem and the same trophic level of food chain vary significantly by several orders of magnitude.

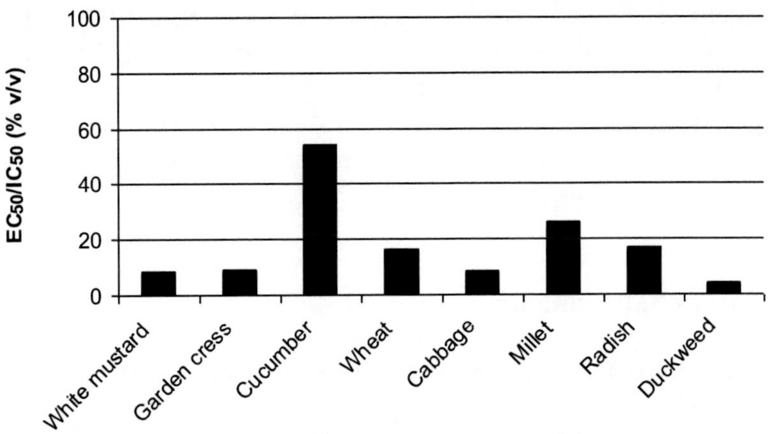

Figure 1. Phytotoxicity of the leachate sampled in November 2011.

Duckweed *Lemna minor* is a species commonly found in fresh water ecosystems in a moderate climate. It is an important source of food for various water birds and fish. It also provides important habitats for invertebrates and thus it is a very important part of the fresh water ecosystem [95]. Duckweed *Lemna minor* is frequently used for assessment of landfill leachate phytotoxicity, because it can be significantly affected by leachates if released untreated into the environment. *Lemna minor* belongs to a sensitive species as confirmed by several studies using a battery of biotests for assessment of landfill leachate toxicity [52, 57]. In our study $168hEC_{50}$ for *Lemna minor* was 4.2 % v/v. Clément and Bouvet (1993) [60] showed similar sensitivity ($120hIC_{50}$ = 4 % v/v) of duckweed *Lemna minor* to landfill leachate as well as Bernard at al. (1996) [52] ($120hIC_{50}$ = 3 – 4.8 % v/v – recalculated from Toxic Units (TU)) and Kalčíková at al. (2011) [57] (168hEC50 = 2.4 % v/v– recalculated from TU). Terrestrial plants have not been used for assessment of landfill leachates phytotoxicity so far and data about their toxicity are scarce.

In our previous study [57] focused on evaluation of hazard of landfill leachate, white mustard *Sinapis alba* was applied. Low polluted leachate (COD = 457 mg L^{-1}, N-NH_4^+ = 70 mg L^{-1}, Cl^- = 325 mg L^{-1}) was almost non-toxic to *Sinapis alba* (72hIC50 = 77 % v/v – recalculated from TU), while more polluted leachate (COD = 5025 mg L^{-1}, N-NH_4^+ = 1444 mg L^{-1}, Cl^- = 2000 mg L^{-1}, 72hIC50 = 8.5 % v/v – recalculated from TU) was comparable to leachate from this study (Table 2) expressing similar phytotoxicity (72hIC50 = 8.7 % v/v) (Figure 1). In another study [64], wheat (undefined species) was used for monitoring the treatment efficiency of a combined treatment process used for stabilized landfill leachate (COD = 2817 mg L^{-1}, N-NH_4^+ = 2000 mg L^{-1}, Cl^- = not determined). Phytotoxicity of raw leachate to wheat was EC_{50} = 9.5 % v/v (duration of the test is not defined, EC_{50} recalculated from TU) which is similar to the phytotoxicity of the leachate to wheat *Tritium aestivum* in our study (72hIC50 = 16.4 % v/v) (Figure 1).

Leachates are complex mixtures of different pollutants and therefore it is very difficult to assess the contribution of every single compound to the final phytotoxic effect. However, there are some underlying surveys about the phytotoxicity of different compounds found in landfill leachates.

It has been proved that high COD expressing non-specific organic pollutants and high concentrations of ammonium nitrogen are associated with increasing toxicity [96] as well as high concentrations of chlorides can cause osmotic stress in plants [97]. It is known that different species within the same trophic level can respond differently as seen in Figure 1. Some of tested species showed a high tolerance to landfill leachate. Such species could be used for the phytoremediation of leachates by the wetland system. Tested millet *Panicum miliaceum* could be a good species for phytoremediation due to its high tolerance to leachate and its ability to grow in wetland and riverine environments [77].

Fenton Oxidation Experiments

Fenton oxidation was applied as one of the advanced oxidation processes (AOPs), which was often successfully employed for the treatment of mature leachates. Experiments were conducted with raw leachate (R1, March 2011, Table 2) at a constant temperature (25 ± 5 °C) and pH (3.0 ± 0.2) with four molar ratios of Fenton reagents $M(Fe^{2+}):M(H_2O_2)$ = 1:3 (T1), 1:5 (T2), 1:8 (T3) and 1:10 (T4).

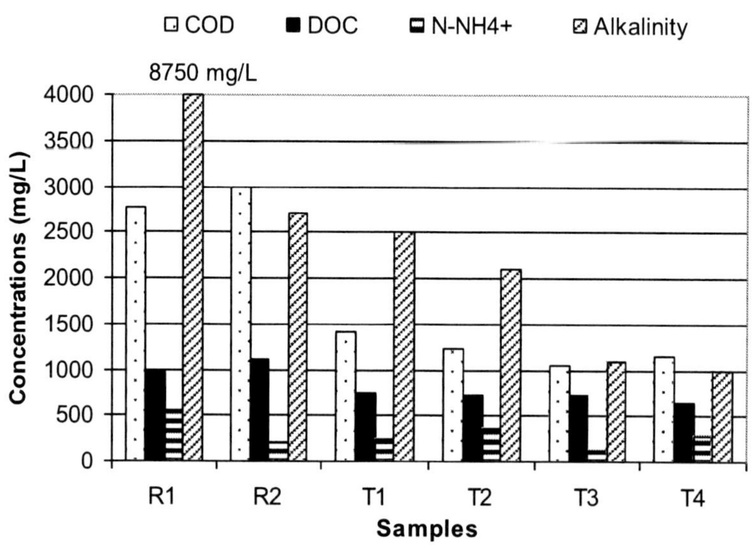

Figure 2. Physico-chemical parameters of raw leachate (R1) sampled in March 2011. Leachate exposed to pH changing and boiling (R2) and leachates treated by Fenton oxidation with different molar ratios of Fenton reagents $M(Fe^{2+}):M(H_2O_2)$ = 1:3 (T1), 1:5 (T2), 1:8 (T3) and 1:10 (T4).

The effect of pH adjusting and boiling of the sample during the Fenton procedure was also investigated (R2). Different parameters such as COD, DOC, alkalinity, concentration of ammonium nitrogen (Figure 2) and toxicity to *Vibrio fischeri* and *Sinapis alba* (Figure 3) were monitored prior to, and after oxidative treatment experiments.

Removal of COD slightly increased with increasing molar ratios of Fenton reagents from 50% (1:3) up to 61% (1:8). Removal of DOC was even lower from 29% (1:3) up to maximal DOC removal of 46% at 1:10 ratio. COD and DOC of the leachate were not significantly affected by changing of pH and boiling (R1, R2). However, alkalinity and ammonium nitrogen contents rapidly decreased during the procedure.

It was assumed that ammonium nitrogen decreased after Fenton oxidation when the ammonium nitrogen (ionic form) in landfill leachate was converted by increased pH to the gas form ammonia [98] and stripped out during boiling. The proportion of volatile NH_3 of total ammonia nitrogen (NH_4^+-N) is a function of pH and temperature [99] (Equation /1/):

$$NH_3(\%) = 1/(1 + 1.82 \times 10^{(9 - pH)}) \quad (1)$$

In Fenton experiments pH was increased up to 12 and in such case 99.8% of ammonium nitrogen is expected to be in the form of volatile NH_3 and thus stripped out of the system due to intensive mixing during boiling. Reduction of ammonium nitrogen was clearly not a consequence of oxidation processes. Differences between the concentrations of ammonia nitrogen after the different experiments (T1, T2, T3, T4) could be explained by a slightly different boiling time of boiling and/or a small difference among the pH of each sample. Alkalinity also significantly decreased (R1, R2: 70% removal) during changing of pH and boiling.

The nonsuccess of Fenton oxidation for removal of organic compounds could be a consequence of a high chloride content, which can act as a catalyst poison and a hydroxyl radical scavenger [46]. Another reason could be a presence of recalcitrant contaminates which are simply so resistant that they cannot be oxidized even by such strong oxidizers as hydroxyl radicals [1]. However, in spite of the different molar ratios investigated we were unable to fulfill legislative limits for the discharge of treated leachate into the environment [53]. COD did not drop below 300 mg L^{-1} (more than 1000 mg L^{-1} at the most efficient T3 ratio, BOD_5 exceed 30 mg L^{-1}, while ammonium nitrogen was never below 50 mg L^{-1} (114 mg L^{-1} in the case of T3). Fenton oxidation was proved to be non-efficient and treated leachates are not allowed to be released into the environment and they have to be additionally treated.

Toxicity of raw landfill leachate B2 to bacteria *Vibrio fischeri* rapidly decreased during Fenton oxidation at all ratios and there was no inhibition observed (< 15% inhibition) just by the changing of pH and boiling. Bacteria *Vibrio fischeri* was sensitive to the high concentration of ammonium nitrogen and after its 60% removal the leachate became practically non-toxic and the effect of Fenton oxidation could not be evaluated. Thus, if physico-chemical parameters and toxicity are compared, ammonium nitrogen and alkalinity are expected to be mostly responsible for toxicity of such concentration of landfill leachate to bacteria *Vibrio fischeri*.

The obtained results are in accordance with several authors [49, 59] who investigated the relationship between physico-chemical parameters and the toxicity of landfill leachate. They found out that high concentrations of ammonium nitrogen and alkalinity are one of the main factors responsible for high toxicity of leachates to many organisms.

In spite of the significant amounts of ammonium nitrogen and alkalinity removed, toxicity of leachate to white mustard *Sinapis alba* even increased slightly after changing pH and boiling (R2). In the first two cases (T1 and T2) toxicity after Fenton oxidation significantly increased to 100% inhibition, there was no germination of seeds observed.

The reason could be formation of phytotoxic by-products during the Fenton process. In the case of T3 and T4 Fenton oxidation inhibition of root growth of *Sinapis alba* slightly decreased from 82% inhibition (R2) to 57% inhibition (T3 and T4) and Fenton process treatment efficiency measured by *Sinapis alba* was only 30%.

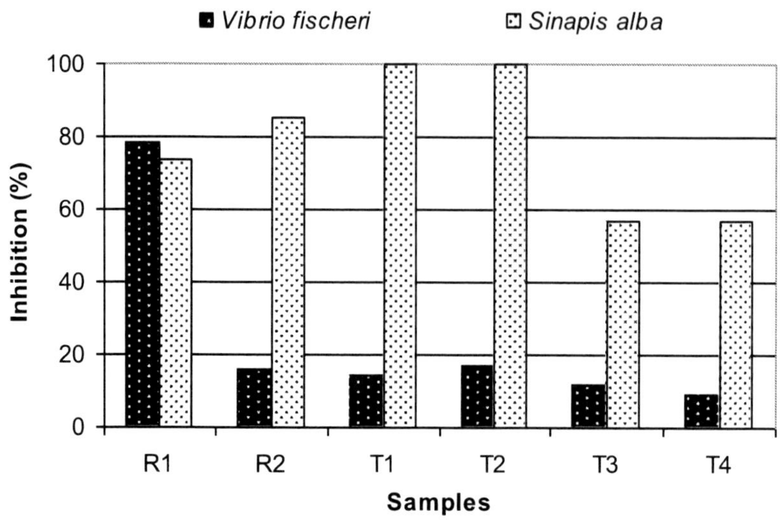

Figure 3. Toxicity of raw leachates (R1), leachate exposed to pH changing and boiling (R2) and treated leachates by Fenton oxidation with different molar ratios of Fenton reagents $M(Fe^{2+})$:$M(H_2O_2)$ 1:3 (T1), 1:5 (T2), 1:8 (T3) and 1:10 (T4).

Biodegradability (expressed as BOD_5/COD ratio) of treated leachate after Fenton oxidation was determined. The BOD_5/COD ratio of the raw sample was 0.33, however biodegradability slightly increased after treatment of T1 and T2 (low molar ratios of Fenton reagents) to 0.41 – 0.42 and could be evidence for formation of some metabolites with a high phytotoxic effect, which was observed after treatment by low molar ratios of Fenton reagents. The BOD_5/COD ratio of treated leachates T3 and T4 (high molar ratios of Fenton reagents) decreased to 0.25 – 0.28. These data confirmed that Fenton oxidation is not an appropriate single step treatment process and is also not suitable as a pretreatment stage, followed by biological treatment, because of reduced biodegradability of the remaining organics.

Another problem related to Fenton oxidation is also the formation of waste iron sludge. Its concentration was also determined in every experiment. Organic carbon content was also determined to indicate the possible adsorption or coagulation removal of organics from the leachate during oxidative treatment. Results showed that concentrations of iron sludge are quite high, approximately 70 g L^{-1}, in comparison to a similar study [31], where the concentration of iron sludge was a few times lower, approximately 20 g L^{-1}. The generation of such amount of iron sludge in real treatment systems would definitively lead to an undesirable increase of the treatment cost of the Fenton process. However, organic carbon was not detected (< 1%) in any waste iron sludge confirming that organic carbon was removed only by oxidation and not by the coagulation process.

CONCLUSION

The aim of our study was to determine phytotoxicity of regional municipal landfill leachate and to monitor phytotoxicity prior to, and after one of the AOPs – Fenton oxidation. Investigated landfill leachate was polluted by persistent organic substances and expressed characteristics of mature landfill leachate. It was toxic to several plant species, especially to aquatic plant duckweed *Lemna minor*.

Leachate was also toxic to several terrestrial plant species. Some terrestrial plants showed a high tolerance to leachate and thus have potential for their application in phytoremediation. Fenton oxidation was not successful for the treatment of investigated leachate. Maximal treatment efficiency was only 58% of COD removal at a higher molar ratio (1:10) and phytotoxicity decreased for only 30%. In some cases, formation of phytotoxic pollutants after Fenton oxidation was identified.

Phytotoxicity tests were confirmed to be a necessary supplement for reliable evaluation of the hazardous impact of landfill leachate and they are also important for confirmation of treatment efficiency of selected treatment methods. Commonly applied toxicity tests with producers and consumers in the ecosystem do not assure enough data on the impact of hazardous landfill leachates to avoid long-term pollution problems.

REFERENCES

[1] Deng, Y. and Englehardt, J.D. (2006). Treatment of landfill leachate by Fenton process. *Water Research*, *40*, 3683-3694.

[2] Neczaj, E., Okoniewska, E. and Kacprzak, M. (2005). Treatment of landfill leachate by sequencing batch reactor. *Desalination*, *185*, 357-362.

[3] Eurostat. [23. 7. 2010]. Available from: http://epp.eurostat.ec.europa.eu.

[4] Statistični urad Republike Slovenije. [3. 1. 2012]. Available from: http://www.stat.si/novica_prikazi.aspx?id=4244.

[5] Ministrstvo za okolje Republike Slovenije. [3. 1. 2012]. Available from: http://www.mop.gov.si/si/delovna_podrocja/odpadki/pogosta_vprasanja_in_odgovori/.

[6] Morcet, M., Aran, C., Bogner, J., Chanton, J., Spokas, K. and Hebe, I. (2003). Methane mass balance: A review of field results from three French landfill case study. *Proceedings Sardinia 2003, Ninth International Waste Management and Landfill Symposium*, S. Margherita di Pula, Cagliari, Italy; 6 - 10 October 2003.

[7] Baderna, D., Maggioni, S., Boriani, E., Gemma, S., Molteni, M., Lombardo, A., Colombo, A., Bordonali, S., Rotella, G., Lodi, M. and Benfenati, E. (2011). A combined approach to investigate the toxicity of an industrial landfill's leachate: Chemical analyses, risk assessment and in vitro assays. *Environmental Research, 111*, 603-613.

[8] Jones, D.L., Williamson, K.L. and Owen, A.G. (2006). Phytoremediation of landfill leachate. *Waste Management, 26*, 825-837.

[9] Sizirici, B. and Tansel, B. (2010). Projection of landfill stabilization period by time series analysis of leachate quality and transformation trends of VOCs. *Waste Management, 30*, 82-91.

[10] Pivato, A. and Gaspari L. (2006). Acute toxicity test of leachates from traditional and sustainable landfills using luminescent bacteria. *Waste Management, 26,* 1148-1155

[11] Marttinen, S.K., Kettunen, R.H., Sormunen, K.M., Soimasuo, R.M. and Rintala J.A. (2002). Screening of physical-chemical methods for removal of organic material, nitrogen and toxicty from low strength landfill leachate. *Chemospere, 46*, 851-858.

[12] Di Iaconi, C., Ramadori, R. and Polez, A. (2006). Combined biological and chemical degradation for treating a mature municipal landfill leachate. *Biochemical Engineering Journal, 31,* 118-124.

[13] Isidori, M., Lavorgna, M., Nardelli, A. and Parrella A. (2003). Toxicity identification evaluation of leachates from municipal solid waste landfills: a multispecies approach. *Chemospere, 52,* 85-94.

[14] Kjeldsen, P., Barlaz, M.A., Rooker, A.P., Baun, A., Ledin, A. and Christensen T.H. (2002). Present and long-term composition of MSW landfill leachate: a review. *Critical Reviews in Environmental Science and Technology, 32*, 297-336.

[15] Žgajnar Gotvajn, A., Tišler, T. and Zagorc-Končan, J. (2009). Comparison of different treatment strategies for industrial landfill leachate. *Journal of Hazardous Materials, 162*, 1446-1456.

[16] Wisznovski, J., Robert, D., Surmacz-Gorska, J., Miksch, K. and Webere, J.V. (2006). Landfill treatment methods: *A review. Environmental Chemistry Letters, 4,* 51-61.

[17] Read, A.D., Phillips, P. and Robinson G. (1997). Landfill as a future waste management option in England: the view of landfill operators. *Resources, Conservation and Recycling, 20*, 183-205.

[18] Renou, S., Givaudan, J.G., Poulain, S., Dirassouyan, F. and Moulin, P. (2008). Landfill leachate treatment: Review and opportunity. *Journal of Hazardous Materials, 150*. 468-493.

[19] Ford D.L. (1992). Toxicity reduction evaluation and control. Techomic Publishing Company, Lancaster, 34-87.

[20] Batarseh, E.S., Reinhart, D.R. and Berge, N.D. (2010). Sustainable disposal of municipal solid waste: Post bioreactor landfill polishing. *Waste management, 30,* 2170-2176.

[21] Blakely, N.C. and Young, C.P. (1990). Leachate Digestion. *Proceedings of the International Conference - Landfill Gas: Energy and Environment '90*. Bournemouth, UK, 16-19 October 1990.

[22] Can landfill leachate be treated by anaerobic digestion? [4.1. 2012]. Available from: http://www.anaerobicdigestion.com/html/anaerobic_ digestion_of_leachat.php.

[23] Kheramand, S., Karimi-Jashni, A. and Sartaj, M. (2010). Treatment of municipal landfill leachate using a combined anaerobic digester and activated sludge system. *Waste Management, 30*, 1025-1031.

[24] Cecen, F. and Aktas, O. (2004). Aerobic co-treatment of landfill leachate with domestic wastewater. *Environmental Engineering Science, 21*, 303-312.

[25] Cecen, F. and Cakiroglou, D. (2001). Impact of landfill leachate on the co-treatment of domestic wastewater. *Biotechnology Letters, 23*, 821-826.

[26] Vossen, W.J., Peperkamp, H. and Folmer, D.T. (2001). Progress of sustainable landfill stabilization after 3 years of leachate recirculation at the Vlagheide landfill (NL). *Proceedings of the 13th International Waste Managemant and Landfill Simposium*, Santa Margherita di Pula, Italy, 2011.

[27] Warith, M.A. (2003). Solid Waste Management: New Trends in Landfill Design. *Emirates Journal for Engineering Research, 8*, 61-70.

[28] Cortez, S., Teixeira, P., Oliveira, R. and Mota, M. (2010). Ozonation of polishing treatment of mature landfill leachate. *Journal of Hazardous Materials, 182,* 730-734.

[29] Prousek, J. (1995). Fenton reaction after a century. *Chemické Listy, 89,* 11-21.

[30] Zhang, H., Choi, H.J. and Huang, C-P. (2005). Optimization of Fenton process for the treatment of landfill leachate. *Journal of Hazardous Materials B*, *125*, 166–174.

[31] Žgajnar Gotvajn, A., Zagorc-Končan, J. and Cotman, M. (2011). Fenton′s oxidative treatment of municipal landfill as an alternative to biological process. *Desalination, 275,* 269-275.

[32] Mahmud, K., Hossain, M.D. and Shams, S. (2011). Different treatment strategies for highly polluted landfill leachate in developing countries. *Waste Management* (In press).

[33] Rivas, F.J., Beltran, F., Carvalho, F., Acedo, B. and Gimeno, O. (2004). Stabilised leachates: sequential coagulation-flocculation+chemical oxidation processes. *Journal of Hazardous Materials, 116*, 95-102.

[34] Vedrenne, M., Vasquez-Medran, R., Prato-Garcia, D., Frontana-Uribe, B.A. and Ibanez, J.G. (2012). Characterization and detoxification of mature landfill leachate using a combined coagulation-flocculation/ photo Fenton treatment. *Journal of Hazardous materials, 205-206,* 208-215.

[35] Lopez, A., Pagano, M., Volpe, A. and Di Pinto, A.C. (2004). Fenton's pre-treatment of mature lendfill leachate. *Chemosphere, 54*, 1005-1010.

[36] Torrades F., Pérez M., Mansilla H.D. and Peral J. (2003). Experimental design of Fenton and photo-Fenton reactions for the treatment of cellulose bleaching effluents. *Chemosphere, 53,* 1211-1220.

[37] Pintar, A., Besson, M., Gallezot, P., Gibert, J. and Martin D. (2004). Toxicity to *Daphnia magna* and *Vibrio fischeri* of Kraft bleach plant effluents treated by catalytic wet-air oxidation. *Water* Research, 38, 289-300.

[38] Oliviero, L., Barbier, J. and Duprez, D. (2003). Wet air oxidation of nitrogen-containing organic compounds and ammonia in aqueous media. *Applied catalysis B: Environmental, 40,* 163-184.

[39] Anglada, A., Urtiaga, A., Ortiz, I., Mantzavinos, D. and Diamadopoulos, E. (2011). Treatment of municipal landfill leachate by catalytic wet air oxidation: Assessment of the role of operating parameters by factorial design. *Waste Management, 31,* 1833-1840.

[40] Lin, S.H. and Ho, S.J. (1996). Catalytic wet-air oxidation of high strength industrial wastewater. *Applied Catalysis B:* Environmental, 9, 133-147.

[41] Lei, L., Hu, X. and Yue, P.L. (1998). Improved wet air oxidation for the treatment of dyeing wastewaters concentrate from membrane separation process. *Water Research, 32*, 2753-2759.

[42] Poznyak, T., Bautista, G.L., Chairez, I., Cordova, R. I. and Rios, L.E. (2008). Decomposition of toxic pollutants in landfill leachate by ozone after coagulation treatment. *Journal of Hazardous Materials, 152,* 1108-1114.

[43] Prousek, J. (1996). Advanced oxidation processes for water treatment. Chemical processes. *Chemické Listy, 90*, 229-237.

[44] Silva, A.C., Dezotti, M. and Sant′Anna Jr., G.L. (2004). Treatment and detoxification of a sanitary landfill leachate. *Chemosphere, 55,* 207-214.

[45] Poblete, R., Otal, E., Vilches, L.F. and Fernandez-Pereira, C. (2011). Photocatalytic degradation of humic acids and landfill leachate using a solid industrial by-product containing TiO_2 and Fe. *Applied Catalysis B: Environmental, 102,* 172-179.

[46] Cho, S.P., Hong, S. C., and Hong S.-I. (2002). Photocatalytic degradation of the landfill leachate containing refractory matters and nitrogen compounds. *Applied Catalysis B: Environmental, 39,* 125-133.

[47] Meerof, D.E., Bloetscher, F., Reddy, D.V., Gasnier, F., Jain, A. S., McBarnette, A. and Hamaguchi, H. (2012). Application of photochemical technologies for treatment of landfill leachate. *Journal of Hazardous Materials, 209-210,* 299-307.

[48] Shrab, E., Brown, K.W. and Donnelly, K.C. (1993). Acute and genetic toxicity of municipal landfill leachate. *Water, Air, and Soil Pollution, 69,* 99–112.

[49] Pablos, M.V., Martini, F., Fernández, C., Babín, M.M., Herraez, I., Miranda, J. Martínez, J., Carbonell, G., San-Segundo, L., García-Hortigüela, P. and Tarazona, T.V. (2011).Correlation between physicochemical and ecotoxicological approaches to estimate landfill leachates toxicity. *Waste Management, 31,*1841-1847.

[50] Montvydienė, D., Marčiulionienė, D., Karlavičienė, V. and Hogland, W. (2008). Phytotoxicity assessment of effluent waters, surface water and sediments. In: Hlavinek, P., Bonacci, O., Marsalek, J. and Mahrikova, I. (Eds.), Dangerous pollutants (Xenobiotics) in Urban Water Cycle (171-180). Dordrecht, The Netherlands: Springer.

[51] Wilke, B.-M., Riepert, F., Koch, C. and Kühne T. (2008). Ecotoxicological characterization of hazardous waste. *Ecotoxicology and Environmental Safety, 70,* 283-293.

[52] Bernard, C., Guido, P., Colin, J. and Du-Delepierre, A.L. (1996). Estimation of the hazard of landfills through toxicity testing of leachates: I. Determination of leachate toxicity with a battery of acute tests. *Chemosphere, 33,* 2303–2320.

[53] Official Gazette of Republic of Slovenia. (2008). Decree on the emission of substance in the discharge of landfill leachate, ULRS 6, 8219–8221, (In Slovene).

[54] Nunes, B.S., Carvalho, F.D., Guilhermino, L.M. and Van Stappen, G. (2006). Use of the genus *Artemia* in ecotoxicity testing. *Environmental Pollution, 144,* 453-462.

[55] Svensson, B.-M., Mathiasson, L., Mårtensson, L., and Bergström S. (2005). *Artemia salina* as test organism for assessment of acute toxicity of leachate water from landfills. *Environmental Monitoring and Assessment, 102,* 309-321.

[56] Devare, M. and Bahadis, M. (1994). Biological monitoring of landfill leachate using plants and luminescent bacteria. *Chemosphere, 28,* 261-271.

[57] Kalčíková, G., Vávrová, M., Zagorc-Končan, J. and Žgajnar Gotvajn, A. (2011). Evaluation of the hazardous impact of landfill leachates by toxicity and biodegradability tests. *Environmental Technology, 32,* 1345-1353.

[58] Ellouze, M., Aloui, F. and Sayadi S. (2009). Effect of high ammonia concentrations on fungal treatment of Tunisian landfill leachates. *Desalination, 246,* 468-477.

[59] Clément, B. and Merlin, G. (1995). The contribution of ammonia and alkalinity to landfill leachate toxicity to duckweed. *Science of The Total Environment, 170,* 71-79.

[60] Clément, B. and Bouvet, Y. (1993). Assessment of landfill leachate toxicity using the duckweed *Lemna minor*. *The Science of Total Environment, 134,* 1179-1190.

[61] Dave, G. and Nilsson, E. (2005). Increased reproductive toxicity of landfill leachate after degradation was caused by nitrite. *Aquatic Toxicology, 73,* 11-30.

[62] Olivero-Verbel J., Padilla-Bottet, C. and De la Rosa, O. (2008). Relationships between physicochemical parameters and the toxicity of leachates from a municipal solid waste landfill. *Ecotoxicology and Environmental Safety, 70,* 294-299.

[63] Bloor, M.C. and Banks, C.J. (2005). Acute and sub-lethal toxicity of landfill leachate towards two aquatic macro-invertebrates: Demonstrating the remediation potential of air stripping. *Environment International, 31*, 1114-1122.

[64] Li, W., Hua, T., Zhou, Q., Zhang, S. and Li, F. (2010). Treatment of stabilized landfill leachate by the combined process of coagulation/flocculation and powder activated carbon adsorption. *Desalination, 264,* 56-62.

[65] Guo, J.-S., Abbas, A.A, Chen, Y.-P., Liu, Z.-P., Fang, F. and Chen, P. (2010). Treatment of landfill leachate using a combined stripping, Fenton, SBR, and coagulation process. *Journal of Hazardous Materials, 178,* 699-705.

[66] Tatsi, A.A., Zouboulis, A.I., Matis, K.A. and Samaras, P. (2003). Coagulation–flocculation pretreatment of sanitary landfill leachates. *Chemosphere, 53,* 737-744.

[67] Gill, S.S. and Tuteja, N. (2010). Reactive oxygen species and antioxidant machinery in abiotic stress tolerance in crop plants. *Plant physiology and Biochemistry, 48,* 909-930.

[68] Sang, N., Han, M., Li, G. and Huang M. (2010). Landfill leachate affects metabolic responses of *Zea mays L.* seedlings. *Waste management, 30,* 856-862.

[69] Cleuvers, M. and Ratte, H.-T. (2002). Phytotoxicity of coloured substances: is *Lemna* Duckweed an alternative to the algal growth inhibition test? *Chemosphere*, 49, 9-15.

[70] Wang, W. and Williams, J.M. (1990). The use of phytotoxicity tests (common duckweed, cabbage, and millet) for determining effluent toxicity. *Environmental Monitoring and Assessment, 14,* 45-58.

[71] Pandard, P., Devillers, J., Charissou, A.-M., Poulsen, V., Jourdain, M.J., Férard J.-F., Grand, C. and Bispo A. (2006). Selecting a battery of bioassays for ecotoxicological characterization of wastes. *Science of Total Environment, 363,* 114–125.

[72] ISO 11269-1 (1993). Soil quality - Determination of the effects of pollutants on soil flora - Part 1: Method for the measurement of inhibition of root growth, Edition 1.

[73] The Guideline of the Department of waste management to determine the ecotoxicity of waste. (2007). Ministry of the Environment of the Czech Republic, Prague, (In Czech).

[74] Wang, X., Sun, Ch., Wang, Y. and Wang, L. (2002). Quantitative structure relationship for the inhibition toxicity to root elongation of *Cucumis sativus* of selected phenols and interspecies correlation with *Tetrahymena pyriformis*. *Chemosphere, 46,* 153-161.

[75] Kalčíková, G., Zagorc-Končan, J., Zupančič, M. and Žgajnar Gotvajn, A. (2012). Variation of landfill leachate phytotoxicity due to landfill ageing. *Journal of Chemical* Technology and Biotechnology (In press).

[76] Fiskesjö G. (1993). Technical methods section. *Allium test1*: A 2-3 day plant test for toxicity assessment by measurement the mean root growth of onions (*Allium cepa* L.). *Environmental Toxicology and water Quality: An international Journal, 8,* 461-470.

[77] Wang, W. and Williams J. (1988). Screening and biomonitoring of industrial effluents using phytotoxicity tests. *Enviromental Toxicology and Chemistry, 7,* 645-652.

[78] Molnárová, M. and Fargašová, A. (2009). Se(IV) phytotoxicity for monocotyledonae cereals (*Hordeum vulgare* L., *Triticum aestivum* L.) and dicotyledonae crops (*Sinapis alba* L., *Brassica napus* L.). *Journal of Hazardous Materials, 172*, 854-861.

[79] Arambašić, M.B, Bjelić, S., Subakov, G. (1995). Acute toxicity of heavy metals (copper, lead, zinc), phenol and sodium on *Allium cepa* L., *Lepidium sativum* L. and *Daphnia magna* St.: Comparative investigations and the practical applications. *Water Research, 29,* 497-503.

[80] U.S. EPA. (1996). Ecological Effects Test Guidelines OPPTS 850.4200 *Seed Germination/Root Elongation Toxicity Test.*

[81] OECD (2003). Guidelines for the testing of chemicals proposal for updating guideline 208 Terrestrial Plant Test: 208: *Seedling Emergence and Seedling Growth Test.*

[82] Gehringer, M.M., Kewada, V., Coates, N. and Downing, T.G. (2003). The use of *Lepidium sativum* in a plant bioassay system for the detection of microcystin-LR. *Toxicon, 41,* 871-876.

[83] Mlakar R. (2011). The impact of municipal landfill leachate on the environment. *Diploma thesis.* University of Ljubljana, Ljubljana, (In Slovene).

[84] ISO 5815–1. (2003). Water Quality – Determination of Biochemical Oxygen Demand after n Days (BOD_n), Part 1: *Dilution and seeding method with allylthiourea addition*, Edition 1.

[85] ISO 6060. (1989). *Water Quality – Determination of the Chemical Oxygen Demand*, Edition 2.

[86] ISO 8245. (1999). *Water Quality – Guidelines for the Determination of Total Organic Carbon (TOC) and Dissolved Organic Carbon (DOC),* Edition 2.

[87] ISO 7150–1. (1984). *Water Quality – Determination of Ammonium*, Part 1: Manual Spectrometric Method, Edition 1.

[88] Standard Methods for the Examination of Water and Wastewater. (2005). 21th ed., American Public Health Association, American Water Works Association, Water Environmental Federation, Washington, DC.

[89] ISO 20079. (2005). Water Quality – Determination of the toxic effect of water constituents and waste water to duckweed (*Lemna minor*) – Duckweed growth inhibition test, Edition 1.

[90] ISO 11348–3. (2007). Water Quality – Determination of the inhibitory effect of water samples on the light emission of *Vibrio fischeri* (Luminescent bacteria test), Part 3: *Method using freeze-dried bacteria*, Edition 2.

[91] Bayo, J., Angosto, J.M. and Gómez-Lópec M.D. (2009). Ecotoxicological screening of reclaimed disinfected wastewater by *Vibrio fischeri* bioassay after a chlorination-dechlorination process. *Journal of Hazardous Materials, 172*, 166-171.

[92] Kalčíková, G., Vávrova, M., Zagorc-Končan, J. and Žgajnar Gotvajn, A. (2011) Seasonal variations in municipal landfill leachate quality. *Management of Environmental Quality, 22,* 612-619.

[93] ISO 6341. (1996). Water quality – Determination of the inhibition of the mobility of *Daphnia magna* Straus (Cladocera, Crustacea) - Acute toxicity test, Edition 3.

[94] Zagorc-Končan, J., Kalčíková, G. and Žgajnar Gotvajn, A. (2011). Comprehensive evaluation of local municipal landfill leachate quality - a useful tool for appropriate treatment management. *Proceedings Sardinia 2011, 13th International Waste Management and Landfill Symposium*, S. Margherita di Pula (Cagliari), Sardinia, Italy, October 7, 2011.

[95] Drost, W., Matzke, M. and Backhaus, T. (2007). Heavy metal toxicity to Lemna minor: studies on the time dependence of growth inhibition and the recovery after exposure. *Chemosphere, 67,* 36-43.

[96] Bernard, C., Colin, J.R. and Le Dû-Delepierre A. (1997). Estimation of the hazard of landfills through toxicity testing of leachates: 2. Comparison of physico-chemical

characteristics of landfill leachates with their toxicity determined with a battery of tests. *Chemosphere, 35,* 2783-2796.

[97] Larcher, W. (1995). *Salt stress in Physiological plant ecology: ecophysiology and stress physiology of functional groups*, Springer, Berlin, 416–420.

[98] Halling-*Sø*rensen, and B. Jørgensen, S.E. (1993). The removal of nitrogen compounds from wastewater. Elsevier science publishers B.V. Amsterdam, The Netherlands.

[99] Erickson, R.J. (1985). An evaluation of mathematical models for the effect of pH and temperature on ammonia toxicity to aquatic organisms. *Water Research, 19,* 1047-1055.

In: Landfills
Editors: G. B. C. Cabral and B. A. E. Botelho

ISBN: 978-1-62257-121-5

Chapter 6

POLYMERS RECYCLING: A REVIEW OF RECENT WORKS

Kotiba Hamad and Fawaz Deri
Faculty of Science, Department of Chemistry,
Laboratory of Materials Rheology, Damascus University, Syria

ABSTRACT

Polymer materials with a range of excellent mechanical properties, low density, durability and low cost, are widely used in the daily needs of contemporary society, ranging from simple packaging to heavy construction. From another side, a continued increase in the use of plastics has led to an increasing amount of plastics ending up in the waste stream, causing in dangerous environmental problems. Recycling is the process for treatment and reusing these wastes. Generally, plastics are made from crude oil so that recycling of plastics helps to conserve this natural resource.

This chapter presents a review of the recent works about the polymers recycling. Also it presents the last developments in the recycling of traditional polymers and their blends and composites using the mechanical and chemical methods.

INTRODUCTION

During last decades, the great population increase worldwide together with the need of people to adopt improved conditions of living led to a dramatical increase of the consumption of polymers (mainly plastics). Materials that appears interwoven with the consuming society where we live so it would be hard to imagine a modern society today without plastics. Plastics have found a myriad of uses in fields as diverse as household appliances, packaging, construction, medicine, electronics, and automotive and aerospace components.

A continued increase in the use of plastics has led to an increasing amount of plastics ending up in the waste stream, which caused in more interesting in the plastic recycling and reusing. Hundreds of works in the last ten years have focused on this object. This chapter concentrates on the reclamation and recycling of plastics. There are several options for how this can be done: reuse, mechanical recycling, and chemical recycling.

- **Reuse:** the most common examples of reuse are with glass containers, where milk and drinks bottles are returned to be cleaned and used again. Reuse is not widely practiced in relation to plastic packaging - plastic products in general tend to be discarded after first use. However, there are examples of reuse in the marketplace. For example, a number of detergent manufacturers market refill sachets for bottled washing liquids and fabric softeners. Consumers can refill and hence reuse their plastic bottles at home, but in all of these cases the reusing of the plastic bottles and containers do not continue for long time epically in the food applications.
- **Mechanical recycling:** also known as physical recycling. The plastic is ground down and then reprocessed to produce a new component that may or may not be the same as its original use [1].
- **Chemical recycling:** the polymer is turned back into its oil/hydrocarbon component in the cases of polyolefin's and functional monomers in the case of polyesters, polyamides.... and for use as raw materials for new polymer production and petrochemical industry [2].

Mechanical recycling and chemical recycling are the most widely practiced of these methods and the most of the studies discussed in this chapter are focused on the two methods. However, from industrial point of view, the mechanical recycling is the most suitable because its low cost and reliability. The aim of this chapter is to discuss the most recent works about the chemical and mechanical recycling of some important and commercial polymers and their systems.

Polylactic Acid (PLA) and Its Systems

Polylactic acid (PLA) is one of the most important biodegradable polyesters, it is derived from renewable sources (mainly starch and sugar). Until the last decade, the main uses of PLA have been limited to biomedical and pharmaceutical applications such as implant devices, tissue scaffolds and internal sutures, because of its high cost, and low molecular weight. Since, the existence of both hydroxyl and a carboxyl group in lactic acid enables it to be converted directly into polyesters via a polycondensation reaction. PLA took a lot of interest in the last 10 years [3-5]. Since PLA is biodegradable, which would significantly reduce environmental pollution associated with PLA waste. The knowledge about the material recycling and changes in the properties of PLA upon its multiple processing is a very important subject. Duigou et al. [6] studied the effect of recycling on mechanical behavior of biocompostable flax/poly(L-lactide) composites which were used as alternative to glass fibre-reinforced petrochemical polymers. The composites were prepared using a single screw extruder and the prepared composites were then molded using injection machine, the recycling process were performed on six injection cycle. The effect of the injection number on the rheological, mechanical and morphological properties of the composite was studied. The results showed that the stiffness of PLA is improved after the addition of the fiber but stress at break and strain at break were decreased. Also it was found that the stiffness of the composites is not effected with the injection number and this behavior were noted the work of Pillin et al. [7] and it was attributed to the to the increase of crystallinity in PLA after processing cycles. The stess at break and strain at break of the composites were decreased after injection cycles. This drop may be caused by fibre damage during recycling, the reduction in fibre length results in more strain concentrations and a higher risk of debonding. Also the reduction of molecular weight during the recycling process causes the decrease of stress at break and strain at break. The rheological properties showed that the viscosity of the composites is more that of pure PLA and the recycling caused in decreasing the viscosity of pure PLA and the composites, this behavior was attributed to the reduction of the molecular weights in the composites.

Tsuneizumi et al. [8] studied the chemical recycling of PLA based polymer blends, the studied blends were polylactic acid/polyethylene (PLA/PE) and polylactic acid/poly(butylenesuccinate) (PLA/PBS). Two routes to selective chemical recycling of PLLA/PE blend were tested:

- ❖ Direct separation of PLLA and PE first by their different solubilities in toluene, followed by the chemical recycling of PLLA using montmorillonite.
- ❖ The selective degradation of PLLA in the PLLA/PE blend by montmorillonite in a toluene solution at 100 °C forming the lactic acid oligomer (LA) with a small molecular weight. The PE remained unchanged and was quantitatively recovered by the reprecipitation method for material recycling.
- ❖ In a similar procedure, chemical recycling of PLLA/PBS blend was also carried out and compared by two routes:
- ❖ The direct separation of PLLA and PBS by solubility in toluene.
- ❖ The sequential degradation of PLLA/PBS blend using a lipase first to degrade PBS into cyclic oligomer, which was then repolymerized to produce a PBS. Next, PLLA was degraded into repolymerizable LA oligomer (Figure 1).

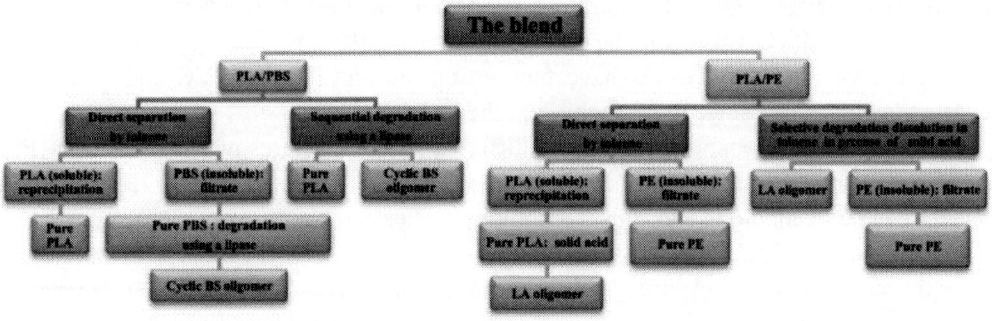

Figure 1. Diagram of chemical recycling of PLA/PE and PLA/PBS blends.

Very recently, the effect of recycling on the properties of PLA/PS polymer blend was studied [9]. A simple blend of (50/50) PLA/PS (PLA50) was prepared using a single-screw extruder and the granules of PLA50 were injected into dog bone-shaped samples and the samples were processed again. Solution viscosity, mechanical properties and rheological properties of the samples were determined.

The results showed that intrinsic viscosity of the samples, which is directly related to the molecular weight, decreases after each processing cycle. Also it was found that the viscosity decreases steadily with increasing the processing number. The compounded blend (PLA50) had a viscosity of 3100 Pa.s, and after each processing cycle, the viscosity of the blend decreased nearly by a factor of 0.15-0.3, while it was found in another study [7] that the zero viscosity of pure PLA decreased by a factor of 0.82 just after one processing cycle, and this difference was attributed to the good thermal stability of PS comparing with PLA. These observations were attributed to the reduction of the molecular weights with the processing cycles. By comparing, the flow activation energy of the samples increased with processing number, also this behavior was attributed to the reduction of the molecular weight with the processing cycle. The relationship between the flow activation energy and molecular weight was reported by Collins and Metzger [10], where they found that as the molecular weight of the polymer increases the influence of the temperature on the viscosity (flow activation energy) decreases.

The mechanical results showed that stress at break and strain at break of the sample decreased sharply after each processing cycle, these results was also attributed to a reduction of the molecular weights after the processing cycles which causes a lower cohesion in the blend. The same behavior was noted in Young's modulus of the sample but in this case the reduction of the mechanical property (Young's modulus) was not sharp after each processing cycle.

Polyvinyl Chloride (PVC) and Its Systems

The low cost and high performance of polyvinylchloride (PVC) products, combined with the wide range of properties that can be obtained from different formulations has contributed to the widespread use of PVC in construction products. There has been a long time-lag between PVC consumption and the amassing of PVC waste arising from the long life of PVC products, which can be up to 50 years. It is obvious that all the PVC that is being produced will become waste some day. The European Association of Plastics Converters (EuPC) have estimated that the PVC waste for the periods between 2010 and 2020 will arise from the following sources. So many works have focused on the recycling of PVC and its systems (blends and composites) since the beginning of the last decade .

Lee and Shin [11] separated PVC from different plastic mixtures with polystyrene (PS), polyethylene terephthalate (PET), polyethylene (PE) and polypropylene (PP) using a triboelectrostatic technology. In this technology, negative and positive charges can be imparted to the particles, of the two polymers in a mixture, then they can be separated by passing through an external electric field. The kind of the charges on the polymer depends on the ability of polymer to the electron loses or gains. The material with a higher affinity for electrons gains electrons will charge negatively, while polymer with the lower affinity loses electrons will charge positively. Figure 2 shows the triboelectrostatic charging sequence of various plastics [12]. For the removal of PVC from the two-component mixed plastics such as PVC/PET, PVC/PP, PVC/PE or PVC/PS, separation results show a recovery of 96-99% with the pure extract content in excess of 90%.

Yarahmadi et al. [13] studied the recyclability of PVC sheets which are used in building floors application. The study has shown that PVC floorings as plastic waste can be mechanically recycled without the addition of new plasticizers.

Augier et al. [14] studied the effect of the wood fibre filler on the internal recycling of PVC based composites. For studying the effect of wood fibre content on the recycling of PVC, the recycling of PVC and wood fibre-reinforced PVC was investigated and compared. The effect of the recycling on the mechanical properties of both PVC and its composites was investigated. The results showed that the addition of wood fiber to PVC improves its recyclability where it was found that up to five cycles, the composite properties remained stable. However, after 10 cycles and especially at 20 cycles, the flexural strength increased, whereas the other mechanical properties remained almost constant. In general it was to say that It is possible to recycle the composite internal waste five times, without adding fresh material, where no significant change appeared until five cycles.

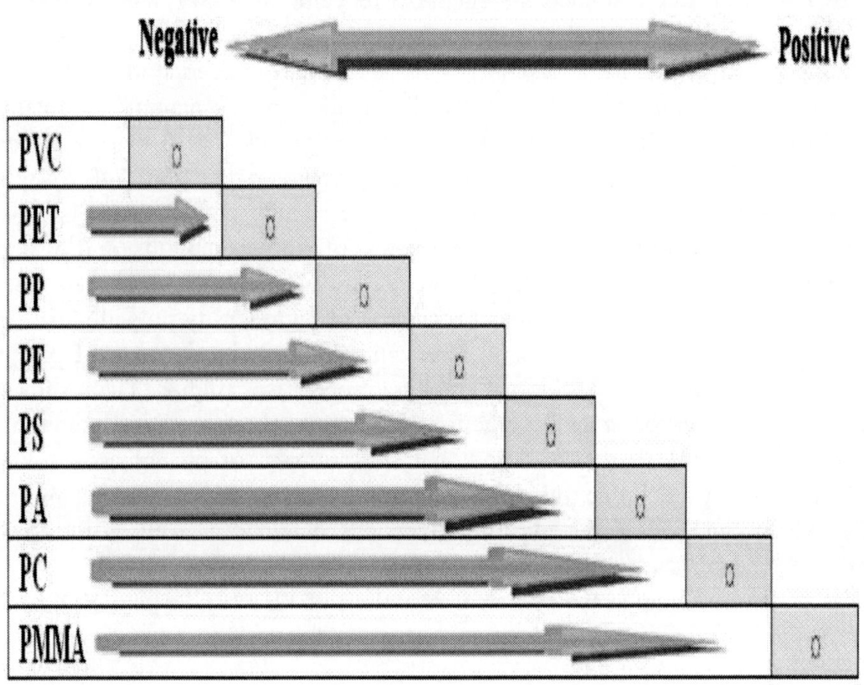

Figure 2 Triboelectrostatic charging sequence of various polymers.

Very recently Petchwattana et al. [15] studied the recycling of PVC/wood composites (wood – plastics composites (WPC)). The effect of recycling on the mechanical and structure properties of the composites. The waste composites were mixed with raw composites and re-processed for seven times. The results showed that the molecular weights of the PVC decrease due to the molecular chain scission during re-processing. Mechanical test results demonstrated that the composites could be recycled as WPC again without critically affecting its mechanical performance, where the impact strength of the composite remained rather constant and closes in value to that of the composite from freshfeed, and the flexural strength of the composite decreased slightly with re-processing cycles.

Polyethylene's (PE's) and Their Systems

Polyethylene (PE) is one of the most widely used plastics. It is characterized by a density in the range 0.918-0.965 g/cm^3 and is very tough and flexible. Its major application is in packaging film although its outstanding dielectric properties means it is also widely used as an electrical insulator. Other applications include domestic ware, tubing, squeeze bottles and cold water tanks.

Puente et al. [16] studied the chemical recycling of LDPE using fluid catalytic cracking (FCC) method at 500 °C, different commercial catalyses were used. The process was done in a solution of LDPE using toluene as a solvent. Results were qualitatively similar in all the catalysts. The contribution is centered mainly in the gasoline fraction, with high aromatic content, although the production of gases is also important, with a high proportion of valuable light olefins C_3–C_4; isoparaffins C_4–C_5 are significant as well.

Hajekova and Bajus [17] studied the chemical recycling of LDPE and PP wastes using the thermal decomposition method in two steps, in the first one the polymer wastes were decomposed individually in a batch reactor at 450 °C and they converted to wax/oil products. In the second step the wax/oil products were dissolved in heavy naphtha to obtain steam cracking feedstock. The selectivity and kinetics of copyrolysis for 10 mass% solutions of wax/oil from LDPE or PP with naphtha in the temperature range from 740 to 820 °C at residence times from 0.09 to 0.54 s were studied. The results showed that it is possible to perform polyalkenes recycling via the copyrolysis of polyalkene oils and waxes with conventional liquid steam cracking feedstocks on already existing industrial ethylene units.

Vicente et al. [18] studied the chemical recycling of high density polyethylene (HDPE). Thermal cracking method was used for the recycling process in presence of phenol as a solvent, the effect of phenol on the thermal cracking process was studied. The results showed that the main products in the cracking reaction are olefins which are very important in the petrochemical industry. Also it was found that the presence of phenol as a solvent can promote the cracking reaction due to its role in favoring random scissions and chain reactions (Figure 3), which can explain the plastic conversions obtained as well as the yields and selectivity's of each hydrocarbon product.

Achilias et al. [19] studied the chemical recycling of PE's (LDPE and HDPE) and PP which are obtained from different application such as packaging film, bags, pipes, food-retail outlets. Tow techniques were used:

- Dissolution/reprecipitation method using different solvents and non-solvents.
- Catalytic pyrolysis using fluid catalytic cracking process (FCC).

In the first technique the polymers were dissolved in xylene and reprecipited using *n*-hexane, the pure polymers were then dried. In the second one the polymers were thermal cracked in presence of FCC catalysts. The pure polymers obtained from the first technique were mechanically tasted and compared with the mechanical properties of the wastes. The products of the second technique were characterized using spectroscopes methods. The first leads to high recovery of polymer with the disadvantage of using large amounts of organic solvents. From the measurements of the tensile mechanical properties of samples before and after recycling, it was found that the pure polymers are almost identical to the virgin polymers. Furthermore, pyrolysis was investigated as a promising technique for thermochemical recycling of these polymers where a series of alkanes and alkenes of different carbon number, which can be used in petrochemical industry, could be recovered using pyrolysis method.

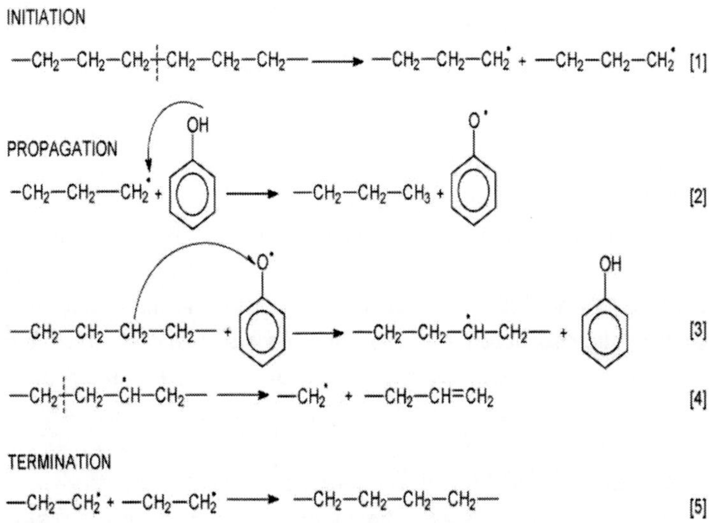

Figure 3. The role of phenol on the thermal cracking of HDPE.

Vallim et al. [20] used recycled HDPE for preparing HDPE/PA6 blend. The blends were prepared using twin-screw extruder. Morphological, mechanical and thermal properties of the prepared blends were studied. The results showed that the mechanical properties and the thermal stability of the blend improve by the using of recycled HDPE which attributed to the decrease of size domains of the recycled HDPE.

Achilias et al. [21] used dissolution–reprecipitation technique to recycle polymers from waste plastic packaging used in different applications (PE, PP, PET, PVC). FTIR was used for characterizing the recovered polymer. The mechanical properties of the polymers were compared before and after recycling process. Very good polymer recoveries were obtained in almost all waste samples examined, while lower values in some samples were attributed to the removal of additives present in the original waste products. The result showed that the mechanical properties of the polymers changed slightly after the recycling process (decreasing and increasing).

Recently, Wei et al. [22] studied the chemical recycling of plastics waste compounded from PP, LDPE and HDPE using fluid catalytic cracking process (FCC), the effect of different catalysts on the yield and selectivity was compared. It was found that the zeolitic catalysts give a higher yield comparing with non- zeolitic catalysts in case of volatile hydrocarbons, different types of catalysts and its yield and selectivity at 360 °C are shown in Table. 1.

Polypropylene (PP) and Its Systems

Aurrekoetxea et al. [23] studied the effect of the recycling on the morphology and properties of PP. In this work, PP samples were injected several times and the effect of the multiprocessing on the properties of PP were investigated. The results showed that the melt viscosity of PP decreases after processing which was attributed to the decreasing the molecular weight of PP after processing. Also it was found that the recycled PP exhibited greater crystallisation rate, higher crystallinity and equilibrium melting temperature than those measured for virgin PP. Elastic modulus and yield stress increase with the number of recycling cycles which attributed to the decrease of crystallization rate of PP after processing. However, due to decreasing of molecular weight of PP, the elongation at break and fracture toughness decrease.

Phuong et al. [24] used recycled PP in preparation of PP/organophilic modified layered silicates nanocomposites. The recycling process of PP was done using a twin – screw at different temperatures. the process was repeated 10 times. PP nanocomposites were prepared using the recycled PP. Rheological and mechanical properties of the composites were determined. The results showed that the melt flow index (MFI) increased with increasing the number of extrusion cycle, this behavior was noted in many works which focused on pure PP recycling and was attributed to the thermal degradation of PP during the extrusion. The mechanical results showed that tensile strength of the composites decreased with increasing the number of the extrusion cycle while the impact strength of the composites was improved.

Bourmaud et al. [25] studied the effect of recycling on the rigidity of PP/vegetal fiber composites. The recycling process was done in this case by multi-extrusion process. The results showed that the rigidity of the composites nearly reminded constant after the processing cycle. This behavior was attributed to good stabilization of the fibre aspect ratio after recycling.

Bahlouli et al. [26] studied the effect of recycling on the properties of PP – based composites (ethylene propylene diene monomer (EPDM)/PP and talc/PP). The recycling process was done by multi-extrusion process of the studied composites. Rheological, mechanical and structural properties of the composites were determined and compared after each extrusion process. The results showed that the melt viscosity of the composites decreases with processing number in the same way of pure PP [27], the same mechanical results were noted where the mechanical properties of the composites decrease with processing number. All noted changes in the composites properties were attributed to the changes in the structural properties of the composites during the processing cycle. The obtained results from this work were useful for optimizing the recycling process and for a better use of the recycled materials in components design.

Polystyrene (PS) and Its Systems

Lee et al. [28] used clinoptilolite as a catalysts in degradation of PS wastes. The clinoptilolites showed good catalytic activity for the degradation of PS with very high selectivity to aromatic liquids.

Brennan et al. [29] studied the effect of recycling on the properties of ABS and high impact polystyrene (HIPS) and the blend ABS/HIPS. Mechanical and thermal properties of recycled ABS and HIPS determined and compared. the results showed that in the two cases (ABS and HIPS) the effect of recycling on the glass transition temperature, tensile strength and tensile modulus were negligible, but the strains at break and impact strength were reduced considerably. The blending of ABS and HIPS had no effect on the properties, and the blend restored significant amounts of ductility, as seen by increases in the strain at break.

Arandes et al. [30] studied the thermal cracking of PS and polystyrene-butadiene (PSB) dissolved in light cycle oil (LCO), mesoporous silica was used as catalyst. The cracking of the PS/LCO blend produces high yields of styrene, whereas the cracking of the PSB/LCO blend produces a stream of products with petrochemical interest.

Elmaghor et al. [31] used Poly(ethylene-*co*-vinyl acetate) and poly(styrene-*b*-ethylene/butylenes-b-styrene) as compatibilizers in the recycling of the ternary blend PS/HDPE/PVC. The compositions were prepared using a single-screw extruder. The extrudates were subjected to gamma radiation to improve the compatibility of the blend. The results showed that the compatibilizers and the irradiation were improved the mechanical properties of the blend where impact strength of the blend was enhanced. Also high elongation at break was achieved.

Williams et al. [32] studied the chemical recycling of PS using fluid catalytic cracking process. Two catalysts were used, zeolite ZSM-5 and Y-zeolite, the effect of the temperature on the yield of the process were studied.

Table 1. Summary of the main products of LDPE/HDPE degradation at reaction temperature of 360 •C over various catalysts

Degradation results	Catalyst type				
	USY	ZSM-5	MOR	ASA	MCM-41
Yield (wt% feed)					
Gaseous	87.5	93.1	90.2	85.6	87.3
Liquid	3.7	3.3	4.3	4.7	5.6
Residue	8.8	3.6	5.5	9.7	7.1
Involatile residue	4.7	2.4	2.9	7.4	5.2
Coke	4.1	1.2	2.6	2.3	1.9

The results showed that the main product from the uncatalysed process of polystyrene was an oil consisting mostly of styrene and other aromatic hydrocarbons. The gases were found to consist of methane, ethane, ethene, propane, propene, butane and butene. In the presence of either catalyst an increase in the yield of gas and decrease in the amount of oil produced was found, but there was significant formation of carbonaceous coke on the catalyst. Increasing the temperature of the Y-zeolite catalyst and also the amount of catalyst in the catalyst bed resulted in a decrease in the yield of oil and increase in the yield of gas.

Achilias et al. [33] studied the chemical recycling of PS. Catalytic and non-catalytic pyrolysis experiments were carried out in a fixed bed reactor using either model polymer or commercial waste products as the feedstock. It was found that the pyrolysis oil fraction could be re-polymerized again to produce raw PS. However, aromatic compounds included in this fraction may act as chain transfer agents, resulting in alterations in the shape of the reaction rate curve and lowering significantly the average molecular weight and the glass transition temperature of the PS produced.

García et al. [34] recycled extruded PS wastes using dissolution–reprecipitation technique. Different solvents were used. The effect of the solvent on the chain stability of PS during the recycling process was studied. The results of this work showed that the used solvent in this work has no effect on the chain during the recycling process.

In another work, Garcia et al. [35] studied the stability of PS wastes in a dissolution–reprecipitation technique for recycling PS. Different solvents were used such as (benzene, toluene, xylene, tetrahydrofuran, chloroform…). The results showed that the action of these solvent does not produce any degradation of PS chains. The solubility of PS in the mentioned solvents at different temperatures has been investigated. The solvent can be easily recycled by distillation

Acrylonitrile–Butadiene–Styrene Copolymer (ABS) and Its Systems

Szabo et al. [36] recycled ABS and ABS/polymethylmethacrylate (PMMA) blend wastes using thermal decomposition method. The results of this work showed that the decomposition temperature of the blend is less than that of the pure ABS, but the additional products derived from ABS detain the direct feedstock recycling of the MMA monomer.

In anothore work Karahaliou and Tarantili [37] studied the stability of ABS during the repeated cycle of extrusion process where it was subjected to 5 extrusion cycles. The mechanical and rheological properties showed that ABS has good stability during the processing cycles.

Boronat et al. [38] studied the effect of multiprocessing cycle conditions (temperature and shear rate) on the properties of ABS. Two grades of ABS were injected and tested (high viscosity grade and low viscosity grade). It was found that each of the two grades show a different behavior upon reprocessing. Low viscosity grade shows a reduction of viscosity with increasing the number of processing cycles, which was attributed to the degradation of this polymer. High viscosity grade, conversely, shows an increase of melt viscosity as the number of injection molding cycles is increased.

Yeh et al. [39] used recycled ABS in preparation ABS/wood composites and compared with virgin ABS/wood composites. The composites with 50 % wood and coupling agent were prepared using a twin-screw extruder. The mechanical properties of the composites were compared. It was found that while the impact strength and ductility of the virgin and recycled polymers were significantly different, the composite properties differed only slightly from each other.

Perez et al. [40] studied the effect of reprocessing on the properties of ABS. ABS was extruded and injected 10 times, mechanical, thermal and rheological properties of ABS were determined after each processing cycle. The result showed that the mechanical properties neither the melt flow index nor the tensile strength were affected by the number of reprocessing cycles, but the impact strength was decreased slightly, so it was to say that ABS has good recyclability and it can be recycled in presence of toughness agents.

Chen et al. [41] used recycled ABS as an additive to virgin ABS. The effect of recycled ABS content on the mechanical properties of the blend was studied. The results showed that there is no obvious effect of recycled ABS content on the tensile strength, elongation at yield, flexural strength, flexural modulus, and impact strength. However, hardness and glass transition temperature of blend increase with increasing recycled ABS content in the blend. Also it was found that melting flow index increases slightly with increasing recycled ABS content.

Recently Bai et al. [42] suited the effect of reprocessing on the mechanical properties of ABS/$CaCO_3$ composites. It was found that at a low content of $CaCO_3$ (less than 10 %) the impact strength of the composites decreased with the number of processing cycle which was attributed to the thermal degradation of the rubber phase in ABS, whereas at a high content of $CaCO_3$ (higher than 15 %) the impact strength of the composites increased with the number of processing cycle.

Very recently Scaffaro et al. [43] used the recycled ABS as an additive to virgin ABS. The prepared blend was then reprocessed. The effect of recycled ABS content and of reprocessing cycles on the physical properties of the prepared blends was evaluated by mechanical tests and by rheological and thermo-mechanical measurements. The results showed that the content of recycled ABS has no effect on the viscosity of the blend. However, at all blend compositions the viscosity of the blends decreases with the number of the processing cycle. In the case of mechanical properties it was found that the tensile strength, tensile modulus, elongation at break and impact strength of the blends decrease with the number of the processing cycle and decrease slightly with increasing recycled ABS content in the blend.

Polycarbonate (PC) and Its Systems

Elmaghor et al. [44] used virgin ABS grafted maleic (ABS-g-MA) anhydride in recycling process of PC wastes. The blend was prepared using a twin-screw extruder. The results showed that mechanical properties (tensile properties and impact strength) of recycled PC greatly improved by blending with ABS-g-MA. The improvement of the toughness of PC by the addition of ABS-g-MA was attributed to the reaction between MA functional group in ABS-g-MA and the end hydroxyl groups of PC.

Fraisse et al. [45] prepared a blend of recycled PC with recycled polyethylene terephthalate (PET). The blends in different ratio were prepared using twin-screw extruder at 270 °C. The obtained granules of the prepared blends were molded using two different methods, compression molding and injection molding, thermal, mechanical and rheological properties of the blends were determined and compared. The recycling process is shown in Figure 4.

It was found that the blend (PC/PET) shows a good mechanical properties which attributed to the interfacial reaction between the blend components (transesterification between PET and PC wastes occurs during blending in the molten state), which caused in good compatibility in the blend. The compatibility between the blend components was detected using the Tg's value.

In another work Maria and Sanchez [46] studied the recyclability of PC/butylene terephthalate (PBT) blend. The effect of the recycling on the mechanical properties of the blend was studied. The result showed that the elongation at break of the recycled blend was very good and both tensile strength and tensile modulus were effected with the recycling process so it was to say that the PC/PBT blend has good recyclability.

Very recently [47], the chemical recycling of PC waste to coproduce bisphenol A (BPA) and carbohydrate carbonates was studied by selecting glycerol (Gly) and glucose (Glu) as model carbohydrates (CHs).

Polyethylene Terephthalate (PET) and Its System

Pingale and Shukla [48] used the microwave irradiation in the chemical recycling of PET wastes in ethylene glycol (EG), zinc acetate was used as catalysts. The yield of the main product (bis(2-hydroxyethyl) terephthalate (BHET)) was nearly same as that obtained earlier by conventional electric heating [49].

However, the time taken for completion of reaction was reduced drastically from 8 h to 35 min. This has led to substantial saving in energy.

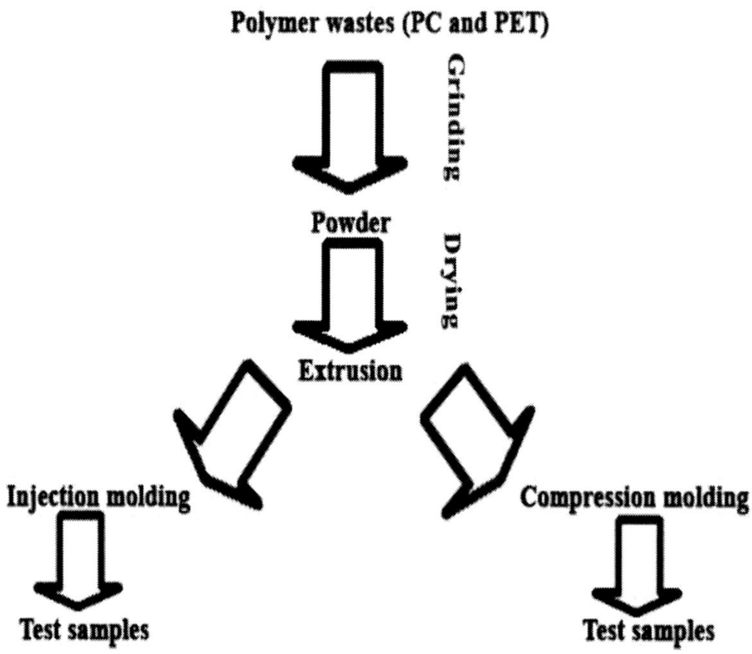

Figure 4 The recycling process of PC/PET blend.

Navarro et al. [50] recycled PET by the addition of virgin HDPE. The blends in different ratio were prepared using extrusion process. The prepared blends were then injected to obtain the test samples.

Thermal and rheological properties of the prepared blends were determined. It was found that the presence of HDPE in the blend decreases the melt viscosity of the blend which means good flow ability of the blend comparing with recycled PET. However, incompatibility between HDPE and PET in the blend was detected at HDPE content higher than 5 % which causes in poor mechanical properties comparing with PET. The results of this work indicated that it will be possible to modify PET wastes using a small amount of virgin HDPE less than 5 %.

Shukla et al. [51] studied the chemical recycling of PET wastes, ethylene glycol and sodium sulfate were used in the recycling process to obtain bis(2-hydroxyethyl) terephthalate (BHET) as a main product of the reaction. BHET was used to prepare hydrophobic disperse dyes for synthetic textiles.

Fonseca et al. [52] used ethylene glycol (EG) in presence of metal salts (zinc acetate, sodium carbonate, sodium bicarbonate, sodium sulphate and potassium sulphate) for PET recycling (Figure 5). The reaction was done at different temperatures. The main product of the reaction was bis(2-hydroxyethyl) terephthalate (BHET). The highest yields of the reaction were obtained with zinc acetate and sodium carbonate as depolymerization catalysts at 196 °C.

Shi et al. [53] used PET wastes in preparing polymer composites with ABS and SiO_2. The composites were prepared using a single screw extruder. SiO_2 was used in this composites to the reduce the interfacial tensions between PET and ABS. The results of this study showed that the mechanical properties of composites improve with increasing the SiO_2 content in the composite.

Achilias et al. [54] depolymerized PET wastes in presence of diethylene glycol (DEG) (glycolysis) under microwave irradiation. The reaction was carried out in a sealed microwave reactor in which the pressure and temperature were controlled. The depolymerization products were characterized using FTIR. The results were compared with that obtained without using microwave irradiation. It was found that the complete degradation of PET can be done at temperature higher than 180 °C. Also the results showed that in the normal condition (without irradiation), the reaction needs more than 4 h to complete the PET degradation, which confirms the importance of the microwave power technique and the substantial energy saving achieved.

Also in another work done by Achilias et al. [55] PET wastes were depolymerized in presence of ethanolamine (aminolytic) under microwave irradiation to enhanced the wastes degradation. The reaction products were characterized using FTIR. The main product of the reaction was bis(2-hydroxyethyl) terephthalamide. The results showed that the complete degradation of PET wastes was done at temperature higher than 260 °C. By comparing with the previous work [54] it could be said the presence of diethylene glycol caused in complete degradation of PET at lower temperature comparing the presence of ethanolamine.

In another work, Fonseca et al. [56] studied Kinetics of catalytic glycolysis of PET wastes. The catalyst used in this reaction was sodium carbonate in presence of ethylene glycol (EG) where the authors found in their previous work [52] that sodium carbonate gives the highest yield when it uses as a catalyst.

Very recently Mantia et al. [57] modify PET wastes by the addition of small amounts of virgin PLA. The blend was prepared using melt mixing technology. The effect of the small amounts of PLA on the mechanical and rheological properties of the prepared blend. The results showed that the viscosity of the blend is less than that of PET, but the blend was more thermal sensitivity comparing with PET. The mechanical results showed that the small amount additives of PLA do not affect on the tensile properties, where the mechanical properties of the blend are similar to those of PET.

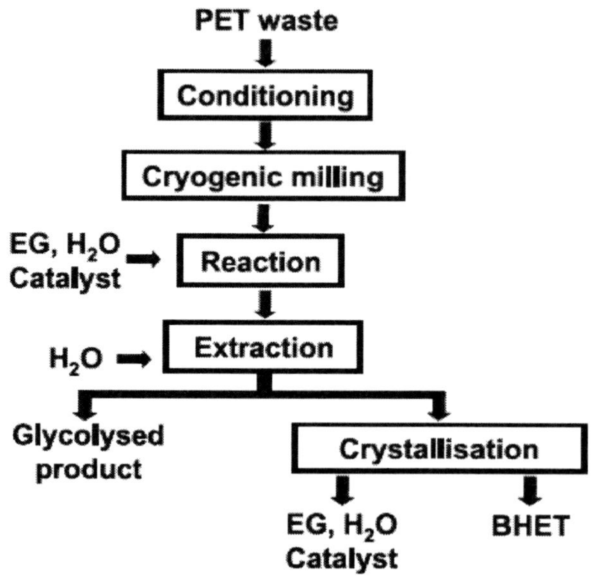

Figure 5 The reaction and analytical procedure for the glycolysis of PET wastes.

Polyamides (PA's) and Their Systems

Su et al. [58] studied the effect of the reprocessing on the mechanical and rheological properties of PA6. PA6 sample was processed 16 times and its properties were compared with of the virgin PA6. The results showed a reduction in the molecular weight and an increase in the molecular weight distribution as a consequence of a decrement in melt viscosity of PA6, but no chemical changes were noted in PA6 structure during the recycling process. The mechanical test showed that the tensile properties increases after each processing cycle while the impact strength decreases.

Bernasconi et al. [59] blended a recycled short glass fibre reinforced PA66 with virgin materials and studied the effect of the recycled materials content on the tensile strength of the blend. Tensile test results showed a decrease of both elastic modulus and tensile strength, whereas the strain at break increased, for increasing content of reprocessed material.

Goitisolo et al. [60] studied the effect of reprocessing on the properties of PA6 nanocomposites. The composite was reprocessed using injection molding the process was repeated 5 times. The properties of the composite was determined after each processing cycle. No chemical changes were observed in the composite during the processing cycles the same results were obtained in pure PA6 recycling [61], but the viscosity of the composite decreased with the number of processing cycle which was attributed to the decrease of the molecular weight. Also it was found that the strain at break of the composite decreased with the number of processing cycle which means less ductility.

Hassan et al. [62] studied mechanical, thermal and morphological properties of irradiated recycled PA and waste rubber powder blends. The results of this blend were compared with that of the non-irradiated blend. The results showed that the incorporation of rubber waste into recycled PA caused in decrease of the mechanical performance because of a too weak interfacial adhesion. However, the irradiation process caused in improve the compatibility of the prepared blend. In another work to Hassan et al. [63] the effect of carbon black (CB) content on the properties of irradiated recycled PA and waste rubber blend. The results showed that the tensile strength, elongation at break and Young's modulus of the blend increase with increasing CB content. Also El-Nemr et al. [64] studied the effect of acrylonitrile butadiene rubber (NBR) content on the properties of irradiated recycled PA6/66 and waste rubber blend. The results showed that the mechanical properties improve by the addition NBR.

Nemade et al. [65] studied the chemical recycling of PA in water at different temperatures and pressure. The effect of the temperature and pressure on the yield of the depolymerization reaction was studied. The results showed that the high yield of the reaction can be obtained at 240 °C and 700 psi.

Dorigato and Fambri [66] used recycled short fiber of PA66 in the reinforcing process of the commercial grade of PA12, the effect of PA66 content on the thermal and mechanical properties was studied. It was found from this study that *Tg* of PA12 increases with increasing the fiber content. Mechanical properties and the morphology of the material indicated good interfacial adhesion between the components.

REFERENCES

[1] J. Cui and E. Forssberg, J. Hazar. Mater B. 99, 243 (2003).

[2] F. Sasse and G. Emig, *Chem. Eng. Tech.* 21, 777 (1998).

[3] G. Stoclet, R. Seguela and J. M. Lefebvre, *Polymer*. 52, 1417 (2011).

[4] K. Hamad, M. Kaseem and F. Deri, *J. Polym. Res.* DOI 10.1007/s10965-011-9586-6 (2011).

[5] M. A. Huneault and H. Li, *J. Appl. Polym. Sci.* 119, 2439 (2011).

[6] Duigou, I. Pillin, A. Bourmaud, P. Davies and C. Baley, *Composites: Part A*. 39, 1471 (2008).

[7] Pillin, N. Montrelay, A. Bourmaud and Y. Grohens, *Polym. Degrad. Stab.* 93, 321(2008).Y.

[8] Tsuneizumi, M. Kuwahara, K. Okamoto and S. Matsumura, *Polym. Degrad. Stab.* 95, 1387 (2010).

[9] K. Hamad, M. Kaseem and F. Deri, *J. Mater. Sci.* 46, 3013 (2011).

[10] E. A. Collins and A. P. Metzger, *Polym. Eng. Sci.* 10, 57 (1970).

[11] J. K. Lee and J. H Shin, *Korean. J. Chem. Eng.* 19, 267 (2002).

[12] J. Brandrup, M. Bittner, and G. Menges, "Recycling and Recovery of Plastics" Hanser Publishers (1996).

[13] N. Yarahmadi, I. Jakubowicz and L. Martinsson, *Polym. Degrad. Stab.* 79, 439 (2003).

[14] L. Augier, G. Sperone, C. Vaca-Garcia and M. Borredon, *Polym. Degrad. Stab.* 92, 1169 (2007).

[15] N. Petchwattana, S. Covavisaruch and J. Sanetuntikul, *Constr. Build. Mater*, 28, 557 (2012).

[16] G. de la Puente, C. Klocker and U. Sedran, *Appl. Cata B: Environ.* 36, 279 (2002)

[17] E. Hajekova and M. Bajus, *J. Anal. Appl. Pyrolysis*. 74, 270 (2005).

[18] G. Vicente, J. Aguado, D.P. Serrano and N. Sanchez, *J. Anal. Appl. Pyrolysis*. 85, 366 (2009).

[19] D.S. Achilias, C. Roupakias, P. Megalokonomos, A.A. Lappas and E.V. Antonakou, *J. Hazard. Mater.* 149, 536 (2007).

[20] M. Vallim, J. Araujo, M. Spinace and M. Paoli, *Polym. Sci. Eng*. 49, 2005 (2009).

[21] D. S. Achilias, A. Giannoulis and G. Z. Papageorgiou, *Polym. Bull.* 63, 449 (2009).

[22] T. Weia, K. Wua, S. Leeb and Y. Lina, *Resou, Conser, Recyc*. 54, 952 (2010).

[23] J. Aurrekoetxea , M. A. Sarrionandia, and I. Urrutibeascoa, *J. Mater. Sci*. 36, 2607 (2001).

[24] N. Phuong and V. Gilbert, *J. Rein. Plast. Compos.* 27, 1983 (2008).

[25] Bourmaud, and C. Baley, *Polym. Degrad. Stab.* 94, 297 (2009).

[26] N. Bahlouli, D. Pessey, C. Raveyre, J. Guillet, S. Ahzi, A. Dahoun and J. Hiver, *Mater. Desig.* 33, 451 (2012).

[27] Z. Gao, T. Kaneko, I. Amasaki and M. Nakada, *Polym. Degrad. Stab.* 80, 269 (2003).

[28] S. Lee, J. Yoon, J. Kim and D. Park, *J. Anal. Appl. Pyrolysis.* 64, 71 (2002).

[29] L. B. Brennan, D. H. Isaac and J. C. Arnold, *J. Appl. Polym. Sci.* 86, 572 (2002).

[30] J. Arandes, J. Erena, M. Azkoiti, M. Olazar and J. Bilbao, *J. Anal. Appl. Pyrolysis*. 70, 747 (2003).

[31] F. Elmaghor, L. Zhang and H. Li, *J. Appl. Polym. Sci*. 88, 2756(2003).

[32] P. Williams and R. Bagri, *Inter. J. Ener. Res*. 28, 31 (2004).

[33] D. Achilias, I. Kanellopoulou, P. Megalokonomos, E. Antonakou, and A. Lappas, *Macro. Mater. Eng*. 292, 923 (2007).

[34] Gracia, M. Garca, G. Duque, A. Lucas, and J. Rodrguez, *J. Mater. Sci.* 11, 2 (20050.

[35] M. Garca, I. Gracia, G. Duque, A. Lucas, and J. Rodrguez, *Waste. Manage.* 29, 1814 (2009).

[36] E. Szabo, M. Olah, F. Ronkay, N. Miskolczi and M. Blazso, *J. Anal. Appl. Pyrolysis.* 92, 19 (2011).

[37] E. K. Karahaliou and P.A. Tarantili, *Polym. Sci. Eng.* 49, 2269 (2009).

[38] T. Boronat, V.J. Segui, M.A. Peydro and M.J. Reig, *J. Mater. Proc. Tech*. 209, 2735 (2009).

[39] S.Yeh , S. Agarwal and R. Gupta, *Compos. Sci. Tech*. 69, 2225 (2009).

[40] J. Perez, J. Vilas, J. Laza, S. Arnaiz, F. Mijangos, E. Bilbao and L. Leon, *J. Polym. Environ*. 18, 71 (2010).

[41] S. Chen, W. Liao, M. Hsieh, R. Chien and S. Lin, *Polym. Plast. Tech. Eng.* 50, 306 (2011).

[42] X. Bai, Z. Wu and N. Feng, *Advan. Mater. Res*. 455-456, 845, (2012).

[43] R. Scaffaro, L. Botta and G. Benedetto, *Eur. Polym. J.* 48, 637 (2012).

[44] F. Elmaghor, L. Zhang, R. Fan and H. Li, *Polymer*. 45, 6719 (2004).

[45] F. Fraisse, V. Verney, S. Commereuc and M. Obadal, *Polym. Degrad. Stab.* 90, 250 (2005).

[46] E. Maria and S. Sanchez, *Polym. Test*. 26, 378 (2007).

[47] K. Hidaka, Y. Iwakawa, T. Maoka, F. Tanimoto and A. Oku, *J. Mater. Cycles. Waste. Manag*. 11, 6 (2009).

[48] N.D. Pingale and S.R. Shukla, *Eur. Polym. J.* 44, 4151 (2008).

[49] S. R. Shukla and A. M. Harad, *J. Appl. Polym*. Sci. 97, 513 (2005).

[50] R. Navarro, S. Ferrandiz, J. Lopez and V.J. Segui, *J. Mater. Proc. Tech.* 195, 110 (2008).

[51] S.R. Shukla, A. Harad and L. Jawale, *Polym. Degrad. Stab.* 94, 604 (2009).

[52] R. Fonseca, I. Ingunza, B. Rivas, S. Arnaiz and J.I. Ortiz, *Polym. Degrad. Stab.* 95, 1022 (2010).

[53] G. Shi, Li. He, C. Chen, J. Liu, Q. Liu and H. Chen, *Advan. Mater. Res.* 150-151, 857 (2010).

[54] D. Achilias, H. Redhwi, M. Siddiqui, A. Nikolaidis, D. Bikiaris and G. Karayannidis, *J. Appl. Polym. Sci.* 118, 3066 (2010).

[55] D. Achilias, G. Tsintzou, A. Nikolaidis, D. Bikiaris and G. Karayannidis, *Polym. Inter.* 60, 500 (2011).

[56] R. Fonseca, I. Ingunza, B. Rivas, L. Giraldo and J. Ortiz, *Chem. Eng. J.* 168, 312 (2011).

[57] F.P. Mantia, L. Botta, M. Morreale and R. Scaffaro, *Polym. Degrad. Stab.* 97, 21 (2012).

[58] K. Su, J. Lin and C. Lin, *J. Mater. Proc. Tech.* 192–193, 532 (2007).

[59] Bernasconi, D. Rossin and C. Armanni, *Eng. Fract. Mech.* 74, 627 (2007).

[60] Goitisolo, J. Eguiazable and J. Nazabal, *Polym. Degrad. Stab.* 93, 1747 (2008).

[61] P. Eriksson, A. Albertsson, P. Boydell, K. Eriksson and J. Manson, *Polym. Compos.* 17, 823 (1996).

[62] M. Hassan, N. Badway, A. Gamal, M. Elnaggar and S. Hegazy, *Nucl. Inst. Meth. Phys. Res B.* 268, 1427 (2010).

[63] M. Hassan, N. Badway, A. Gamal, M. Elnaggar and S. Hegazy, *Nucl. Inst. Meth. Phys. Res.* 268, 2527 (2010).

[64] K. El-Nemr, M. Hassan and M. Ali, Polym. Advan. Tech. 21, 735 (2010).

[65] M. Nemade, S. Mishra and V. S. Zope, *J. Polym. Environ.* 19, 110 (2011).

[66] Dorigato and L. Fambri, *Polym. Compos.* 32, 786 (2011).

In: Landfills
Editors: G. B. C. Cabral and B. A. E. Botelho

ISBN: 978-1-62257-121-5

Chapter 7

VALORISATION OF REGENERATED LDPE BY BLENDING WITH OTHER THERMOPLASTIC POLYMERS

N. Belhaneche-Bensemra, R. Irinislimane, S. A. Kabdi, C. C. Chouiref and N. Guerfi
Laboratoire des Sciences et Techniques de l'Environnement,
Ecole Nationale Polytechnique, Alger, Algeria

ABSTRACT

Polyethylene (PE) is the highest volume thermoplastic consumed in the world, constituting the dominant component of plastics wastes. It is becoming the most recycled polymeric material. Polymer blends are becoming commercially important because they offer a property balance not found in either homopolymer. Furthermore, blending technology is used to enhance polymer waste recycling.

This chapter is intended to focus on some recent scientific research in this field. Regenerated low density polyethylene (rLDPE) coming from various neat or dirty wastes sources (bags, ustabilized agricultural films, carbon black stabilized agricultural films, milk pouches) were used. They were obtained after milling, washing and extrusion in a conventional recycling plant. They were first characterized in terms of physico- chemical (density, melt flow index, water absorption, melting temperature and structure by Fourier transform infrared spectroscopy (FTIR) and mechanical properties (tensile properties and shore D hardness)).

These rLDPE samples were recycled in the form of blends with ethylene- propylene-diene-monomer (EPDM) in the presence of peroxide, with polyvinyl chloride (PVC) in the presence of chlorinated polyethylene (CPE) as compatibizer, with polystyrene (PS) and polypropylene (PP) in presence of maleic anhydride functionalized elastomeric ethylene copolymer as comptiilizer. Blends of various compositions were realized and characterized physically (density, water absorption) and mechanically (tensile properties and shore D hardness).

The experimental data showed an improvement of mechanical properties. Thus, from the point of view of practical applications, this kind of blending is an effective way of recycling LDPE wastes.

INTRODUCTION

The use of plastics is increasing steadily in the world. Annually, a large quantity of non biodegradable industrial and domestic plastics is disposed of. Among solid wastes, plastics are increasing more rapidly especially from packaging. The quantity of municipal solid waste (MSW) is a significant problem [1]. The growing volume of plastics contained in municipal wastes and the demand for solutions for the social and environmental problems deriving therefrom is a current concern. Although plastics account only for a few parts of such wastes, since they are very visible they are subject to harsh criticism. As an alternative, incineration is an increasingly popular means of disposal. Incineration, however, may not be the optimal solution. Materials being incinerated can be have their by-products become toxic gases, airborne particulates, volatile metallic oxide fumes, and/ or toxic ash [2].

In the past few years there has been a growing interest in the recycling of polymeric materials. The driving force comes from both economic and environmental reasons [3-12]. The mechanical recycling of post- consumer plastics remains one of the preferred recycling options for ecology and energy reasons, as far as this way is economically profitable. Not all plastics are recyclable in good economic conditions. In spite of enlarging scientific research budget in the recycling domain, recycling techniques for mixed plastics are still few (at industrial scale) [13].

Polyolefins occupy the largest share of the plastics market [14] and are becoming the most recycled polymeric materials. In 2008, 46% of the plastics processed in Europe were polyolefins, corresponding to a volume of 22.3 Mt/a [15]. Low density polyethylene (LDPE) is a widely used commodity thermoplastic and it represents an important weight percentage of the polyolefins found in waste stream. Recycling and reusing of polyethylene is getting more important. Because of its simple structure polyethylene surpasses the most other materials concerning its ability to be reused [16]. However, LDPE alone is not really interesting for particular applications because its mechanical properties are influenced by the aging of the product [17].

Polymer blends are becoming commercially important because they offer a property balance not found in either homopolymer [18]. A blending technique is an efficient way for the modification of general plastics and the treatment and recycling of plastic waste. The mechanical recycling of post- consumer plastics remains one of the preferred recycling options for ecology and energy reasons, as far as this way is economically profitable. This chapter is intended to focus on some recent scientific research in this field. Regenerated low density polyethylene (rLDPE) coming from various neat or dirty wastes sources (bags, ustabilized agricultural films, carbon black stabilized agricultural films, milk pouches) were used.

They were obtained after milling, washing and extrusion in a conventional recycling plant. They were first characterized in terms of physico- chemical and mechanical properties. These rLDPE samples were recycled in the form of blends with ethylene- propylene-diene-monomer (EPDM) in the presence of peroxide, polyvinyl chloride (PVC) in the presence of chlorinated polyethylene (CPE) as compatibizer, polystyrene (PS) and polypropylene (PP) in presence of maleic anhydride functionalized elastomeric ethylene copolymer as compatibilizer.

Regenerated LDPE/EPDM Blends

It is well known that thermoplastic elastomers (TPEs) combine the elastic and mechanical properties of thermoset crosslinked rubbers with the melt processability of thermoplastics. Indeed, a particular family of TPE, called thermoplastic vulcanisates (TPVs), consists in dispersing an uncrosslinked rubber phase into a thermoplastic matrix by melt blending, and to dynamically crosslink that rubber phase in the melt [19].

As a result, the products obtained consist of crosslinked rubber particles dispersed in a thermoplastic matrix. The thermoplastic matrix explains the melt processability of TPVs, while the crosslinked elastomer particles, "glued" together by thermoplastic inter-layers, explain the elasticity. Most commercial TPVs are based on blends of polypropylene (PP) and EPDM terpolymers [19, 20]. In this work, rLDPE is blended with small quantities of EPDM in the presence of peroxide. The addition of a crosslinking agent in the blend may form semi-interpenetrating or interpenetrating networks and co-crosslinked products during the melt-mixing process [*21*]. The co-crosslinked products that are formed at the interlayer of two component polymers may act as an in situ compatibilizer. Also, the formation of semi-interpenetrating networks can enhance the mechanical properties of the matrix phase [22]. Outdoor resistance may be ameliorated too because of EPDM stability against high temperatures, oxygen and ozone, thus giving to the corresponding blends good heat, oxidation and ozone resistance [23, 24]. The special combination of properties offered by such blends opens up a number of possible applications in such sectors, as the automotive industry, the shoe industry, sanitary ware, flexible filter elements, flexible pipes and tubing and cable drums.

In the present work rLDPE is blended with small quantities of ethylene-propylene-diene monomer (EPDM). The optimization of the peroxide content required for the crosslinking of the LDPE/EPDM blends was due by measuring torque and tensile strength. Once the peroxide content was optimized, different blends were obtained by varying the EPDM content. Then they were characterized physically (density, water absorption) and mechanically (tensile properties and hardness). Finally, the blends behavior under the conjugated effect of heat and water was determined at 70°C for 7 days.

Regenerated LDPE/PVC Blends

Poly (vinyl chloride) (PVC) is among the highest volume thermoplastics consumed in the world, constituting an important component of plastics wastes. As it is well known, PVC/PE blends are immiscible, giving rise to a phase- separated system with a very low degree of adhesion between components [18]. This lack of adhesion causes poor mechanical properties which limits the practical use of these blends.

The use of a third polymer to "compatibilize" the blend and to obtain better mechanical properties is a key technology to obtain polymer blends with desirable properties. Many studies have been done on the compatibilisation of this pair of polymers [25-28]. In this work, rLDPE is blended with PVC in the presence of chlorinated polyethylene (CPE) as compatibilizer. This latter is a commonly used impact modifier of PVC and its compatibility depends upon the chlorine content and the distribution of the chlorine atoms on the PE backbone [29]. CPE with 36 % chlorine is the optimum composition for obtaining the necessary features of impact, processing and strength [30-32] and it is used in this work as the third component of rLDPE/ PVC blends.

Regenerated LDPE/PP Blends

Blending of different plastic resins has long been practiced to tailor blends for specific processing and performance requirements. Blends of PE and polypropylene (PP) are among the binary systems that have attracted much attention [33-39]. However, preparation of LDPE/PP blends is hindered by the low compatibility of this polymer pair [40, 41]. In this work, the preparation, characterization and compatibilization of rLDPE/ PP blends was carried out.

Regenerated LDPE/PS Blends

Polystyrene (PS) is one of the most widely used plastics in the world and represents an important weight percentage of the polymers found in waste stream. PS possesses high modulus and high tensile strength. However, it is brittle and easy to crack while PE shows high toughness and good impact performance even at low temperature [42]. Blending is an attractive way to produce a new toughened PS material. However, owing to the great differences in chemical structure, solubility and cristallizability, PS is incompatible with PE. Blends of these compounds consist of mixtures of phases of the two components, either dispersed or continuous.

Typically, these blends exhibit weak adhesion between the PS and PE phases which manifests as inferior mechanical properties in the final composite. Many studies have focused on compatibilization of these two polymers [42-45]. Almost all compatibilizers reported have been block copolymers consisting of at least two blocks, with one block similar in structure of chemistry to PS and another similar to PE. Thus, the diblock copolymer, as it were, ties the two phases together. The effect of compatibilization on the morphology of the blend [43, 44, 46, 47] and its mechanical properties [44, 45, 48, 49] has been studied. Compatibilization promotes the formation of an interlocking structure in PS/ PE blends, which allows more equal sharing of imposed stresses and might therefore improve the mechanical properties of the blends.

One use for recycled PE is extrusion into shapes resembling lumber. These products are used as building materials in a variety of applications, including decking, landscaping, signs and outdoor furniture. One drawback to some plastic lumber products, especially those based on PE, is their tendency to creep. One route to reduce creep consists in blending the low-creep-resistance (or high-creep-speed) PE with a lower –creep-speed plastic such as PS [48]. These blends may possess mechanical properties intermediate between PS and PE. Therefore, they may be tough, but not as brittle as PS alone. The objective of this work was to investigate the feasibility of improving the performance of two regenerated LDPE samples coming from dirty wastes as products for the manufacture of building materials by blending with PS in presence of a compatibilizer.

EXPERIMENTAL

1. Materials

Four regenerated LDPE coming from different sources (Table 1) and obtained after milling, washing and extrusion in a conventional recycling plant were used. Virgin LDPE (B24/2) and EPDM (NORDEl IP 4640) were commercial products from ENIP (Algeria) and UMAC-MIDWEST (USA), respectively. The crosslinking agent was di-tertio butyl peroxide (Luperox F40 MGEVT) from ATOFINA (France). Stearic acid (Loxiol G20) from HENKEL (Germany) was used. PVC (VICIR S 1200) was a commercial product from CIRES (Portugal). CPE with 36 % chlorine was supplied by Aldrich.

Table 1. Description of the LDPE samples used

Designation of rLDPE	Origin of the LDPE scraps	Nature of the scraps	Final aspect of the granules
R1	Stabilized agricultural films	Neat	Green
R2	Milk pouches	Dirty	Grey
R3	Bags and unstabilized agricultural films	Dirty	Yellow
R4	Carbon black stabilized agricultural films	Dirty	Black

The PVC additives were dioctyl phtalate (DOP) from SGP (Tunisia) as plasticizer, Interstable from AKZO (Italy) as lead based thermal stabilizer and stearic acid from SOGIS-SPZ (Italy) as lubricant. PS (Lacqrene 7240 from Atofina, France) and PP (Sunoco, USA) were commercial products. Maleic anhydride functionalized elastomeric ethylene copolymer (Exxelor VA 1801), a commercial product from Exxon Mobil (France) was used as a compatibilizer.

2. Characerization of rLDPE, Virgin LDPE, PVC, PP and PS

The density and water absorption were measured, respectively, according to the ISO/ R1183-1970 (F) and ISO 62-1980. The melt flow index (MFI) was measured at 190°C according to the ISO 1133-1981(F). The melting temperature (Tm) was evaluated with a Dupont TA instruments type 910 differential scanning calorimeter at a heating rate of 10°C/min. The level of oxidation (R_{ox}) was evaluated by FTIR analysis using a Shimadzu FTIR 8300 apparatus according to the following relation [50]:

$$Rox = A_{C=O} / A_{CH2} \quad (1)$$

where $A_{C=O}$ and A_{CH2} correspond to the absorbances of the carbonyl band at 1720 cm^{-1} and the methylene band at 729 cm^{-1}, respectively [50]. The tensile properties and shore D hardness were measured according to the ISO 527- 1993 and ISO 868- 1978 (F), respectively.

Tensile test was performed on a Zwick/ Roell tensile instrument at a crosshead speed of 250 mm/ min. The results of the tensile test and shore D hardness were obtained by averaging the results of five measurements.

3. Preparation of rLDPE Blends

3.1. Regenerated LDPE/EPDM Blends

The peroxide content for crosslinking was optimized as follows: for two different blends of LDPE/EPDM (98/2 and 90/10 % by weight), peroxide content was varied as follows: 0.5; 1; 1.5; 2 and 3% by weight). Thus, the optimum value of peroxide will fit to all blends where the EPDM varies from 2 to 10%. This operation was applied on the three types of rLDPE. Tensile strength and rheometer measurements were used for this optimization.

Mixing of different ingredients (LDPE, EPDM, peroxide and stearic acid) has been carried out by means of a two roll mill type A80 LESCUYER (France). First, LDPE, EPDM and stearic acid are introduced at a temperature of 110°C for 10 min. Then, peroxide is added for 5 min at the same temperature.

The obtained samples (raw samples) have been subjected to an oscillating disc rheometer (ODR100) supplied by MONSANTO according to ISO 3417 (1991) to determine the maximum torque, temperature and time of vulcanization. The resulted blends were then, reticulated and compression moulded into flat workable sheets using a temperature-controlled compression moulder type FONTIJNE (Holland) and a hardened steel plaque mould (210 × 210 × 2 mm) at 170°C for 7 min (as medium value for all blends) and a pressure of 300 kN. The specimens were cut from the obtained plates for the mechanical characterization. Water resistance was estimated according to ISO 1817-1975 using a MEMMERT type thermostated water bath. It consists on immersing LDPE/ EPDM blends specimens at 70°C during 7 days. Then, the immersed specimens are subjected to a tensile test.

3.2. Regenerated LDPE/PVC Blends

rLPDE/PVC Blends

Blends of variable composition from 0 to 100 wt % were prepared in the presence of 3 wt % of heat stabilizer, 1 wt % of lubricant and 30 wt % of plasticizer. The amounts of the three additives were added according to the amount of PVC in the blend. Melt mixing was performed at 125°C on a two-roll mill for 20 minutes. The blends were then pressed in a hydraulic press at 170°C and 300 kN for 5 minutes. The specimens were cut from the obtained plates for the mechanical characterization.

rLDPE/PVC/CPE Blends

rLDPE/ PVC with a weight proportion of 50/50 was used as the basic blend in this study. The amount of CPE for ternary blends was 5, 10, 15 and 20 wt % based on the total amount of the ternary blend. The amounts of the three PVC additives were as mentioned in the previous section. Concerning the pure materials, they were treated in the same way as the blends in order to have the same themomechanical history.

3.3. Regenerated LDPE/PS

rLDPE/PS blends blends of variable composition from 0 to 30 wt% PS were prepared. Melt mixing was performed at 170 °C on a two- roll mill for 20 minutes. The blends were then pressed in a hydraulic press at 180°C and 300 kN for 5 minutes. The specimens were cut from the obtained plates for the mechanical characterization.

rLDPE/PS/Compatibilizer Blends

rLDPE/PS with a weight proportion of 80/ 20 was used as the basic blend in this work. The amount of compatibilizer was 2.5, 5, 10, 15 and 20 wt%. Ternary blends and pure materials were treated in the same way as the binary blends in order to have the same thermo mechanical history.

3.4. Regenerated LDPE/PP Blends

rLDPE/PP Blends

rLDPE/PP blends of variable composition from 0 to 100 wt % PP were prepared by injection moulding.

rLDPE/PP/Compatibilizer Blends

rLDPE/ PP with a weight proportion of 50/50 was used as the basic blend in this work. The amount of compatibilizer was 5, 10, 15 and 20 wt %. Ternary blends and pure materials were treated in the same way as the binary blends in order to have the same thermo mechanical history.

3.5. Characterization of rLDPE Blends

The density and water absorption were measured, respectively, according to the ISO/ R1183-1970 (F) and ISO 62-1980. The melt flow index (MFI) was measured at 190°C according to the ISO 1133-1981(F). The tensile properties and shore D hardness were measured according to the ISO 527- 1993 and ISO 868- 1978 (F), respectively. Tensile test was performed on a Zwick/ Roell tensile instrument at a crosshead speed of 250 mm/ min. The results of the tensile test and shore D hardness were obtained by averaging the results of five measurements. The morphology of the blends was characterized using a Philips XP. 20 Scanning Electron Microscope.

RESULTS AND DISCUSSION

1. Preliminary Characterization of the rLDPEs

The physico- chemical properties (density, water absorption, MFI and R_{ox}) of rLDPEs, virgin LDPE, PVC, PS and PP samples are given in Table 2. The densities of the rLDPE samples are comparable to that of virgin LDPE and to literature [*5,* 51]. The plasticized PVC, PS and PP densities are also conform to literature [51]. LDPE, PS and PP are not totally waterproof [52] and the results confirm it. However, as expected, water absorption for all LDPE samples was lower than that of PVC [52]. Melt flow index increases directly with fluidity and inversely with viscosity. An increased value of MFI means a reduced molecular weight (M_w). According to results of table 2, all rLDPE samples presented lower values of MFI in comparison with virgin LDPE which means that their M_w are higher than that of virgin LDPE. This can be explained by the fact that the initial LDPE scraps were probably subjected to thermo and photo-oxidative degradation during the fabrication process or the use which leads to some crosslinking.

The flows of R1 and R3 were so long and took more than 10 min so the load was changed following the same standard and used the next load after 2.16 kg which is 5 kg. According to results of Table 2, R1 and R3 have the highest Mw followed by R2 and then by R4. It can be suggested that R3 is the more crosslinked since it comes from unstabilized agricultural films that were subjected to photo and thermo-oxidative degradation. On the other hand, R4 possesses the lowest Mw which may be due to the fact that it comes from stabilized agricultural films that have been subjected to chain scission reactions during their initial utilization. The occurrence of oxidation reactions is confirmed by FTIR analysis (R_{ox}). Hence, all rLDPE samples showed a certain level of degradation. Table 3 shows the mechanical properties of rLDPE, virgin LDPE, PVC, PS and PP. Strains at break values of rLDPEs are lower than that of virgin LDPE. This can be explained by a decrease in tensile properties of the regenerated samples. These properties that are good criteria for quality can tell us about deterioration levels of these materials. R4 possesses the lowest stress and strain at break (9.098 MPa and 450%).

Table 2. Physico- chemical properties of rLDPE, virgin LDPE, PVC, PS and PP

Sample	Density (g/cm^3)	Water Absorption at 23°C (%)	Melt flow index (g/10min)	R_{ox}
R1	0.900	0.04	0.30 (**)	0.60
R2	0.901	0.05	0.44 (*)	0.35
R3	0.905	0.07	0.40 (**)	0.40
R4	0.893	0.06	0.50 (*)	0.70
Virgin LDPE	0.902	0.06	0.70 (*)	0.00
PVC	1.270	0.13	/	/
PS	0.978	0.03	2.28**	/
PP	0.906	0.04	/	/

(*) nominal load M= 2,164 Kgf ; (**) nominal load M= 5Kgf.

Table 3. Mechanical properties of rLDPE, virgin LDPE, PVC, PS and PP

Sample	Strain at break (%)	Stress at break (MPa)	Shore D hardness
R1	800.00	14.037	53.0
R2	800.00	12.714	53.0
R3	750.00	12.762	53.0
R4	450.00	9.098	54.0
Virgin LDPE	900.00	10.690	52.0
PVC	300.00	23.174	61.0
PS	31	16.100	72.0
PP	6.50	39.780	70.0

This concords with previous results which showed that the Mw for R4 is the lowest due to chain scission reactions. R2 that comes from milk pouches and contains hardener (for its first application) showed a high stress (12.714 MPa). Concerning R1, it has the highest elongation at break (800%) and stress at break (14.034 MPa) that means the higher molecular weight as it comes from stabilized agricultural films. The shore D hardness values of the rLDPEs are relatively higher than those of virgin LDPE. This result confirms that they were subject to some reticulations during their initial utilization which leads to an increase in their superficial rigidity.

2. Regenerated LDPE Blends

2.1. Regenerated LDPE/ EPDM Blends

Peroxide Optimization for the Crosslinking of the rLDPE/EPDM Blends

Table 4 shows the rheometer results obtained with rLDPE/EPDM blends (98/2 and 90/10% by weight) in presence of various peroxide levels (0.5, 1, 1.5, 2 and 3% by weight). It can be seen that the torque increases with the amount of peroxide and there is no optimum for the blends considered.

This means that the more we add peroxide the best is the torque; 3% of peroxide is too much, and if we exceed it the torque will increase too (this was tested and proved). So this method can not give information's about the needed peroxide content (optimum). In these conditions, the optimum can't be determined but we could conclude from the rheometer results that time and temperature were 7 min and 170°C (all the blends have been subjected to torque and have crosslinked at a temperature of 170°C during 7 min). Stress at break as a function of peroxide concentration is presented in Figures 1-3 for R2/EPDM (98/2, 90/10), R3/EPDM (98/2, 90/10) and R4/EPDM (98/2, 90/10) respectively.

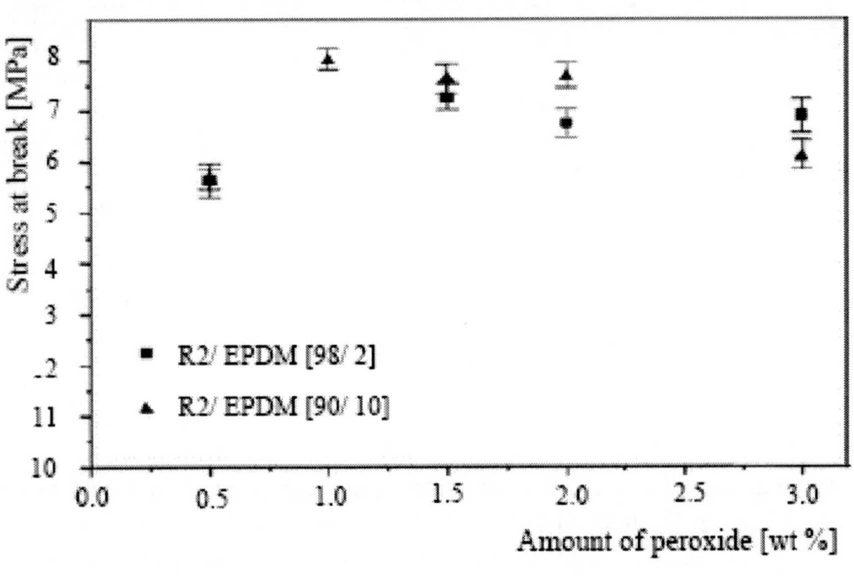

Figure 1. Variation of stress at break as a function of the level of peroxide in the case of R2.

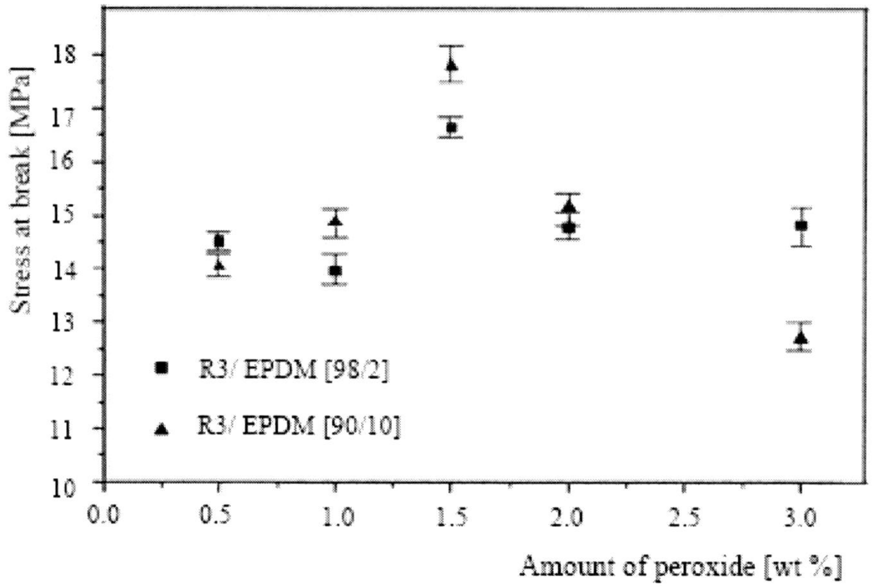

Figure 2. Variation of stress at break as a function of the level of peroxide in the case of R3.

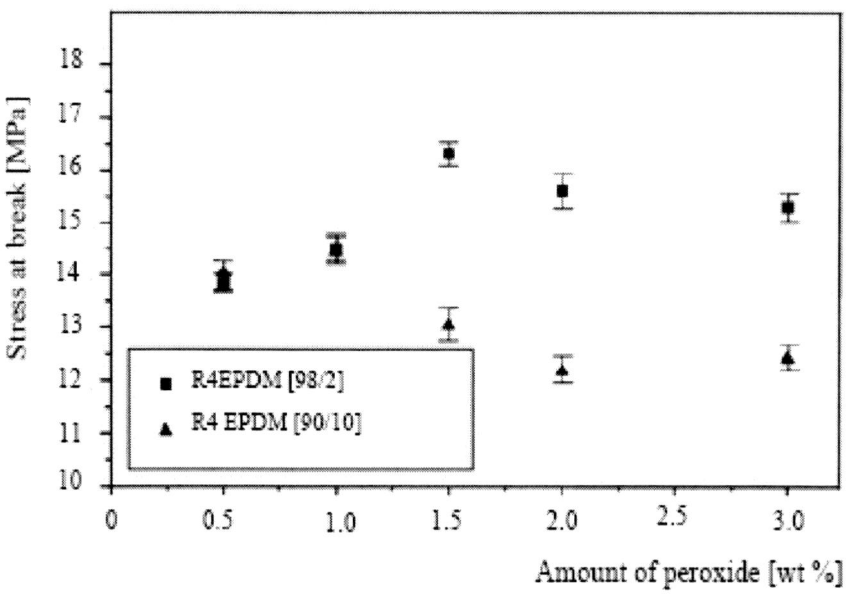

Figure 3. Variation of stress at break as a function of the level of peroxide in the case of R4.

Figure 1 shows that for the two blends considered, the optimum peroxide value is 1.0 % and will be so for all blends of R2/ EPDM in the interval [98/2, 90/10]. Figure 2 shows that for the two blends considered, the optimum peroxide value is 1.5 % and will be so for all blends of R3/EPDM in the interval (98/2, 90/10). Concerning R4, Figure 3 shows that the optimum peroxide value is 1.5% for the blend 98/2, and 1% for the blend 90/10. So for all blends R4/EPDM of composition between [98/2, 90/10], 1% is chosen as an optimum value for the two following reasons:

- At 1.5% peroxide there is a big difference between the two stress values, while at 1 % these values are superposed.
- For economic reasons.

Characterization of Crosslinked rLDPE/EPDM Blends

After peroxide content optimization, the rLDPE/EPDM blends were crosslinked and characterized. Water absorption presented in Figure 4 shows a regular decrease of the absorbed water content as a function of the EPDM added.

Table 4. Rheometer results

Peroxide [%]	Maximum torque [dNm]					
	R2/EPDM		R3/EPDM		R4/EPDM	
	98/2	90/10	98/2	90/10	98/2	90/10
0.5	16.11	16.09	11.32	13.87	11.16	11.69
1.0	21.44	22.32	18.15	20.72	18.87	21.02
1.5	29.80	33.01	27.90	31.41	25.81	28.76
2.0	35.25	42.00	34.21	40.50	32.21	39.54
3.0	46.98	57.26	48.66	60.39	44.45	54.93

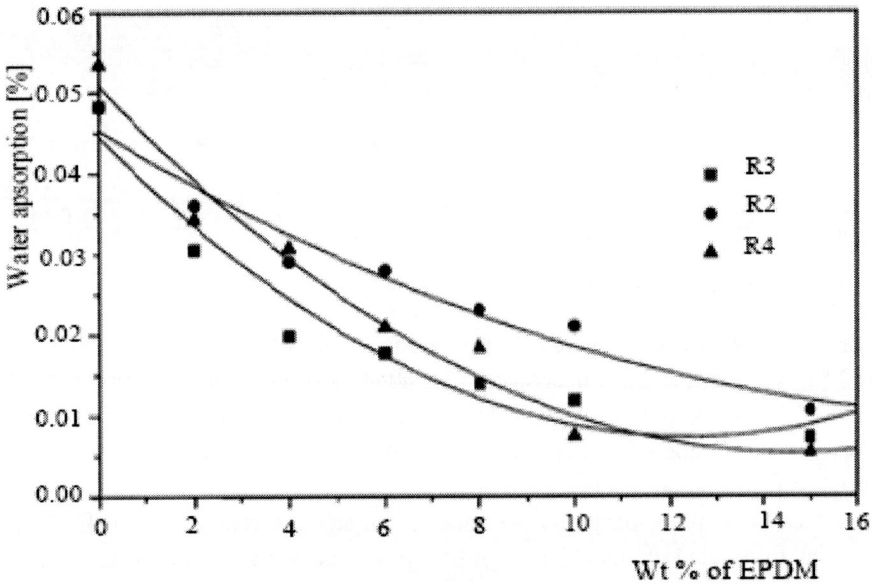

Figure 4. Variation of water absorption with blend composition.

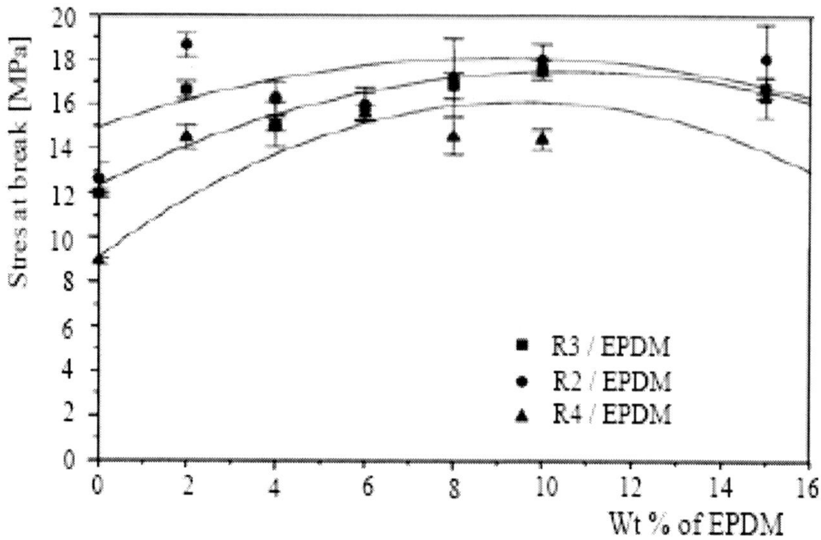

Figure 5. Variation of stress at break with blend composition.

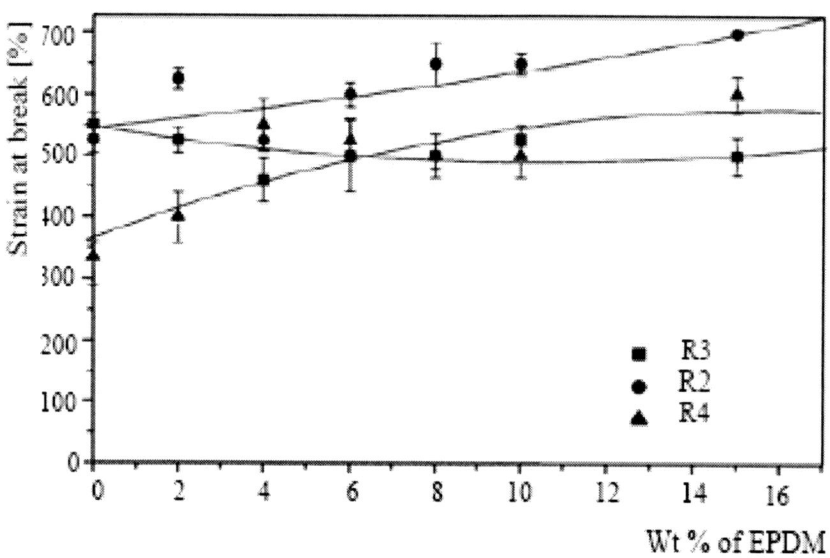

Figure 6. Variation of strain at break with blend composition.

The three rLDPEs have presented an absorption level of 0.0534 % (R4), 0.0483 % (R3) and 0.0482 % (R2) that has been decreased till 0.0054 % (R4), 0.0072 % (R3) and 0.0106 % (R2) at 15% of EPDM which shows the good water resistance of the prepared blends. These results were expected since EPDM is a well resistant polymer against water and moisture so it has imparted its resistance to the prepared blends.

Figures 5 and 6 show respectively the variation of stress and strain at break as a function of the amount of EPDM in the blends. Both stress and strain at break are considerably ameliorated. In literature, thermoplastic elastomers based on PP/EPDM present an average elongation's of 340–400% up to 600% [53].

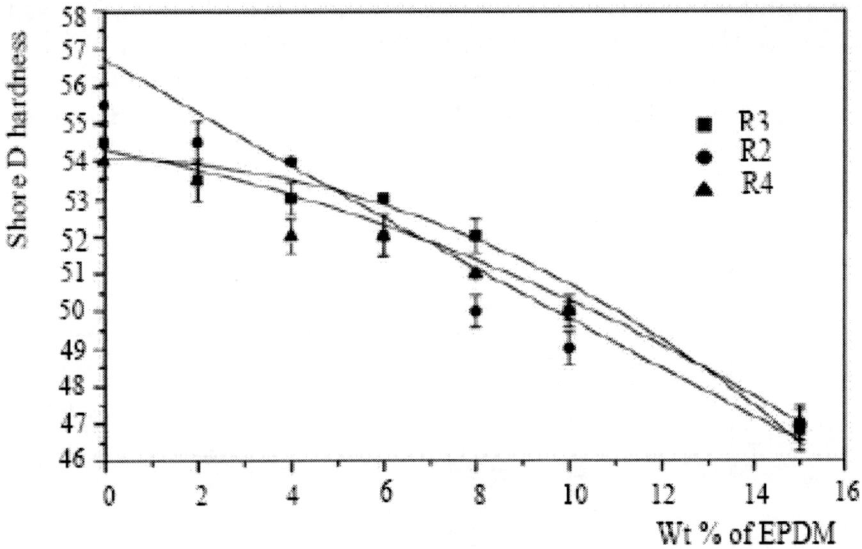

Figure 7. Variation of shore D hardness with blend composition.

Thus, the obtained values according to this valorization are similars and appreciable. The hardness shore D decreases with EPDM content (Figure 7). It can be said that LDPE has gained from EPDM elasticity since the later belongs to the elastomers family and so it has reduced hardness. So EPDM incorporation is a pertinent factor to decrease the hardness.

Effect of Heat and Water on the Crosslinked rLDPE/EPDM Blends

The realized blends represent appreciable valorization for the regenerated products especially since EPDM is incorporated in small quantities (from an economic point of view). In order to evaluate the blends resistance to outdoor conditions, humidity and heat, a water resistance test at 70°C was carried out on samples from R3/EPDM. After this test, the samples were subjected to a tensile test. The variation of stress at break is given in Table 5. It shows that the variation level of stress is 8% for R3/EPDM (100/0) and only 4.5% for the blend R3/EPDM (94/6). So the realized blends exhibit high resistance to the conjugated effect of heat and humidity.

Table 5. Variation of stress at break (%) at 70°C in distilled water

Sample	Stress at break (MPa)		
R3/EPDM (wt %)	0 day	7 days	% of variation
100/0	12.00	12.96	+08.00
94/6	16.90	17.66	+04.50

2.2. Regenerated LDPE/ Plasticized PVC Blends

Characterization of rLDPE/PVC Blends

Figure 8 shows the variation of density as a function of the amount of PVC in the blends. Since the density of plasticized PVC is higher than that of LDPE, it is obvious that the addition of plasticized PVC increases the density of rLDPE/PVC blends. The variation of density with blend composition is quasi- linear. The water absorption of rLDPE/PVC blends versus PVC content is plotted in Figure 9. It can be seen that water absorption increases with the amount of PVC until a certain threshold (50 wt % of PVC for rLDPE samples coming from dirty scraps and 75 wt % of PVC for rLDPE coming from neat scraps) and then decreases. It seems that the binary blends undergo a phenomenon of water absorption due to the immiscibility of the two polymers which favors the presence of micro-cavities and facilitates the water penetration until their saturation.

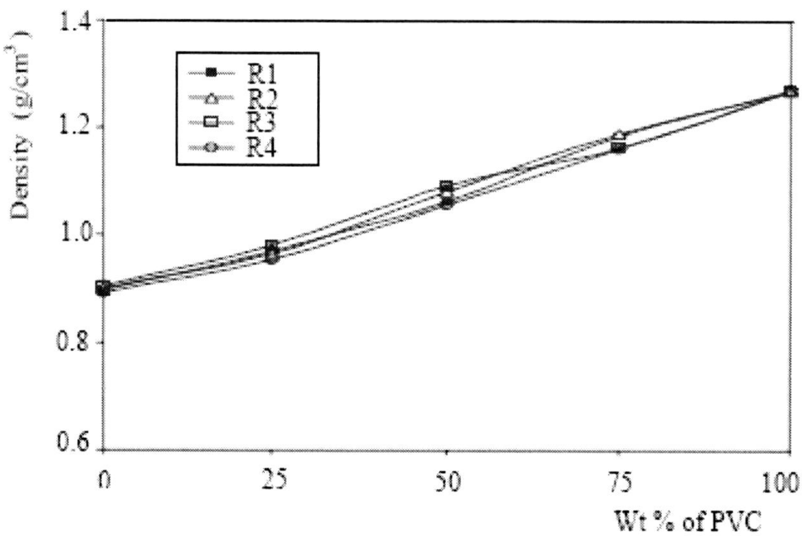

Figure 8. Variation of density with blend composition.

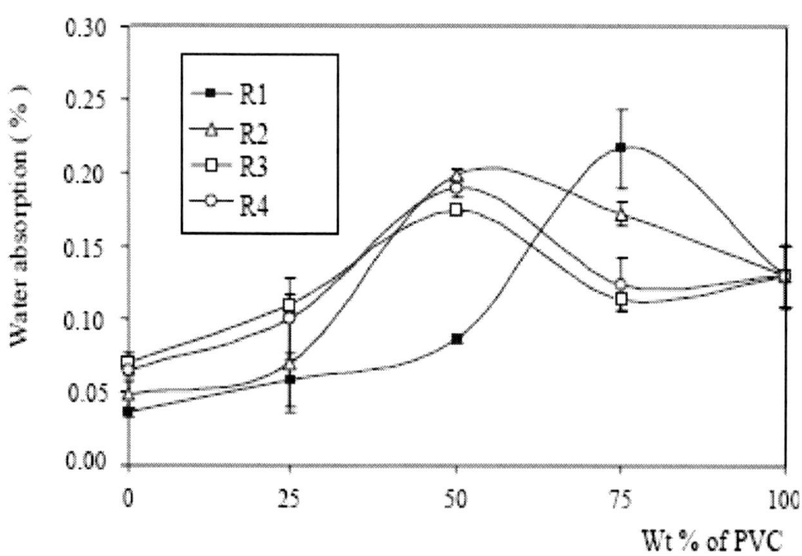

Figure 9. Variation of water absorption with blend composition.

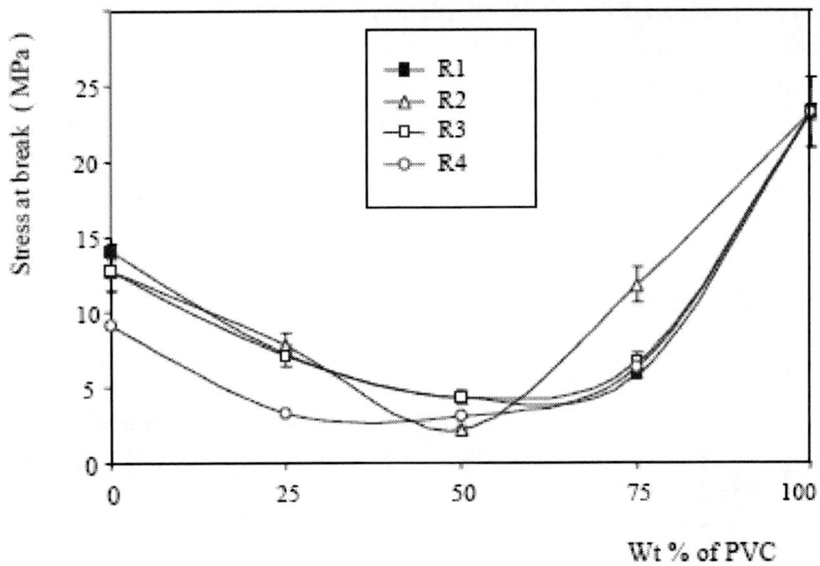

Figure 10. Variation of stress at break with blend composition.

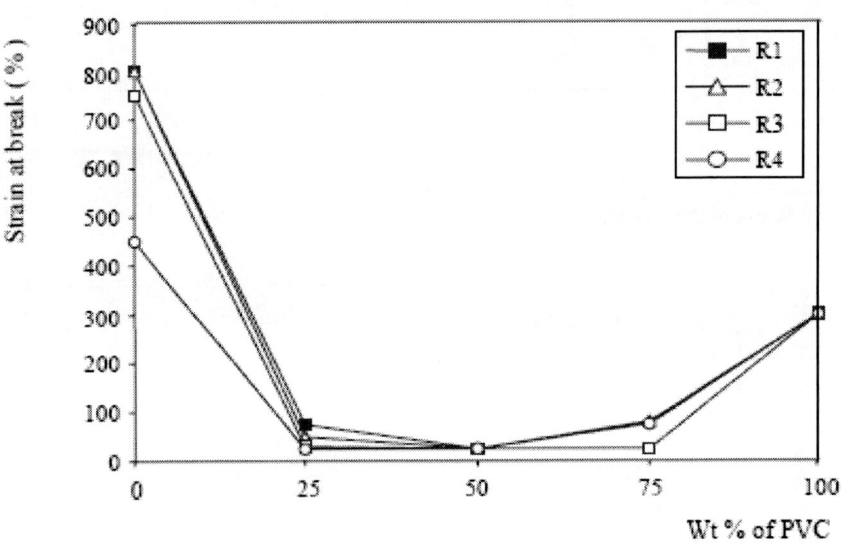

Figure 11. Variation of strain at break with blend composition.

This water penetration leads to the migration of the PVC additives (plasticizer, heat stabilizer, lubricant) outside the polymer matrix and then explains the observed decrease in water absorption [54-56]. Figures 10 and 11 show, respectively, the variation of stress and strain at break with blend composition. It can be seen that all the blends failed at lower values than that of rLDPE and plasticized PVC alone. These results confirm that the binary blends of rLDPE and plasticized PVC are typically immiscible with a poor phase structure and mechanical properties [18]. Therefore, the incorporation of a third component is necessary to enhance the compatibility of this pair of polymers.

Characterization of rLDPE/PVC/CPE Blends

The density of rLDPE/ PVC/ CPE blends versus CPE content is plotted in Figure 12. All the densities of the ternary blends are lower than that of the corresponding binary blends. Furthermore, the density decreases for a level of CPE of about 5 wt % and then increases slowly and seems to reach a plateau. These observations are related to the presence of the compatibilizer which increases the interfacial adhesion between the two incompatible polymers and then reduces the micro- cavities in the blend.

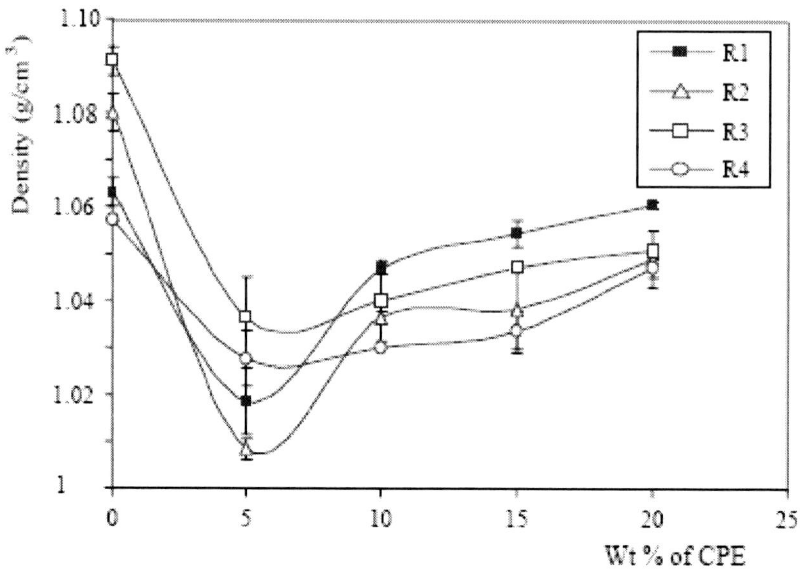

Figure 12. Variation of density with the level of CPE.

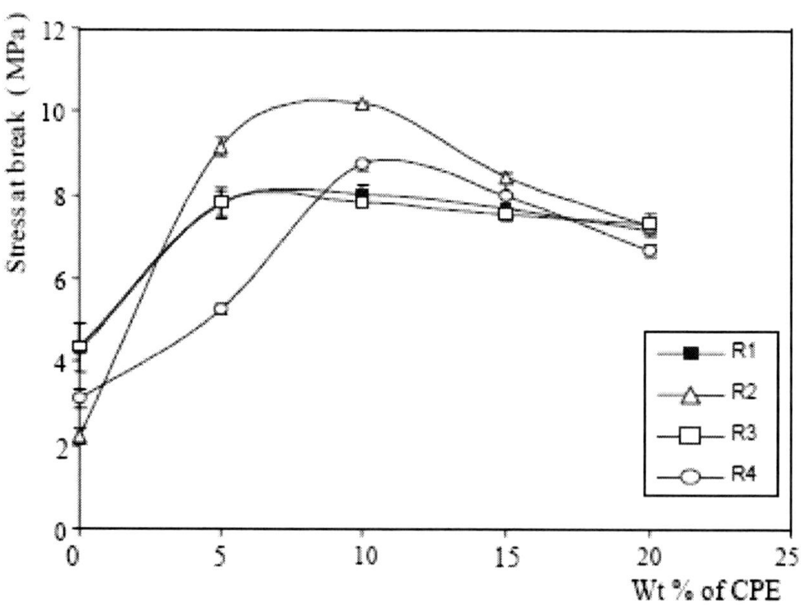

Figure 13. Variation of stress at break with the level of CPE.

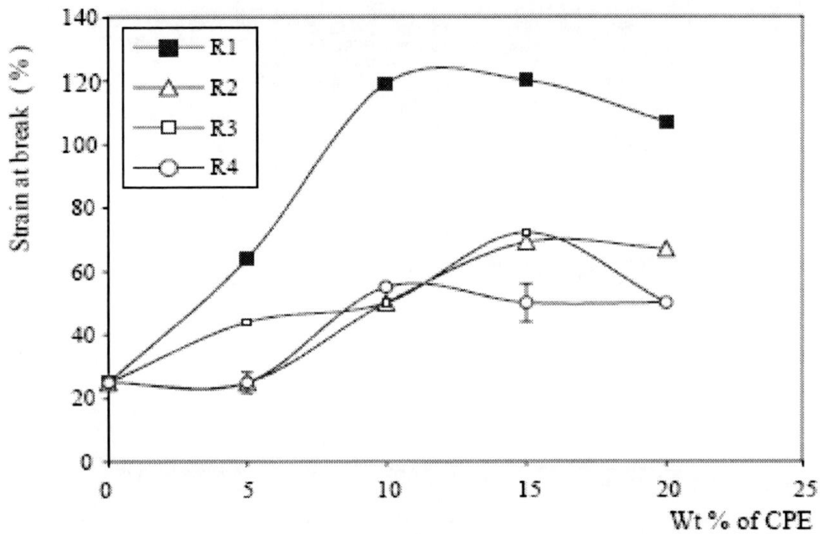

Figure 14. Variation of strain at break with the level of CPE.

Figures 13 and 14 show, respectively, the variation of stress and strain at break with CPE content. The effect of the addition of CPE on the tensile properties of 50/ 50 (wt) rLDPE / PVC blends is very noticeable in our case. It can be seen that all the ternary blends underwent broke at higher stresses and elongations than the binary blends. Both stress and strain at break increased with increasing CPE content and reached a maximum value for about 8- 15 wt % of CPE.

These results can be explained by the fact that CPE addition increases adhesion via interactions with the blend components. However, it seems that the increase of CPE level reduces the adhesion with rLDPE samples which explains the observed decrease in tensile properties after the maximum reached. Scanning electron micrographs of R1/ PVC (50/ 50), R1/ PVC/ CPE (45/ 45/ 10) and R1 / PVC/CPE (42.5/ 42.5/ 15) are shown in Figure 15. They reveal a morphological change when CPE is added to the binary blend. Indeed, as can be seen in Figure 15.a, in the case of non- modified R1 / PVC blend, the association ability of PVC and rLDPE particles is weak which leads to poor mechanical properties. However, when CPE is incorporated (Figures 15.b and 15.c) a much better dispersion is obtained with a better cohesion and a decrease of the micro- cavities size which explain the improvement of the mechanical properties.

2.3. Regenerated LDPE/PS Blends

Characterization of rLDPE/PS Blends

The variation of stress and strain at break with blends composition is given in Figures 16 and 17, respectively. It can be noted that stress at break shows a maximum for low concentrations of PS (10% for R3 and up to 5wt % for R4) and then decreases in both cases while strain at break decreases with blend composition. In both cases, strain at break reached a minimal value up to about 20wt % of PS and remained constant for higher compositions. These poor mechanical performances are related to the immiscibility of the two polymers.

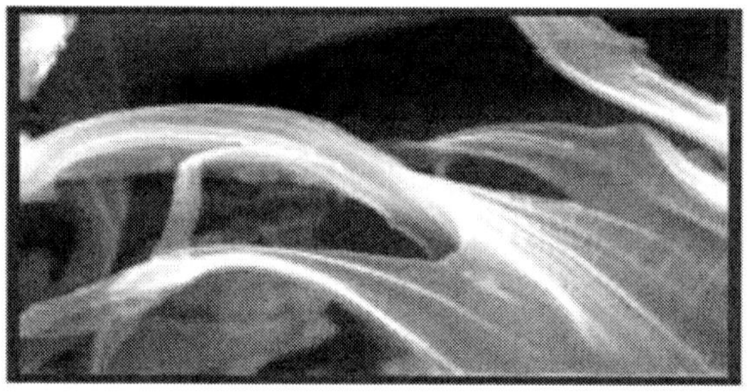

a- R1 /PVC (50/ 50)

b- R1 / PVC /CPE (45 / 45/ 10)

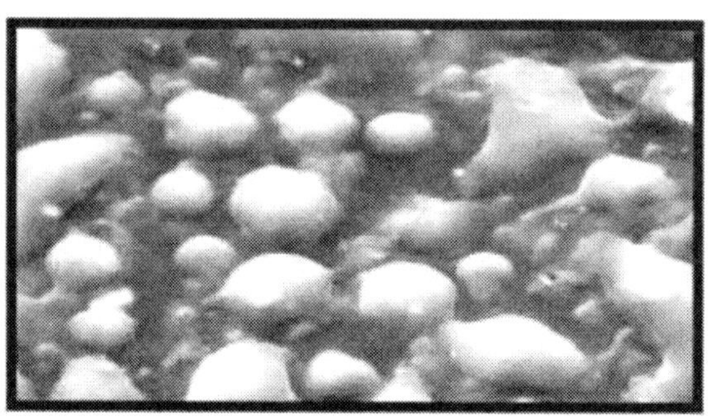

c- R1/PVC /CPE (42.5/ 42.5/ 15)

Figure 15. Scanning electron micrographs of binary and ternary blends (Grx 2500).

Therefore, the incorporation of a third component is necessary to enhance the compatibility of this pair of polymers. In order to investigate the effect of the compatibilizer, the rLDPE/PS (80/20) which exhibited the lowest tensile properties was chosen as the basic blend.

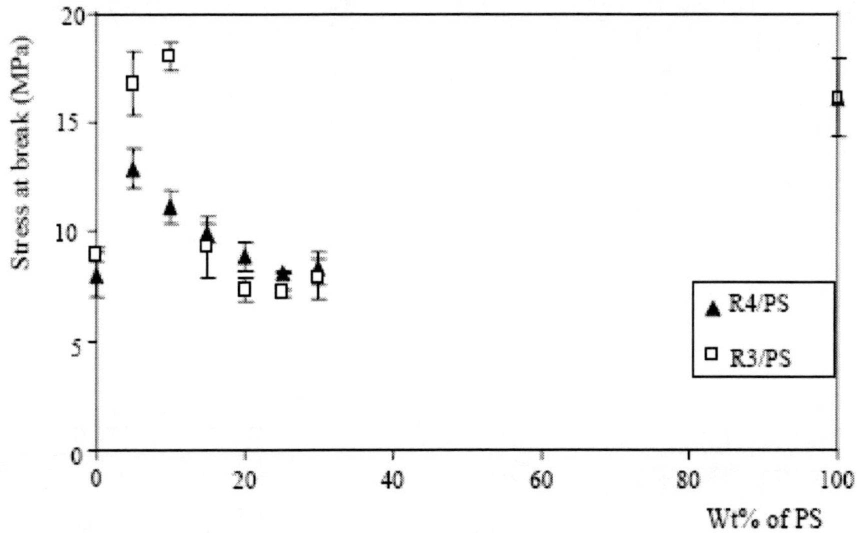

Figure 16. Variation of stress at break with blend composition.

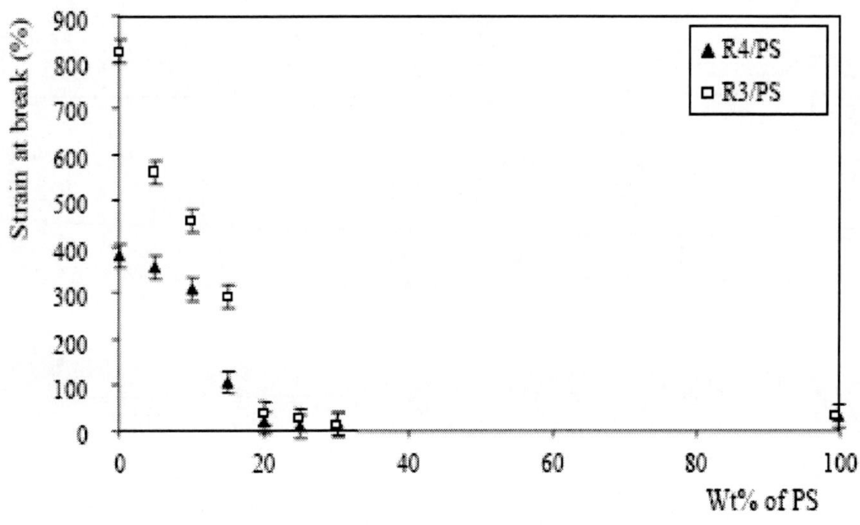

Figure 17. Variation of strain at break with blend composition.

Characterisation of rLDPE/PS/Compatibilizer Blends

In this step, the effect of a compatibilizer (maleic anhydride functionalized elastomeric ethylene copolymer) on the properties of LDPE/PS in a weight proportion of 80/20 was considered. For that purpose, physical and mechanical characterizations were carried out. Figure 18 shows that the density increased slightly with the amount of compatibilizer up to a level of about 5 wt% and then decreased slowly and seems to reach a plateau. The increase can be explained by the fact that the compatibilizer increased the interfacial adhesion between the two incompatible polymers which reduced micro cavities in the ternary blends. However, it seems that the increase of the compatibilizer level reduces the adhesion with rLDPE samples which explains the observed decrease in density after the maximum reached.

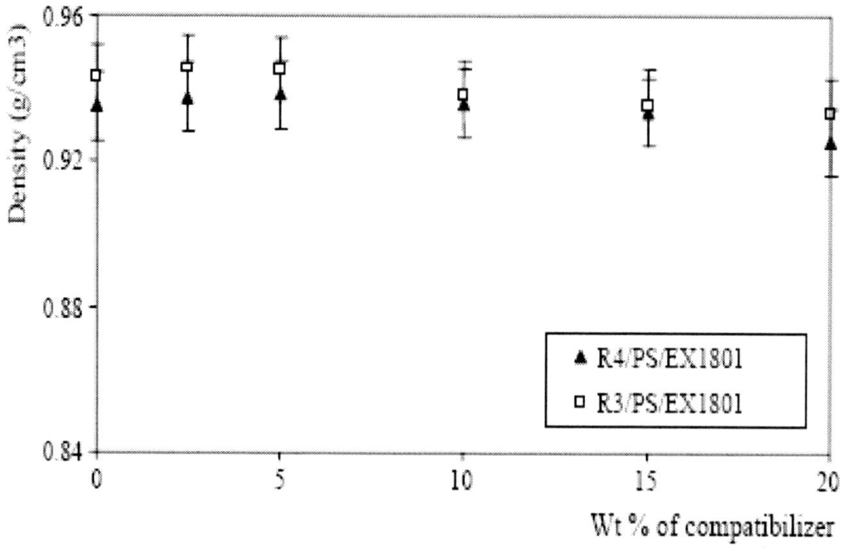

Figure 18. Variation of density with the level of compatibilizer.

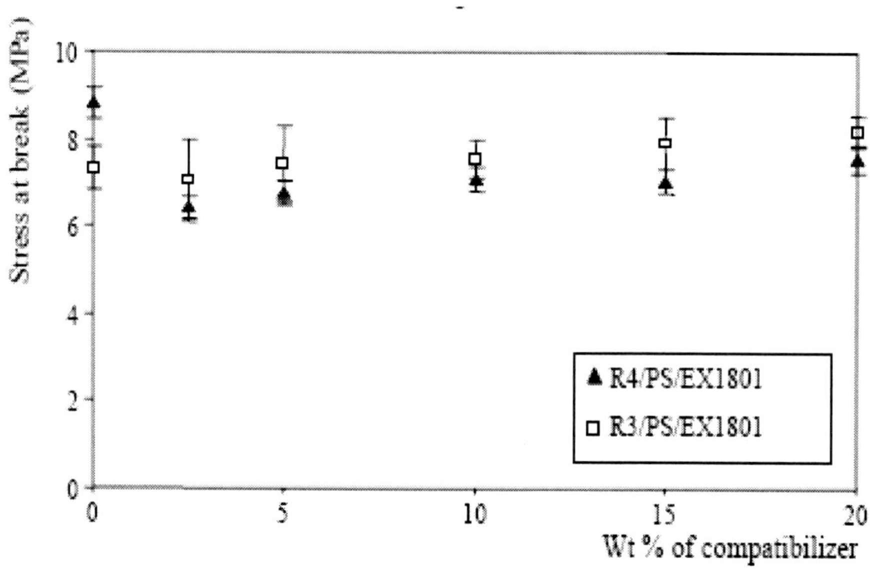

Figure 19. Variation of stress at break with the level of compatibilizer.

The variations of stress and strain at break with the level of compatibilizer are given in Figures 19 and 20, respectively. The addition of the compatibilizer to R4/PS blend hasn't ameliorated stress at break in the case of all the considered concentrations while a weak amelioration is observed with R3/PS blends (Figure 20). A noticeable increase of strain at break is obtained when the compatibilizer is added to R4/PS blend up to 20 wt% while only a weak increase is obtained with R3/PS containing 10 wt % of compatibilizer (Figure 20). In this last case, the augmentation of the compatibilizer concentration has no effect.

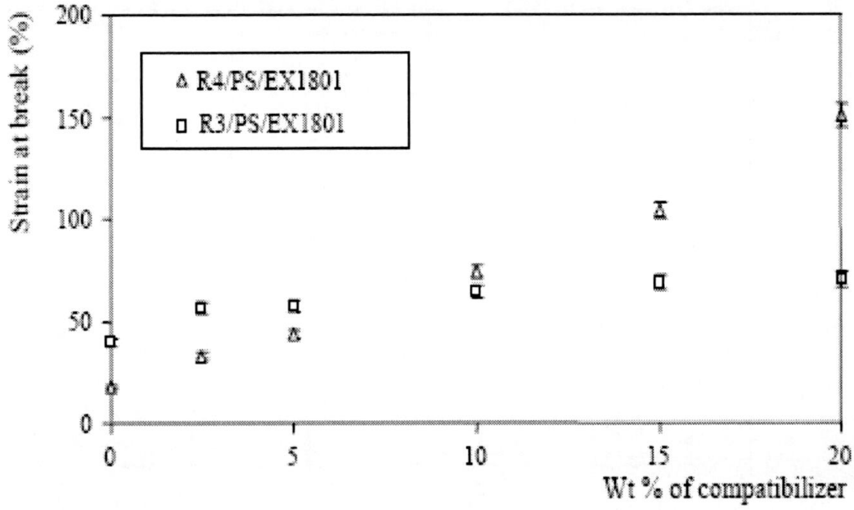

Figure 20. Variation of strain at break with the level of compatibilizer.

2.4. Regenerated LDPE/PP Blends

Characterisation of rLDPE/PP Blends

Density decreased with the level of PP (Figure 21). This is obvious since PP is more light than LDPE. Water absorption (Figure 22) increased with the level of PP. This is due to the immiscibility of this pair of polymers which favours the presence of cavities in the blends and facilitates the wáter penetration. This result was already observed by Kabdi and Belhaneche-Bensemra [57].

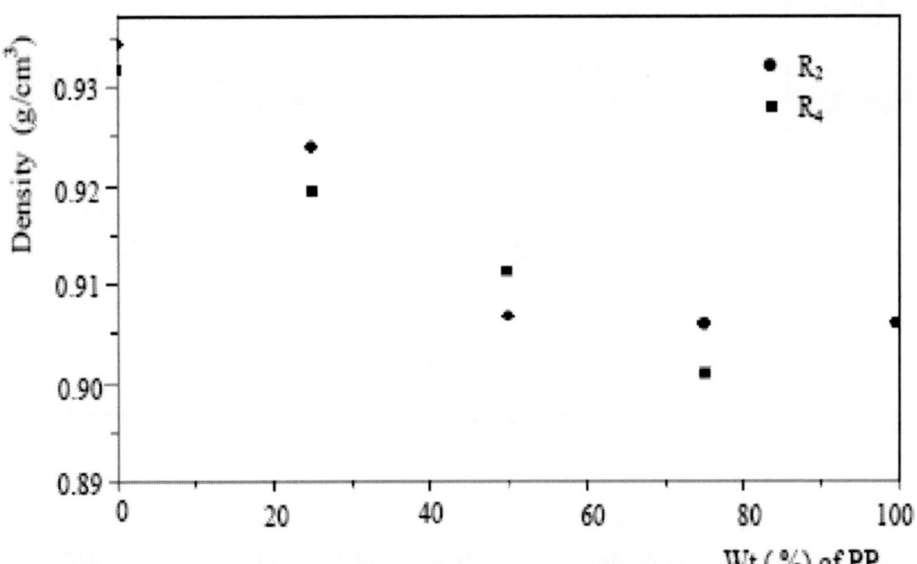

Figure 21. Variation of density with blend composition.

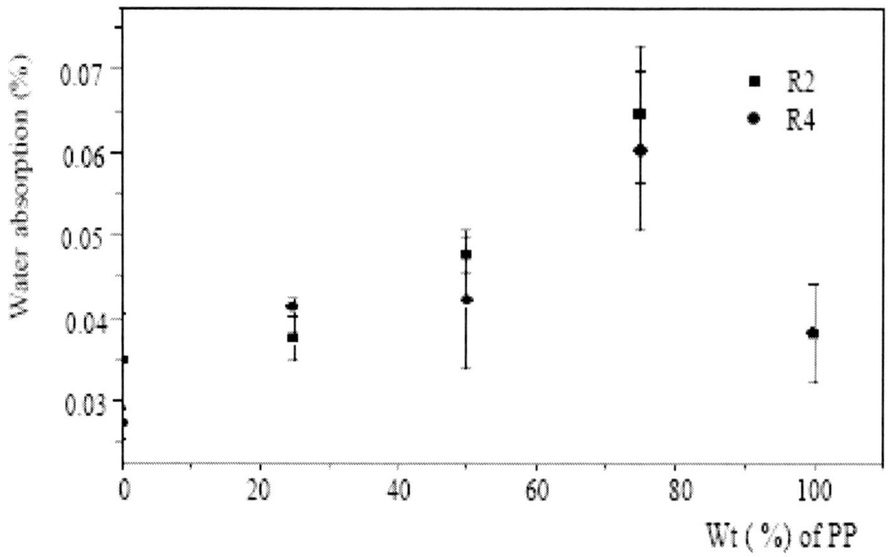

Figure 22. Variation of water absorption with blend composition.

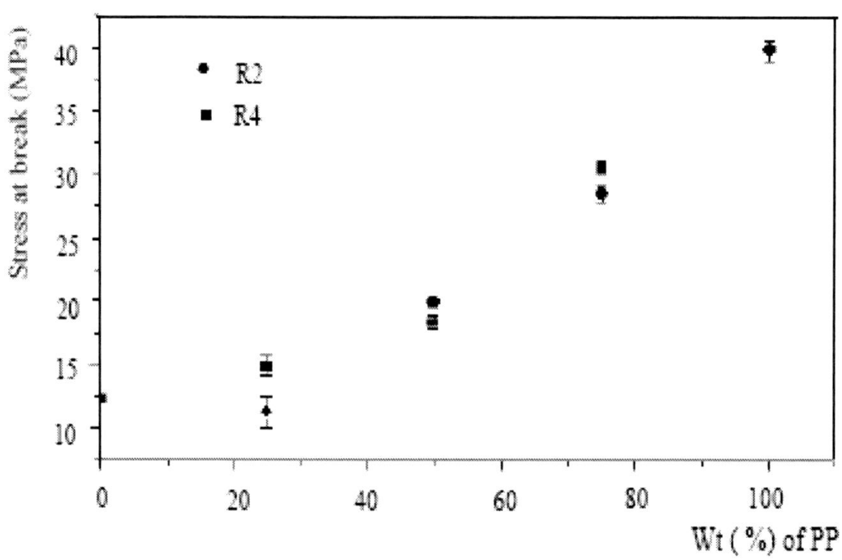

Figure 23. Variation of stress at break with blend composition.

The variation of stress and strain at break with blends composition are given in Figures 23 and 24, respectively. It can be noted that stress at break increased in both cases while strain at break decreased with blend composition. In both cases, strain at break reached a minimal value up to about 20wt % of PP and remained constant for higher compositions. These poor mechanical performances are related to the immiscibility of the two polymers. Therefore, the incorporation of a third component is necessary to enhance the compatibility of this pair of polymers. In order to investigate the effect of the compatibilizer, the rLDPE/PP (50/50) which exhibited the lowest elongations at break was chosen as the basic blend in the next step.

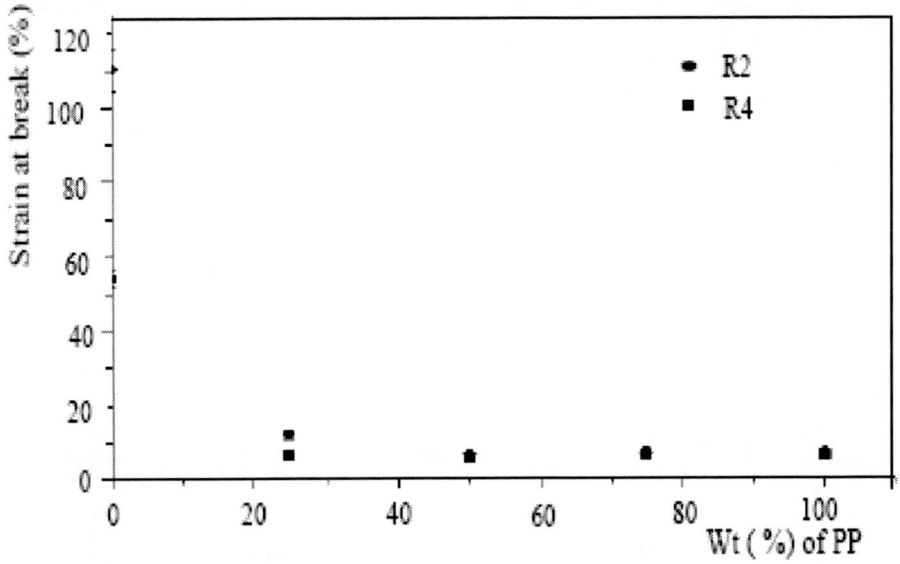

Figure 24. Variation of strain at break with blend composition.

Characterisation of rLDPE/PP/Compatibilizer Blends

Figure 25 shows a decrease of density for about 5 wt % of compatiblizer which denotes a certain amelioration of interfacial adhesion between the two polymers. Figure 26 shows also a decrease of the water absorption for the same level of compatibilizer. These results can be explained by the fact that the compatibilizer increased the interfacial adhesion between the two incompatible polymers which reduced micro cavities in the ternary blends.

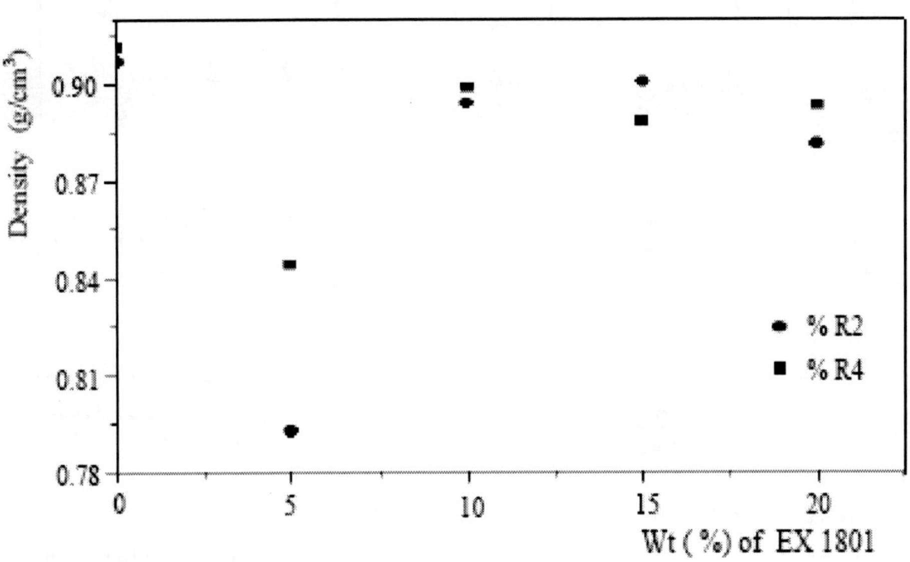

Figure 25. Variation of density with the level of compatibilizer.

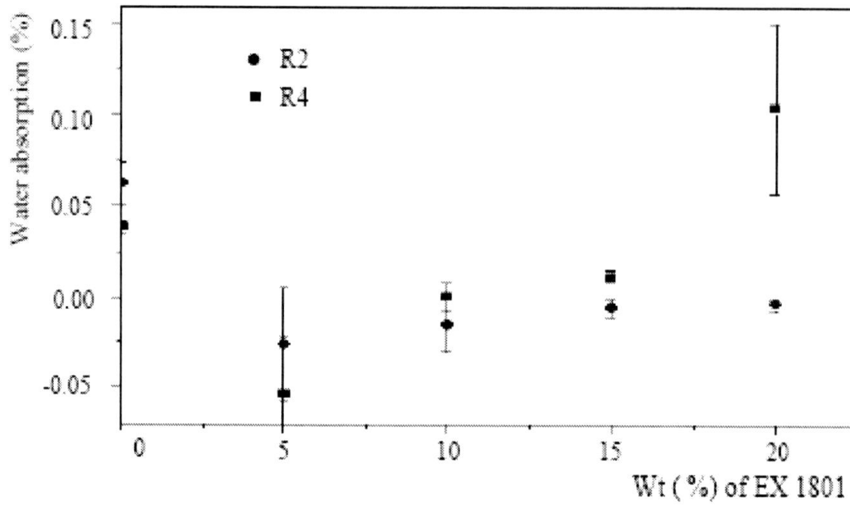

Figure 26. Variation of water absorption with the level of compatibilizer.

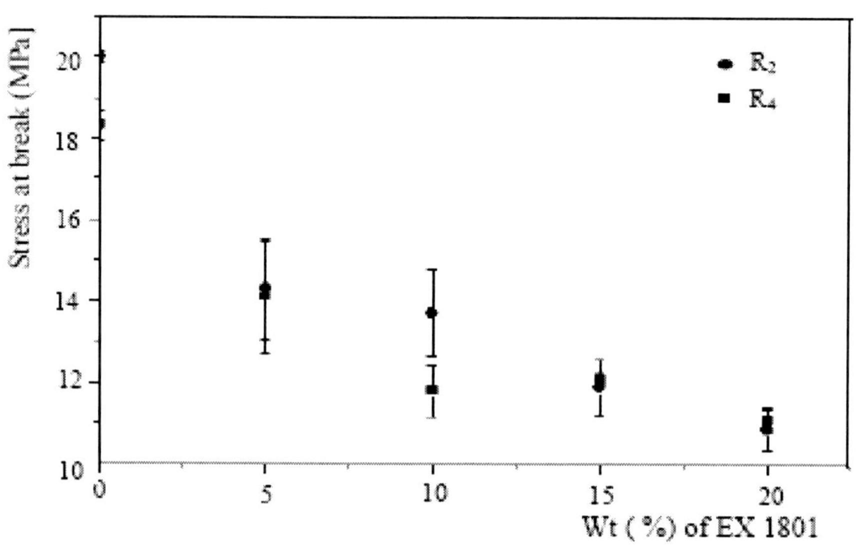

Figure 27. Variation of stress at break with the level of compatibilizer.

The variations of stress and strain at break with the level of compatibilizer are given in Figures 27 and 28, respectively. Figure 27 shows clearly that strain at break decreased with the level of compatibilizer while Figure 28 shows that the strain at break increased. These results indicate the amelioration of mechanical properties by incorporation of the compatibilizer. Figure 29 shows the morphology of the realized blends with R4. In the case of the blend R4/PP, the presence of microcavities indicates the weak association capacity between the particles of rLDPE and PP and explains the weak performances obtained in tensile properties. A change in the morphology is obtained when the compatibilizer is added. A decrease in the microcavities size and a better cohesion between the blends components is obtained. These effects are more pronounced with the level of compatibilizer.

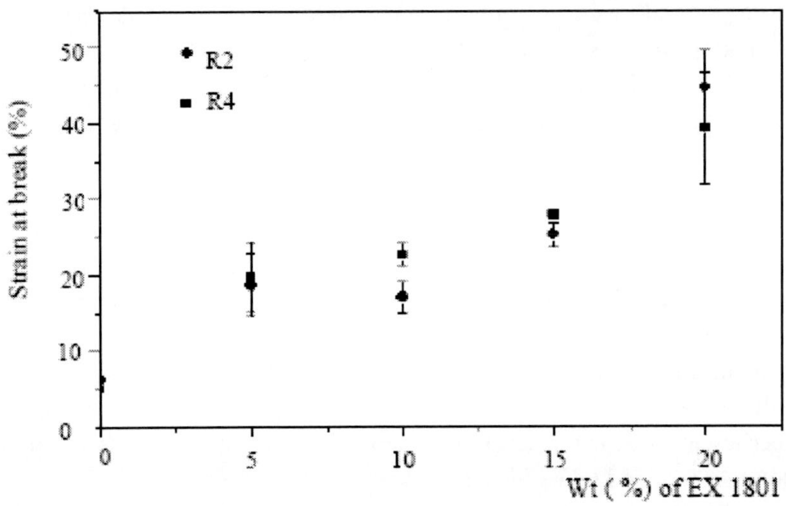

Figure 28. Variation of strain at break with the level of compatibilizer.

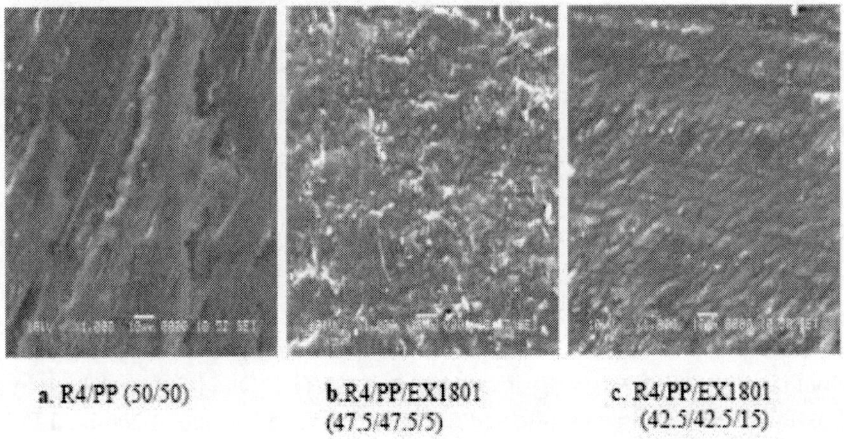

Figure 29. Analysis of R1/PP with and without comptibilizer by scanning electron microscopy. (Gr ×1000).

Then, the amelioration of tensile properties is due to a better interfacial adhesion between these two incompatible polymers.

CONCLUSION

Based on the results of these different studies, the issues to be noted are the following:

- The preliminary characterization of the rLDPE samples showed that their physico-chemical (density, water absorption, MFI, Rox) and mechanical properties (tensile and shore D hardness) depend on the level of degradation of the corresponding wastes.

- Tensile test allowed to optimize the amount of peroxide necessary for the crosslinking of the rLDPE/EPDM blends. The characterization of crosslinked rLDPE/EPDM blends showed an improvement of both stress and strain at break and a decrease in shore D hardness. The realized blends exhibited good behaviour under the combined effect of heat and water.
- The characterization of rLDPE/PVC blends showed that this pair of polymers is immiscible with a poor phase structure and tensile properties. The addition of CPE to binary polymer blends resulted in a significant increase of tensile properties (stress and strain at break) which is due to a better interfacial adhesion as evidenced by scanning electron microscope.
- The characterization of rLDPE/ PS blends showed that this pair of polymers has poor tensile properties due to their incompatibility. A relative amelioration of the tensile properties is obtained when maleic anhydride functionalised elastomeric ethylene copolymer is added to the blends.
- rLDPE/ PP blends exhibited also poor tensile properties due to their incompatibility. The addition of maleic anhydride functionalised elastomeric eythylene copolymer allowed to improve the tensile properties of the blends.

Finally, from the point of view of practical applications, the croslinking of rLDPE/EPDM blends and the addition of a compatibilizer (CPE for PVC/rLDPE blends or maleic anhydride functionalised elastomeric ethylene copolymer for rLDPE/PP or rLDPE/PS bends) are effective ways of recycling LDPE wastes.

ACKNOWLEDGMENTS

The authors wish to thank CABEL (Gué-de-Contantine), INJELEC (Rouiba), TP1/R ENPC (Rouiba), TP1/G ENPC (El-Harrach) and CATEL(Oued-Smar) which made this research possible. The authors are indebted to Dr Belhaneche Hacene of National Polytechnic School (Algiers) for his help.

REFERENCES

[1] F. Monfort-Windels. *Eur J Mech and Env Eng*. 1997, 42(1), 32- 40.

[2] D. Marco, B. Caballero, A. Torres and M.F. Laresgoiti. *J. Chem. Technol. Biotechnol.* 2002, 77, 817- 824.

[3] C. N. Kartalis, C.D. Papaspyrides, R. Pfaendner, K. Hoffmann and H. Herbst. *Polym. Eng. Sci.* 2001, 41, 771- 781.

[4] Z. Dolnzal, V. Pacakova and J. Kovarova. *J. Anal. Appl. Pyr.* 2001, 57, 177- 185.

[5] Pawlak, J. Morawiec, F. Pazzagli M. Pracella and A. Galeski. *J. Appl. Polym. Sci.* 2002, 86, 1473-1485.

[6] Fortelny, D. Michalkova and Z. Krulis. *Polym. Degr. and Stab.* 2004, 85, 975- 979.

[7] G. Grause, W. Kaminsky and G. Fahrbach, *Polym. Degr. and Stab.* 2004, 85, 571- 575.

[8] F. Awaja and D. Pavel. *Eur. Polym. J.* 2005, 41, 1452-1477.

[9] F. Fraisse, V. Verney, S. Commereuc and M. Obadal, *Polym. Degr. and Stab.* 2005, 90, 250- 255.
[10] B. G. Girija, R. R.N. Sailaja and G. Madras. *Polym. Degr. and Stab.* 2005, 90, 147- 153.
[11] C.Lorenzetti, P. Manaresi, C. Berti and G. Barbirol. *J. Polym. Env.* 2006, 14(1), 89- 101.
[12] H. Zhang, W. Guo, Y. Yu, B. Li and C. Wu. *Eur. Polym. J.* 2007, 43, 3662- 3670.
[13] J. Leidner. *Plastics waste: recovery of economic value*. Marcel Dekker, New York, 1981.
[14] R. D. Deanin. *Polymer blends and alloys*. Marcel Dekker, New York, 1999.
[15] M. Gahleitner. *eXPRESS Polm. Lett.* 2011, 5(11), 936.
[16] S. Bertin and J. J.Robin. *Eur. Polym. J. 2005,* 38, 2255- 2264.
[17] K. Berdjane, Z. Berdjane, D.R. Rueda, D. Benachour and F.J. Balta-Callja. *J. Appl. Polym. Sci.* 2003, 89,2046- 2050.
[18] D. R. Paul, S. Newman. *Polymer blends*. Academic Press Inc, New York, 1978.
[19] A.V.Machado and M. Van Duin. *Polymer.* 2005, 46, 6575- 6581.
[20] M. Van Duin and A.V. Machado. *Polym. Degr. and Stab.* 2001, 90, 340- 348.
[21] Z. Fang and M. Zeng. *J. Appl. Polym. Sci.* 2001, 82, 2947- 2955.
[22] G. Ma and Z. Fang. *J. Appl. Polym. Sci.* 2003, 88, 1296-1302.
[23] N.R. Legge, G. Holden et al. *Thermoplastic elastomers.* Hanser Publishers, Munich, Vienna, New York, 1987.
[24] E. Martuscelli, R. Palumbo et al. *Polymer blends: processing, morphology and properties*. Plenum Press, New York, 1980.
[25] C.S. Ha, H.D. Park and W.J.Cho. *J. Appl. Polym. Sci.* 2000, 76, 1048-1053.
[26] Z.P. Fang, G. W. Ma, B.Q. Shentu, G. P. Cai, and C. W. Xu. *Eur. Polym. J.* 2000, 36, 2309- 2311.
[27] G. Ma, Z. Fang and C. Xu. *J. Appl. Polym. Sci.* 2003, 88, 1296- 1303.
[28] Zarraga, M. E. Munoz, J. J.Pena and A. Santamaria. *Polym. Eng. and Sci.* 2001, 41(11), 1893- 1902.
[29] D. J. Walsh, J. S. Higgis and A. Machomachie. *Polymer blends and mixtures*. Martinus Nijhof, Boston, 1985.
[30] Siegmann, L. K. English, E. Baer and A.Hiltner. 1984, *Polym. Eng. Sci.*, 24, n°11, 877- 885.
[31] J. Klaric, N. Stipanelov Vrandecic and U. Roje., *J. Appl. Polym. Sci.* .2000, 78, 166- 172.
[32] B. L. Shurman, U. Niebergall, N. Severin, Ch. Burger, W. Stocker and J. P. Rabe. *Polymer*. 1998, 39, 5283- 5291.
[33] C. M. Tai, R. K. Y. Li, C.N. Ng. *Polym. Test.* 2000, 19, 143- 154.
[34] M. Yang, K. Wang, L. Ye, Y. W. Mai and J. Wu. *Plastics, Rubber and Composites.* 2003, 32(1), 27- 31.
[35] R. Strapasson, S. C. Amico, M.F.R. Pereira and T.H.D. Sydenstricker. *Polym. Test.* 2005, 24, 468- 473.
[36] S. M. Lai, F. C. Chiu and T.Y. Chiu. *Eur. Polym. J.* 2005, 41, 3031- 3041.
[37] M. Munaro and L. Akcerlrud. *Polym. Degr. and Stab.* 2008, 93, 43- 49.
[38] Z. I. Ali, H. A. Youssef, H. M.Said and H.H. Saleh. *Thermochim. Acta.* 2002, 438, 70 - 75.

[39] G. Radonjic and N. Gubeljak. *Macromol. Mater. and Eng.* 2002, 287, 122- 132.
[40] M. Yang, K. Wang, L. Ye, Y. W. Mai and J. Wu. *Plastics, Rubber and Composites.* 2003, 32(1), 21- 26.
[41] S.A. Xu, J. W. Xie and M. Jiang. *Polym. International.* 1999, 48, 1121-1128.
[42] H. F. Guo, S. Packrisamy, R. S. Mani, C. L. Aronson, N. V. Gvozdic and D. J. Meier. *Polymer. 1998,* 39(12), 2495- 2505.
[43] C. Harrats, R. Fayt and R. Jerome. *Polymer.* 2002, 43, 863- 873.
[44] M. F. Diaz, S. E. Barbosa and N. J. Caplati. *Polymer.* 2007, 48, 1058- 1065.
[45] D. Bourry and B. D. Favis. *Polymer.* 1998, 39(10), 1851-1856.
[46] K. Wilczynski, A. Tyszkievicz and Z. Szymaniak. *J Mater Process Technol.* 2001, 109, 320- 323.
[47] B. Xu, J. Simonsen and W. E. Rochefort. *J. Appl. Polym. Sci.* 2000, 76, 1100- 1108.
[48] S. A. Xu and S. C. Tjong. *J. Appl. Polym. Sci.* 2000, 77, 2024- 2033.
[49] G. Akay, T. Tincer and H. E. Ergoz. *Eur. Polym. J.* 1980, 16, 601- 605.
[50] M.A. Chabou, N. Belhaneche-Bensemra. *Ann. Chim. Sci. Mat.* 2002, 27, 93- 105.
[51] D.W.Van Krevelen. *Properties of Polymers.* Elsevier, Amsterdam, 2003.
[52] M. Reyne. *Les plastiques, transformations et applications*, Hermes, Paris, 1991.
[53] J. R. White. *Degradation (Weatherability),* CD-ROM Polymeric Materials Encyclopedia, CRC Press Inc., 1996.
[54] N. Belhaneche-Bensemra, C. Zeddam and S. Ouahmed. *Macromol. Symp.* 2002, 180, 191- 202.
[55] D. Atek and N. Belhaneche – Bensemra. *Eur. Polym. J.* 2005, 41, 707- 714.
[56] M. O. Boussoum, D. Atek and N. Belhaneche-Bensemra. *Polym. Degr. and Stab.* 2006, 91, 579- 584.
[57] S. A. Kabdi and N. Belhaneche- Bensemra. J. Appl. Polym. Sci. 2009, 110(3), 1750 - 1755.

In: Landfills
Editors: G. B. C. Cabral and B. A. E. Botelho

ISBN: 978-1-62257-121-5

Chapter 8

MANAGEMENT OF ELECTRONIC WASTE IN BASQUE COUNTRY: EXPERIENCES OF COLLECTING AND TREATMENT WITH NEW TECHNOLOGIES*

A. Renteria[1#] ***and E. Alvarez***[2†]
[1] TECNALIA, Spain
[2] Faculty of Engineering, University of Deusto, Spain

ABSTRACT

The success of a recycling solution as an alternative to disposal, in order to become economically feasible, depends mainly on two aspects: the responsibility of consumers to take back their appliances (reverse logistics) and the costs of handling and disassembly. The first issue is being dealt by several legislative initiatives, e.g.: European Directives on waste. For the second, a wide spectrum of technological solutions has been suggested and developed.

This chapter will cover the first experiences of collecting the waste of electric and electronic appliances in the Basque Country, a region in Northern Spain. Enforcing laws, marketing strategies, related to logistics, etc, will be explained. The results over the first years of the experience applied in main Basque towns were positive, therefore the collection scheme was expanded to the whole region and other areas of Spain. Main outcomes regarding the economic impact, social awareness and influences over the environment will also be covered.

The main problem concerning the disassembly of devices is their high diversity. A good technical approach should offer an intelligent combination of manual and automated operations, optimized logistics, flexible plant layout, etc. This chapter offers a review of the state of the art of industrial solutions, together with their shortcomings and future trends. In addition, some new optimization techniques will be explained, with the

---

*The proposed contents of the chapter are the experience of collecting and treatment of waste of electric and electronic appliances in the Basque Country area (Northern Spain), results of this initiative, expansion to other areas of Spain, technologies for identification, separation and recycling of this kind of waste, impact (economic, social, environmental..), Case-Studies, statistics, etc.

arantxa.renteria@tecnalia.com.

† esther.alvarez@deusto.es

aim to increase the benefits of the recycling companies, promoting in this way the correct treatment of the electronic waste and avoiding the disposal in landfills.

In order to solve the problem of a growing generation of electronic waste, the promotion of the reuse of devices and components is considered to be one of the most desirable solutions. The concept of reutilization of electronic equipment goes from the reuse of the complete device for a second life, to the alternative of making the most of its valuable components. In order to show specific examples of these solutions, a case study will be explained showing the recycling of TV sets and monitors. A local recycler of TV sets and monitors has been involved in the definition and planning of the recycling facility. This company has carried out a statistical study of the received TV sets and monitors, collecting technical data about the devices. Taking into account these records and other technical information, a recycling scheme was proposed in order to handle, separate, recycle and reuse them. The recycler defined some requirements to be fulfilled by the process in order to design the recycling facility properly. The components and materials obtained from the separation process have been reused in new devices (e.g. electronic components) and recycled as raw materials (e.g. glass) for new usages.

1. Introduction to the Recycling of Waste of Electric and Electronic Equipment

The production of electric and electronic appliances is one of the fastest growing areas, resulting in an increase of their waste. Electric and electronic waste includes an increasing scope of electronic appliances of different nature, from large household devices (e.g. refrigerators, computers, washers, or music equipment) to cellular phones, lamp bulbs, etc.

The technological innovation and expansion of markets allow a fast upgrading of equipment. People no longer take a malfunctioning electronic device to a repair shop because the replacement of the object is often easier and cheaper than repair. As a consequence of this practice and electronic devices' shorter life cycles, there is an ever increasing amount of waste that is turning into is a serious environmental problem. Electronic waste already constitutes from 2% to 5% of the United States (US) municipal solid waste stream and is growing rapidly. European studies estimate that the volume of electronic waste is rising from 3% to 5% per year – almost three times faster than the municipal waste stream [Zoeteman, 2010].

Statistics from the US point out that a percentage from 2 to 5% of solid urban waste is electronic scrap, with a yearly increase of 8%. In Europe, this rise is between 3 and 5%, three times the increase of traditional waste [Hischier, 2005]. The population of Europe, 463 millions of people, generates an average amount of 10 to 12 kg of WEEE per person a year; about 5.1 millions of tons [Zonneveld, 2007]. As an example, 500 million PCs contain approximately 2,872,000 t of plastics, 718,000 t of lead, 1363 t of cadmium and 287 t of mercury [Widmer et al., 2005]. These quantities are growing fast because the market for PCs is far from being saturated and the average lifetime of a PC is decreasing rapidly (6 years for a CPU in 1997, 2 years in 2005). But return rates are still low and the cost of processing a return can be 2 to 3 times that of an outbound shipment [Mangan et al., 2008].

Old computers, televisions and other e-scrap contain valuable materials and components that are recyclable and reusable (see Table 1). But there are hazardous materials in the electronic waste too. The cathode ray tubes (CRTs) used in computer monitors, television sets, and other displays contain significant concentrations of lead (Pb) and other heavy metals. Other dangerous materials used in computers and other electronic devices include cadmium, mercury, hexavalent chromium, plastic and brominated flame retardants. Other materials may change their properties and became hazardous if they are burned. All these materials make electronic recycling particularly hazardous to workers, as well as to the environment.

The European Union (EU) has recognized the urgency of the e-waste problem, and approved two directives dealing with this important issue. The two main laws are "Waste from Electrical and Electronic Equipment," (the WEEE Directive) and a Directive on the "Restrictions on the Use of Certain Hazardous Substances in Electrical and Electronics Equipment" (ROHS). These directives are linked and intended to decrease the amount of hazardous substances found in electrical and electronic equipment, and to reduce the amount of waste going to landfill.

Table 1. Recyclable materials recovered from TV and monitors

Material	Quantity[1]
Iron	756
Copper	425
Aluminium	8
Non-ferrous metals	66
CRT glass	6.628
Other valuable materials	938
Total weight of valuable materials	8.821

1 (Unit: tons).
Source: Sony Corporation CSR Report 2003.

The WEEE Directive requires that producers supply systems for the treatment of WEEE. The goals of the Directive are to:

- Prevent e-waste
- Improve the re-use, recycling and other forms of recovery to reduce e-waste
- Improve the environmental performance of all economic operators involved in the life cycle of electrical and electronics equipment

The Directive also requires the components and materials of e-waste to be labelled. The ROHS, on the other hand, tried to prevent the use of hazardous substances in the production of electrical and electronics equipment by 2008. Due to its danger to human health and the environment, lead is particularly targeted by this legislation. The European Union makes industry responsible for assisting in the solution on several levels. First of all, industry is encouraged to find better, less toxic ways to manufacture products, trying to decrease the risks of the equipment in the future. The Directives also place full financial responsibility on producers to set up collection, recycling and disposal systems, and contain effective and feasible goals for recycling. The cost of this legislation is only an additional 1% to 3% of retail prices.

In order to achieve the targets specified in the Directive it is necessary to set up efficient collection schemes. Following the subsidiary principle of the European Directives, this Directive only defines general requirements to comply with mandatory collection and recycling objectives. The way to define the collection logistics and the organization of the take-back schemes are left to the will of the Member States. Prior to the establishment of the WEEE Directive, some European countries (e.g. Belgium, the Netherlands, Sweden, or Denmark) had already stated national regulations and organized management schemes for WEEE. These systems were adapted to very different national situations and philosophies. Some other countries had to adapt their national laws when implementing the WEEE Directive.

Out of the potential 5 million tons of WEEE that go for disposal each year in the EU, the volume collected for European households is thought to have been around 1 million tons last year. An estimated 60% of this volume was processed complying with WEEE requirements. However, a worrying 40% was sent beyond the EU as untreated WEEE, most notably to Asia and Africa. According to some studies, this problem will grow in the years to come, with an estimated 4.3 million tons improperly treated by 2020 (Commission of the European Communities, 2008, see Figs. 1 and 2).

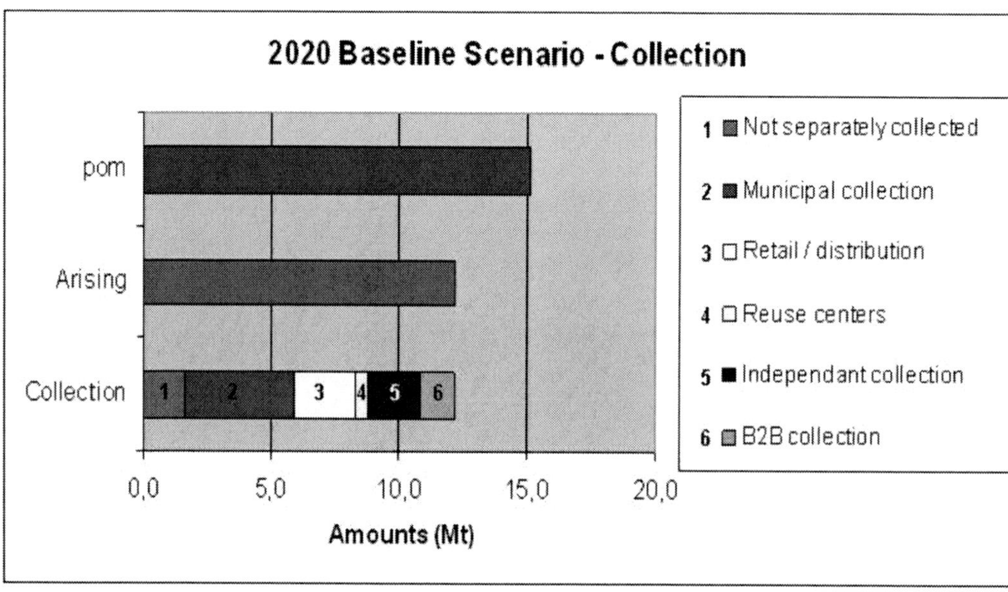

Figure 1. Predicted destination and collection of WEEE in 2020 (pom: put on market).

Apart from the activities carried out by Member States, one of the most prominent EU wide system is the European Recycling Platform (ERP), an entity supported, among others, by Hewlett Packard, Sony, Electrolux and Braun, with the aim to develop Pan –European compliance structures. The idea is to design and manage all aspects of the compliance process, (by now the activity remains only at planning level).In order to work effectively, the ERP will establish national schemes in several countries and will obtain legal coverage to operate. The platform will not send WEEE outside of the country of origin, but needs to develop pan-European agreements with networks of providers.

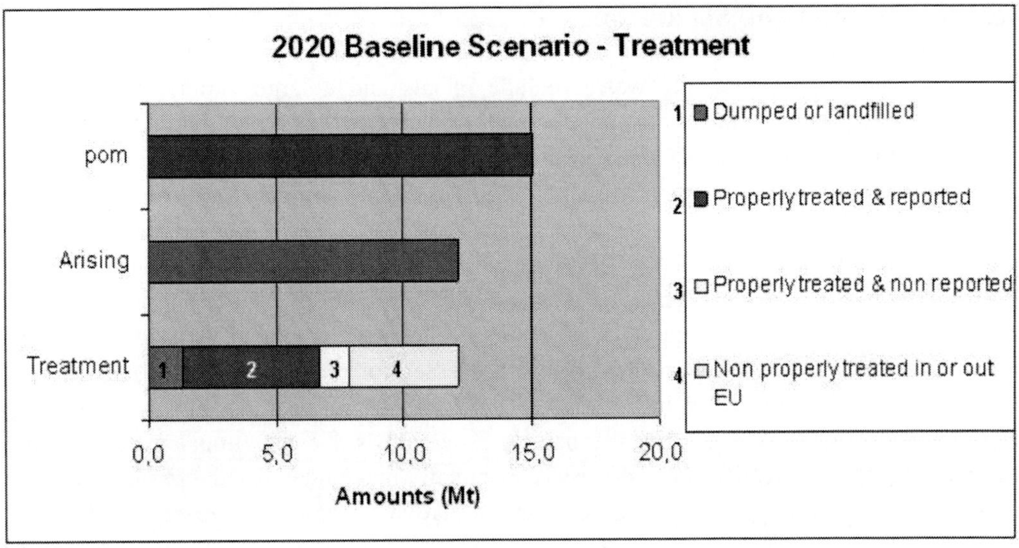

Figure 2. Predicted destination and collection of WEEE in 2020.

The companies involved in the ERP think this platform is an important opportunity to develop alternatives to the national schemes and to create competition, which in turn will stimulate efficiency and cost reductions. But some Member States and producers remain sceptical. As a matter of fact, it is difficult to define collective systems and the associated logistics, at least in the short term, on such scale and taking into account the differences in legislations on each country.

2. New Technologies for the Treatment of WEEE

In general, the treatment of waste of electric and electronic equipment follows these steps:

- Selective tests on the appliance. The aim is to recover those equipments which are still able to be used.
- Identification, disassembly and separation of appliances, with the following goals:
    - Recovery of components for new appliances.
    - Recovery of recyclable material to be used as raw material of other usages: metals (aluminium, copper, iron,..), plastics, paper.
    - Material separation for a specific treatment in special facilities.
    - Shredding of recyclable material as raw material for other usages with the aim to separate into corresponding fractions.

2.1. Identification and Separation

Usual contents of electronic waste include metals and a large variety of resins and plastics, which can be coated, mixed with additives, and made flame retardant (Cui and Forssberg, 2007). Regarding plastics recycling, a main concern is the need to identify and separate plastics found in electronics (mainly additives and contaminants). Several automated solutions have been proposed to separate mixture of plastic and other elements, such as machine vision for image analysis (Renteria and Alvarez, 2007a) and vertical vibration systems (Mohabuth and Miles, 2005). After a correct separation, their properties (physical and mechanical) must meet those of virgin resins if they are intended to be used in high-end products. Nowadays, three types of recycling processes for plastics are being used (Prada, 2011). Firstly, chemical recycling processes use waste plastics as raw materials for petrochemical processes. Secondly, mechanical recycling is a conventional method, which uses a shredding and identification process to eventually make new plastic products. Thirdly, thermal recycling, where plastics are used as an alternative fuel (Cui and Forssberg, 2003).

There are easier solutions to separate other types of materials. A magnetic separation technique can be used to remove ferrous metals from plastics waste stream. If strong magnets are used, many grades of stainless steel can also be removed. Among the non-ferrous metals, aluminium can be separated by Eddy current process. Air classification processes are used to separate light fractions, such as paper, by controlling air velocity. A combination of machine vision and vacuum can be used to identify and separate light fractions of shredded components characterized by colour, shape, etc. as explained in Renteria et al. (2008).

There are five methods for the separation of the two parts of the cathode ray tube:

- Diamond circular saw: it cuts the area of union between panel and funnel. See Fig. 3a.
- Water jet cutting: this method has the following advantages [Geskin, 2002]:
    - Small cutting hole.
    - Low cutting forces.
    - No vibrations in the robot arm.
    - No thermal problems.
- Laser cutting: the material is partially melted and evaporated. The laser beam is focused over or below the surface of the material by means of a lens. Some commercial systems based on CO_2 [Proventia, 2008] already exist. But they present some drawbacks such as the price and the potential problems which may arise from thick glasses.
- Hot air: by heating the *frita* (adhesive material to join together the panel and funnel glass), it is possible to avoid tensions and uncontrolled break of the TV tube. In this way the panel and funnel parts can be separated, but it is time consuming.
- Thermal shock with electric resistances: the aim is to separate the panel from the funnel by consecutively heating and cooling, thus causing the breakage of the

tube in the area where previously a notch has been done. But the main problem is that a well controlled break in the desired area is difficult to achieve.

The traditional method of treatment for printed circuit boards was shredding, but more valuable materials may be obtained if the functional components are extracted. Several methods exist for this alternative:

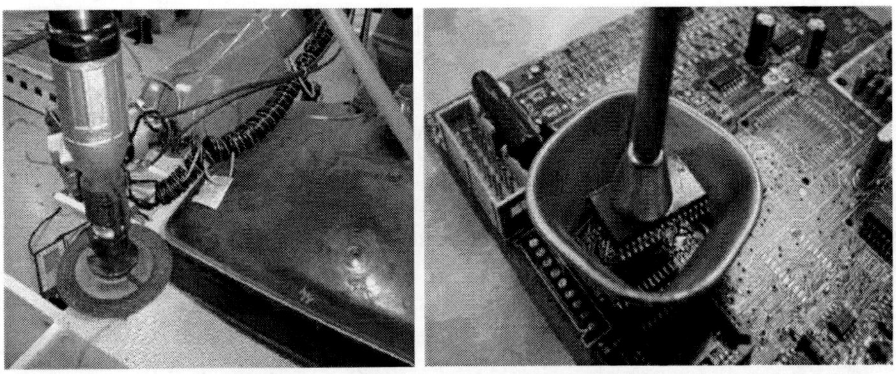

a b

Figure 3a. Diamond saw cutting, 3b: Desoldering of components in printed circuit boards.

- Desoldering by melting in vapour phase: This method has been applied to surface mounted devices (SMD), using the heat generated by the vapour phase of an inert liquid, a common technology applied in the electronic industry to solder components. Heat causes the soldered joints to melt. There is no damage in the components or PCBs but the cycle time is high.
- Desoldering by heating using infrared radiation: This method can be applied to both SMD and thru-hole components. The infrared radiation is focused on the side of the PCB containing the soldered joints in the case of thru-hole elements. In order to minimize the risk of damage to the components, several parameters must be carefully controlled (transmitted power, type of infrared lamps, distance between transmitter and PCB, exposure time). Several drawbacks have been detected: the radiation may cause damage to the layers of the PCBs, (depending on their composition), and also to the electronic part of the SMD components (covered by the packaging). This solution demands a fine tuning of the above mentioned parameters, and consequently makes the final system more complex and expensive.
- Desoldering by heating using hot air jet (Fig. 3b): This technology used to be applied to SMD components, but it also works properly with thru-hole elements. The process consists of the positioning of the vacuum gripper over the components, application of a heated air jet over the soldered joints and extraction of the released component. Similar parameter control as with infrared method must be taken (air flow and temperature, exposure time, distribution of air over the connection pins). As results from the realized tests, no damage has been done to the electronic part of the components due to the short exposure time. The desoldering times are shorter than the ones obtained using the infrared method.

An important issue here is the right diffusion of the hot air over the connection pins which must be homogeneous and focused on the soldered pins, avoiding the packaging part of the SMD component.

2.2. Recycling/Reuse

According to the available data of the main TV set manufacturers [Thomson, 2008], the average amount of plastics in a TV set is 18-15% (similar percentages may be applied to computer monitors). The majority of plastics present in TV sets and monitor housings are HIPS (high impact polystyrene) and ABS (acrilonitryl butadiene styrene), together with the mixture ABS+PC (polycarbonate), including several additives such as stabilizers and flame retardants. In many cases this plastic can comply with technical and legal specifications required by secondary markets and European directives, and can be re-processed to be used in new products.

In computer monitors the ABS is the most frequent polymer and can be recycled for high added values applications. However, monitor housings are painted sometimes, which makes their recycling more difficult. If there are halogenated flame retardants they may cause health problems by the smoke coming form their burning.

The situation has changed significantly during the last years. The discarded electronic equipments contain less metals and more plastic materials. Table 2 shows the variation on the distribution of materials recovered form TV sets recycled by Sony Corporation in Japan [Sony, 2002; Sony, 2003; Sony 2007]. The average composition, in weight, of a monitor is:

- 50% glass.
- 20% ferrous metals.
- 15% plastics.
- 15% other materials, included non-ferrous metals.

Table 2. Recovered materials in TV sets (Sony, Japan)

Material	2001 T and % with respect to total processed	2003 T and % with respect to total processed	2006 T and % with respect to total processed
Iron	704 / *8%*	833 / *7%*	2.264 / *10,7%*
Copper	471 / *5,3%*	469 / *4%*	829 / *4%*
Non-ferrous components	--	73 / *0,6%*	58 / *2,7%*
Tube glass	5.355 / *60%*	7.306 / *61%*	8.093 / *38,5%*
Other materials	575 / *6,4%*	1.034 / *8,6%*	4.106 / *19,5%*
Total weight of recovered materials	7.106	9.724	15.360
Total weight of processed TV sets	8.851	12.008	21.009

The identified secondary markets for the recovered materials from the treatment of monitors and TV sets are:

- Mixed glass (the whole tube with panel and funnel): melting agent for metallurgy.
- Plastics in housing, without banned additives: mechanical recycling.
- Other plastics (housings with banned additives…): energy recovery in a cement facility.
- Metals (Fe, Cu, Al…): metallurgy.
- Panel glass:
    - Filling material for construction [Batayneh, 2007].
    - Insulating material for construction [Andreola, 2007].
    - Melting agent in foundries of lead and copper. In these cases glass is used as an alternative to silica or sand.
    - Manufacturing of asphalt.
- Electronic components: they can be reused as refurbished components for new or used appliances, provided that they have passed all quality controls.

3. Treatment of WEEE in the Basque Country (Northern Spain)

Reverse logistics is the collection of end-of-use appliances, coming from consumers, thus extending supply chains to closed-loop ones [Dyckoff et al., 2004]. It covers the time elapsed since the consumer discards a product, until it is transformed into a value-added product which is again introduced in the market.

In the world of recycling of electronic equipment the economies of scale represent the availability of big volumes of appliances to be recycled in a specific geographical area, in order to organise the take back, transport and processing profitable. They are important because:

- Without big volumes of appliances the cost associated to collecting and transport may be too high to justify the investment. The environmental impact related to the transport (use of petrol and material) is not even justifiable if a small amount of waste is collected.
- Without big volumes of appliances the total capacity of treatment facilities and technologies could be under-used, with the subsequent economic disadvantages (treatment cost per kilo would be too high, and the obtained recycled material would be too low), and also environmental drawbacks. A separation process tested in a labouratory may be correct, but at industrial scale the cost can be out of range.

3.1. Developed Scheme

Between October 1994 and December 1996 the Department of Environment of the Basque Government promoted the implementation of the "Take back and treatment of electric-electronic appliances in their end of life" in the metropolitan area of Bilbao (41,3 km^2, population 375.000). The goals were to acquire experimental data in order to establish a schema to be applied to other municipalities of the autonomous region of the Basque Country, to create a mixed experience with WEEE coming from domestic and industrial sources and to sensitize the citizens. In this project several partners were involved: the Basque Government, the Bilbao City Council, a company that collects waste (CESPA) waste, a large retail chain (Eroski), and a recycling company (Indumetal Recycling, S.A.) [Gobierno Vasco, 1996]. The Bilbao City Council provided three centres for selective collection ("Bilbogarbis"), already installed in the city at that moment. The retail chain organized some collecting points in four shopping centres. There were three types of collecting systems, depending of the source of the waste: domestic, industrial and market flows.

The aim was also to carry out a socio-economic and environmental analysis, covering the following issues:

- Amount of materials collected, classified by type of product and recovery rates.
- Degree of response from the public.
- Economic costs of the recovery.
- Optimization of each process: collection, transport and recovery.
- Environmental and economic benefits.

Table 3. TV sets and monitors collected in Bilbao

Source		Collected appliances	
		Units	tons
Municipal	Colour TV set	2,550	77.83
	B&W TV set	879	15.64
	Monitor	685	8,18
	Total	4,114	101.65
Retail chain	Colour TV set	441	14.08
	B&W TV set	23	0.40
	Monitor	9	0.10
	Total	473	14.60
Industrial	Colour TV set	48	1.62
	B&W TV set	10	0.15
	Monitor	7,436	101.90
	Total	7,494	103.68
TOTAL colour TV sets		3,039	93.54
TOTAL B&W TV sets		912	16.20
TOTAL monitors		8.130	110,19
TOTAL		12.081	219,94

After that first attempt, the experience was expanded to the whole area of the province of Bizkaia, installing centres of selective collection of waste (Plan for Waste Management in Bizkaia, 1997-2001). There were infrastructures for voluntary contribution, accessible by car, for the selective and separate collection of small amounts of domestic waste, industrial and commercial waste, waste from construction and demolition, etc.

3.2. Results of the Initiative

The experiment carried out in Bilbao and Bizkaia has been reinforced in the following years, with an important increase in the collected amounts of waste. In addition, the area of influence covers the whole Basque Country (with a population of 2.2 million people). At present several agencies and recycling companies are in charge of the recycling activities:

- ECOEMBES, for light containers
- ECOVIDRIO, for glass
- Facilities for the recovery of paper and cardboard
- ZABALGARBI, for incineration and energy recovery
- INDUMETAL RECYCLING, for domestic electronic appliances
- BEROHI, for textile products
- RAFRINOR for oil and edible fats
- RECYPILAS, for batteries
- AMBILAMP, for fluorescent lamps
- EMAUS and BERZIKLATU, for bulky waste (furniture...)

The flow of domestic electric/electronic appliances is classified in 3 categories:

- White line: fridges, washing machines...
- Brown line: smaller appliances, such as TV sets, music equipments...
- Grey line: computer components (monitor, keyboards...)

In 2009 a total of 4999 t of domestic electronic appliances were collected in the Basque Country region (1,603 t corresponding to brown line, 802 to grey line and 553 to white line). Part of them is recovered and, after amending and upgrading, they are re-introduced in the market by means of initiatives such as Ekorrepara, Ekocentres and Emaus. From the total amount of collected domestic appliances, 0.3 % of them are reused, 68.4% is recycled, 8.4 % are burnt with energy recovery and 23% are discarded into a landfill [IHOBE, 2011].

The amount of selectively collected household appliances can be added to the amount of fluorescent lamps to cover the definition of WEEE by the European Directive. This gives a total of 5,095 t of WEEE, 2.3 kg/person/year. This is still far from the objective set by the Directive by 2006 (4 kg/person/year). However, considering the fact that there is a non-selective collection of WEEE, the actual data of generation of WEEE per person and year would be closer to the goal.

Another consequence of this scheme for collection and recycling waste is that a planned treatment plant, aimed to burn waste, will not be built due to the increasing recycling rates, so this new plant is no longer needed.

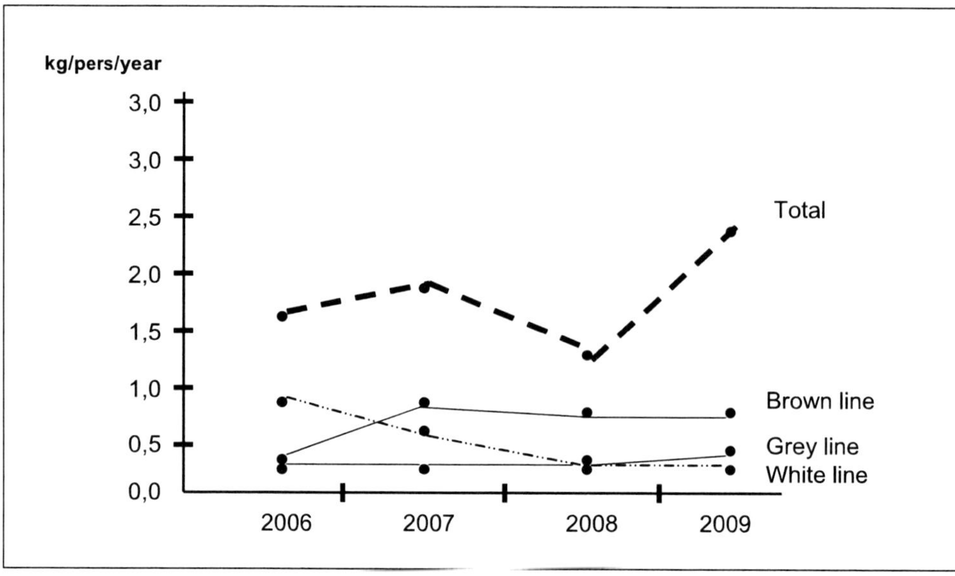

Figure 4. Selective collection of WEEE in the Basque Country, by type, in 2006-2009 (kg/person/year).

3.3. Future Strategies and Trends

The first experience of WEEE collection in the Basque Country was limited to products belonging to brown and grey line, where the amount of collected waste was increasing through the years. In order to fulfil the goal set by the European Directive (and corresponding Spanish laws), the promotion of the recycling of white line products was required. In addition, the Basque Government has authorized the creation of eight management bodies, corresponding to the treatment of electric and electronic equipment, air-conditioners, computer systems, mobile telephones, lamps and general lighting systems.

The plan of the Basque Government for years 2005-2020 [Diputación Foral de Bizkaia, 2005] comprises two clear objectives:

- To establish by 2012 a degree of urban waste generation per person equivalent to what was collected in 2001.
- To reduce by 2012 the urban waste allocated to landfills to a 45% (and 30% by 2020).

3.4. Impact

Apart from the statistics regarding the collection of WEEE and their diversion from the landfills, one of the main impacts of this treatment of WEEE is the social one, by means of the generation of jobs for people at risk of social discrimination. These groups of people are affected by:

- Labour factors: unemployment, under-employment, temporariness, job insecurity, lack of social security, lack of labour expertise….
- Social factors: insufficient or non-regular incomes, debts, homelessness.
- Cultural factors: ethnic minorities, foreigners, illiteracy.
- Personal factors: age, genre, mental and physical disabilities, illness, alcoholism, drug addiction, violence, exile, refugee status, lack of family relationships…

About 7% of companies and agents working with the least favoured groups in society are related to the recycling of WEEE, by means of the collecting sites named Garbigunes.

In the case of the Basque Country a new program was created, Berziklatu, promoted by the local government and two companies working in social initiatives, Rezikleta and Emaus. The goals are:

- To guarantee the reuse and recycling of bulky waste (including WEEE).
- To promote the social and labour insertion of people at risk of social exclusion.
- To promote the environmental consciousness of the citizenship.
- To reduce the social costs in employment and use of raw materials.
- To contribute to the entrepreneurship of the region.

In addition, the companies set up a service to repair WEEE, mainly discarded domestic appliances, and offered these refurbished products to people with lesser incomes for a lower price.

People working at Berziklatu follow a 2 year Plan of Insertion with several stages, in order to develop professional competences, improve qualification, know the ways to a better access to jobs, develop personal resources (autonomy, responsibility…). Therefore, instead of the costs in social subsidies, there are new incomes to the Social Security.

Table 4. Program Birziklatu: data from 2007

<table>
<tr><th>Waste input</th><th>Total kg</th><th></th><th>Waste output</th><th>Total kg</th></tr>
<tr><td>Garbigune (wood)</td><td>627,050</td><td rowspan="8"></td><td>Reuse</td><td>25,749</td></tr>
<tr><td>Garbigune (white line)</td><td>42,400</td><td>Recycling</td><td>685,045</td></tr>
<tr><td>Garbigune (bulky)</td><td>63,160</td><td rowspan="2">Secondary use (co-generation)</td><td rowspan="2">653,360</td></tr>
<tr><td>Street collection by municipality</td><td>940,022</td></tr>
<tr><td>Home collection by municipality</td><td>110,590</td><td>Energy valorization</td><td>314,420</td></tr>
<tr><td>Temporary storage site of WEEE</td><td>320,929</td><td rowspan="2">Management of hazardous waste</td><td rowspan="2">997</td></tr>
<tr><td>Others</td><td>8,200</td></tr>
<tr><td>Total</td><td>2,112,351</td><td>Total</td><td>1,679,571</td></tr>
</table>

4. Case-Study: Reuse of Glass coming from Cathode Ray Tubes

As an example of new technologies applied to the separation of WEEE, we have proposed a new methodology for the definition and implementation of a recycling line for TV sets and computer monitors. The objective of the system is to achieve the highest economic benefit through the sale of the materials contained in TV and monitors. In addition, in this case there is a special emphasis on the recovery of glass from the cathode tube (CRT) of TVs and monitors.

Glass and plastics, unlike iron, copper and aluminium, are materials which are very difficult to extract from the appliance, and reuse is not yet well achieved in new applications. This glass may be redirected to a secondary market. If its proportions of heavy metals or other contaminants are high, the glass is very welcome by manufacturers of low-quality glass, as it can be melted at lower temperatures. Sometimes it is necessary to separate panel and funnel parts of the CRT, since the two fractions are reused in different ways due to their respective content of lead and other components, which restricts its future reuse.

There are some experiences in the reuse of recycled crushed glass for Portland cement and concrete (Shi and Zheng, 2007), although main concern is the expansion and cracking caused by glass aggregates. For this work a potential final user of the obtained glass has been identified (manufacturer of glazed tilcs). Data used for this implementation have been obtained from the actual production information of a recycling company, located close to the potential final user of the obtained glass. Both companies are located in the Northeast area of Spain, a highly populated region where an effective collection of WEEE was established in 1999.

Figure 5. Cathode ray tube.

Graphically, separation tasks applied to the recycling of TV sets and monitors from PCs are represented in Fig. 6. They also comply with the European Directives and national regulations in force in Spain, concerning the recycling of electronic and electric equipment.

Next steps are the design of treatment stations or cells, together with the simulation and optimization of the disassembling operations. The proposed process in previous phases is represented by means of a "virtual prototype" (Fig. 7b), where the operations are represented by 3D models and their function is analyzed using simulation techniques. The objective is to assess the technical solution proposed in the previous phase, taking into account all the data, constraints and conclusions obtained. As a result, an optimized recycling process definition is obtained, in compliance with the required objectives of material recovery.

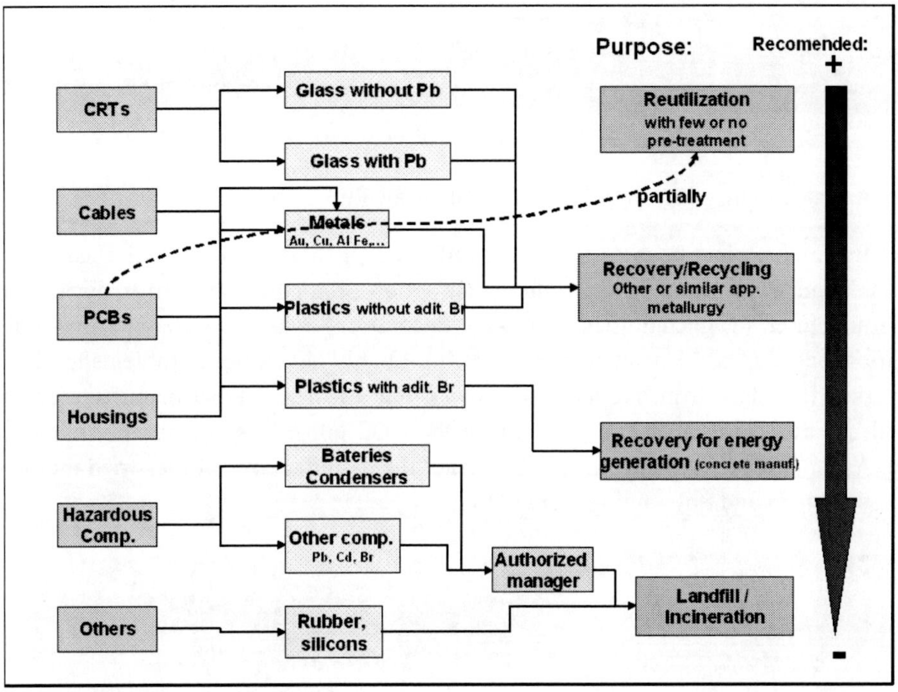

Figure 6. General diagram of components of TV and monitors, and their destination depending on their environmental impact.

Once the most optimized sequence, machines and operations are selected, the recycling plant is configured accordingly (Fig. 7a). In consecutive workstations, internal components, cables, boards and screens are dismantled. These stations can be manual, semi-automatic or fully automatic.

Operation in this cell is as follows: a transportation belt is used to enter CRTs into this cell. The first task is performed by a system which identifies the presence of lead in the panel glass. Detection is based on an X-ray camera; the result of its analysis determines further treatment of the CRT, sending appropriate commands to the robot and other machines. If the panel is lead free, the CRT must be separated into two parts: funnel and panel; otherwise it can be shredded without previous separation. The robot uses a vacuum gripper to handle the CRT and transport it to the next workstations, where a rotating saw cuts the CRT along the joining line between panel (unleaded) and funnel. The final step of the treatment is cleaning of glass fragments (coming from funnel and panel). This is done by means of a vibrating sand bath. Results fulfil the requirements of the manufacturers of glass (for new CRTs or other new applications).

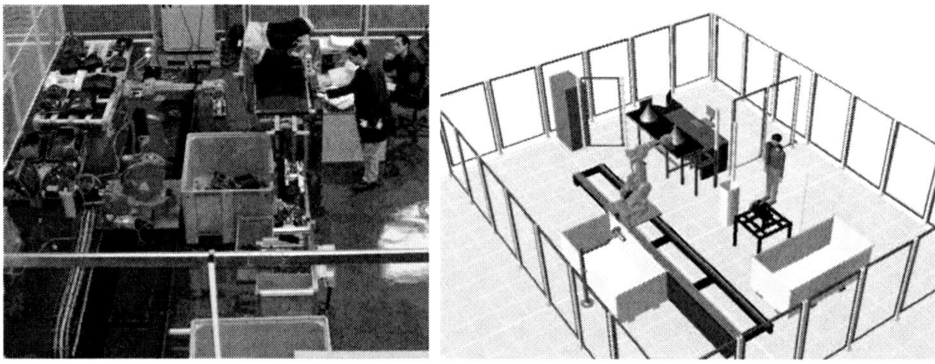

a b

Figure 7a. Actual recycling line, 7b: Virtual model of the facility.

Specific research has been carried out for the use of the recycled panel glass in several applications and fields. Among them the construction sector has been an important one. A local manufacturer of glazed tiles (for wall decoration) was contacted to find out new possibilities for the glass. This manufacturer will probably have future problems to obtain the raw material (flat glass from recycled windows), due to the fact that manufacturers of flat glass will use this material in order to reduce the CO2 emissions and comply with Kioto´s protocol. As an alternative source of raw material, the manufacturer of tiles tried the recycled panel glass (Fig. 8a and 8b) coming from CRTs.

a b

Figure 8a. Obtained glass from panel, 8b: manufactured tiles using recycled glass.

After the initial production of tiles using the recycled glass, some tests have been carried out regarding the obtained quality. The first results are promising, although some problems have to be fixed, mainly those related to the obtained colour of the tiles, impurities and risks of internal tensions. It seems that one of the influential factors may be the different level of granulation of the pounded glass.

CONCLUSION

As general conclusion of the treatment of WEEE and recycling technologies, research in the following areas is recommended (IHOBE, 2004; IPTS, 2000; Martin, 2006):

- Eco-design (design for the environment or for the recycling), applied to new products
- New technologies for the dismantling operations and design of appliances focused on dismantling
- Substitution of hazardous components
- Design and test on pilot plants, and export results to a great scale, to study their environmental impact.
- Development of Life Cycle Analysis models for products and components
- Better identification and characterization of materials, to optimize the ways to separate them and their environmental impact (the electronic waste is heterogeneous and complex in terms of the type, size, and shape of components and materials).
- New developments for information interchange about the products and their recyclability. Waste electric and electronic equipment has been taken into consideration not only by the government but also by the public. With the climate change being of concern, mechanical/physical processing will play an essential role in upgrading of WEEE.
- More accurate economic and financial models. Up to now, the recycling of electronic waste has been mainly focused on the recovery of precious metals (from personal computers and printed circuit boards). However, it is also important to take into account the electronic scrap that contains very low-grade precious metals, such as brown goods.
- Apart from the economic and technological point of view, other issues regarding employment should be taken into account. According to some impact assessment studies the recycling of 10000 t of waste creates around 240 jobs, while incinerating creates 20 to 40 jobs, and land filling about 10.

Recently the European parliament has approved a new Directive introducing more restrictive goals for take back and recycling of WEEE. Now consumers are able to take back their old small electronic appliance without the need of purchasing a new one. On the other side, manufacturers will have less administrative work. The new Directive says that, by 2016, the majority of EU countries will have to take back 45 t of electronic waste per 100 t of appliances introduced in the market previously.

In addition, more constraints will be applied to avoid illegal shipments of WEEE to developing countries, where their treatment is dangerous for workers and the environment.

There will be a growth in the consolidated collection and recycling services to allow more efficiencies and economies of scale. Household WEEE will be recycled in larger sites because volumes of collected equipment will increase. This fact will be the key for the developments of new technologies for treatment, so cost is expected to decrease. However, handling, transport and sorting are important components of the WEEE recycling costs, which are difficult to optimize, so they are expected to remain steady. With respect to organization issues, some new key contractors will appear at national and European level (with high capacity to collect and treat WEEE), and they will absorb the smaller stackeholders.

References

Andreola, F.; Barbieri, L.; Corradi, A.; Lancelloti I. *CRT glass state of art. A case study: Recycling in ceramic glazes*. J. Eur. Ceram. Soc. 2007, 27(2-3), 1623-1629.

Batayneh, M.; Iqbal, M.; Asi, I. *Use of selected waste materials in concrete mixes.* Waste Manage. 2007, 27(12), 1870-1876.

Commission of the European Communities. *Commission Staff Working Paper accompanying for the Proposal of a Directive of the European Parliament and of the Council on waste electrical and electronic equipment (WEEE);* Brussels, BE, 2008.

Diputación Foral de Bizkaia. II *Plan Integral de gestión de residuos urbanos de Bizkaia.* Departamento de Publicaciones: Bilbao, ES, 2005.

Dyckoff, H.; Lackes, R.; Reese, J. *Supply Chain Management and Reverse Logistics;* Springer: Berlin (DE), 2004.

Geskin, E.S.; Goldenberg, B.; Caudill, R. Development of advanced CRT disassembly technology. In Proceedings of the IEEE International Symposium on Electronics and the Environment; IEEE: San Francisco, CA, 2002; pp 249-253.

Gobierno Vasco. *Recogida y tratamiento de equipos eléctricos y electrónicos en su fin de vida.* Flujo doméstico + Flujo ofimático: Una experiencia mixta. Gobierno Vasco. Viceconsejería de Medio Ambiente: Vitoria, ES, 1996.

Cui, J.; Forssberg, E. *Mechanical recycling of waste electric and electronic equipment: a review.* J. Hazard. Mater. 2003, 99(3), 243-263.

Cui, J.; Forssberg, E. *Characterization of shredded television scrap and implications for materials recovery.* Waste Manage. 2007, 27(3), 415-424.

Hischier, R.; Wäger, P.; Gauglhofer, J. *Does WEEE recycling make sense from an environmental perspective?* The environmental impacts of the Swiss take-back and recycling systems for waste electrical and electronic equipment (WEEE). Environ. Impact Asses. Rev. 2005, 25 (5), 525-539.

IHOBE (Agency for Environment of the Basque Government). *Monografía de aparatos eléctricos y electrónicos. Vitoria: Gobierno vasco.* Viceconsejería de Medio Ambiente: Vitoria, ES, 2004.

Ihobe. *Inventario de residuos urbanos de la CAPV 2009.* Vitoria: Ed. Gobierno vasco. Viceconsejería de Medio Ambiente, 2011.

IPTS (Institute for Prospective Technological Studies). *Towards a European solution for the management of waste from electric and electronic equipment.* European Commission – Directorate-General Joint Research Centre: Brussels, BE, 2000.

Mangan, J.; Lalwani, C.; Butcher, J. *Global logistics and supply chain management;* John Wiley and Sons: Chichester, UK, 2008.

Martin, I. *Europe en route to a 'recycling society'.* Recycl. Int. 2006, 38-43.

Prada G. *Evaluación tecnológica de alternativas para el reciclado de RAEE.* Master Thesis, Universitat Politécnica de Catalunya. 2011.

Proventia Automation. [Last accessed: may 2008]. Access: *[http://www.headinvest.fi/sivu/fi/yritys/uutiset/?tila=yksi&uid=3&tyyppi=0]*

Rentería, A.; Álvarez, E. *A review of automation alternatives for recycling plants of electronic appliances.* In Proceedings of FAIM – 17th International Conference on Flexible Automation and Intelligent Manufacturing; Qiu, R.; Philadelfia State University: Malvern, PH, 2007; pp 996-1003.

Rentería, A.; Álvarez, E.; López, J.M. *A simulation based methodology to define the reuse process of printed circuit board's components.* In Proceedings of FAIM – 18th International Conference on Flexible Automation and Intelligent Manufacturing; De Vin, L.J.; University of Skövde: Skövde, SW, 2008; pp 397-403.

Sony Corporation. *Social & Environmental Report* 2002. 2002, 35.

Sony Corporation. *Corporate Social Responsability* Report 2003. 2003, 58.

Sony Corporation. *Corporate Social Responsability* Report 2007. 2007, 63.

Thomson, D. *Views on an emerging issue of concern - the collection and recycling of electronic products, by Electronic Manufacturer's Coalition for Responsible Recycling.* [Last access: november 2008].]. Access: http://www.electronicsrecycling.org/NCER/UserDocuments/Thompson.doc.

Widmer, R.; Oswald-Krapf, H.; Sinha-Khetriwal, D.; Schnellmann, M. and Böni, H. *Global perspectives on e-waste.* Environ. Impact Assess. Rev. 2005, 25, 436-458.

Zoeteman, B.C.J.; Krikke, H.R.; Venselaar, J. *Handling WEEE waste flows: on the effectiveness of producer responsibility in a globalizing world.* Int J Adv Manuf Technol. 2010, 47, 415-436.

Zonnelveld, N. *A treatise on the (illegal) export of WEEE.* In Proceedings of 6th International Electronics Recycling Congress; Geerts, F.; ICM: Hamburg, DE, 2007; pp 17-19.

In: Landfills
Editors: G. B. C. Cabral and B. A. E. Botelho
ISBN: 978-1-62257-121-5

Chapter 9

TOXICITY OF LANDFILLS ASSESSED BY PLANT CYTOGENETIC APPROACHES

Larissa Fonseca Andrade-Vieira *

Departamento de Produção Vegetal
Centro de Ciências Agrárias, UFES
Alegre-ES, Brazil

ABSTRACT

Every year, millions of tons of municipal solid waste are deposited in landfills. Upon rainfall, the production of leachates represents a serious environmental problem. Household waste dumped in landfills is mostly not correctly handled and treated. As a result, landfill leachates can pollute underground water. In this context, it is important to set up simple and accurate methods for monitoring the toxicity of landfill leachates. Many assays are available for this evaluation, employing plants, microbes, aquatic animals, mammals and human peripheral blood lymphocytes as test models. Among the bioassays used to detect toxicity and mutagenicity of environmental pollutants, plant systems have proved to be sensitive, cheap and effective alternatives, finding a wide range of applications as indicators of cytogenetic and mutagenic effects of contaminants. Some of these models evaluate mutations, e.g. the Stamen Hair Mutation (SHM) assay in *Tradescantia*, while others apply cytogenetic parameters to assess genomic modifications, e.g. the *Allium cepa* test. These species have been employed in studies of landfill leachate impact under natural and experimental conditions. The present chapter offers a review of these essays and their efficiency in evaluating cytogenotoxic and mutagenic effects of landfill leachates. Emphasis will be given to their applications, methodologies and endpoints.

* Departamento de Produção Vegetal, Centro de Ciências Agrárias – UFES, Alegre-ES 29500-000, Brazil, Tel: +55 28 3552-8621, Fax: +55 28 3552-8933, E-mail: larissa.cca.ufes@gmail.com.

1. Problems and Toxicity of Landfill Leachates

The exacerbated production of solid litter represents a major problem in the management and handling of urban wastes (Bortolotto et al., 2010). Bakare et al. (2005) described that estimated 0.5 to 4.5 kg of solid residues are generated per person per day, with the production varying in different regions of the world. The options available for disposal of these solid residues are either municipal solid waste (MSW) landfills or incinerators, with the first representing the most widely practiced method of solid waste disposal worldwide (Sang et al., 2010; Bardena et al., 2011). As much as 95% of the total MSW collected worldwide is disposed of in landfills (El-Fadel et al., 1997).

The main pollution threat associated with landfill sites is the production of both potentially explosive gases and liquid leachate (Sang et al., 2010). Leachate emissions from landfill sites are of growing concern, as they may contaminate soil, ground or surface waters. This has been recognized as one of the greatest issues concerning the maintenance of municipal landfills, in view of the risks to environment and human health (Bardena et al., 2011; Kwasniewska et al., 2012). Landfill leachates are liquid effluents, mainly generated by the penetration of precipitated water into the mass of residues undergoing biodegradation at these sites (Bernard et al., 1997; Renou et al., 2008). The infiltration and percolation of rain, ground or flood water through the waste layers of active or abandoned landfill sites, combined with physical, chemical and microbial processes in the waste, transfers pollutants from the waste material to the percolating water, resulting in a solution that may be harmful to a number of organisms, including humans, or to environmental receptors (Bardena et al., 2011; Kjeldsen et al., 2002). Because these leachates can reach distances over 100 m far from the landfill, they can lead to the pollution of subterranean aquifers, as well as surface water in rivers, lakes and other water bodies (Christensen et al., 2001). In addition, even small amounts of landfill leachate can pollute large volumes of groundwater, rendering it improper for domestic and various other uses (Bakare et al., 2000). Another primary issue of pollution by landfill leachate is the potential of landfill sites to generate leachate for hundreds of years following landfill closure (Jones et al., 2006).

Landfill leachates are highly heterogeneous, containing a wide range of organic and inorganic compounds that pose risks to public health and the surrounding environment (Kwasniewska et al., 2012). Leachate composition can vary across one landfill and among distinct ones, mainly depending on waste degradation time. Other factors that affect leachate composition are waste type and landfill technology (Kjeldsen et al., 2002). Despite variability in composition, leachates contain a few basic components from four pollutant groups: dissolved organic matter, inorganic salts, heavy metals and xenobiotic organic compounds (Christensen et al., 1994). Within these groups, over 200 xenobiotic organic compounds have been identified, including hydrophobic, volatile, aromatic and aliphatic organic substances in MSW leachates (Benfenati et al., 1999, 2007; Paxeus, 2000; Schwarzbauer et al., 2002; Bardena et al., 2011; Matejczyk et al., 2011). All these compounds constitute potential hazard when the leachates are released into the environment (Benfenati et al., 2007; Matejczyk et al., 2011). However, in spite of being environmentally toxic, accurate determination of toxicity potential and available treatments to solve the issue is cumbersome, owing to the lack of information from physical-chemical analyses.

The necessity of a more complete database and for more information about the compounds present in leachates, with respect to their identity, concentration and toxicity, has been established as an urgent matter (Matejczyk et al., 2011). The monitoring of landfill leachate quality has historically relied on chemical analyses (Kwasniewska et al., 2012). However, not all organic compounds present in landfill leachates can be identified, owing to the limitations of traditional chemical analyses, in which the number of chemicals that can be routinely detected is still limited, and the very low concentration of each compound (Matejczyk et al., 2011). Moreover, the high number of contaminants in landfill leachates, though appraisable with chemicals methods, does not offer insights into the biological effects of the toxicants, nor into the joint effects of mixtures containing known and unknown compounds on the biota (Kwasniewska et al., 2012). Furthermore, the lack of knowledge about their environmental properties, eco- and toxicological parameters and potential biological effects hinder any attempt to estimate the risk of leachates to humans and environmental receptors (Kwasniewska et al., 2012). In this context, some authors have proposed studies combining physical-chemical examination and (geno)toxicological approaches to improve the assessment of risks associated with landfill leachates (Christensen et al., 2001; Kwasniewska et al., 2012).

Evaluation of landfill leachates by biological toxicity tests (biotests) provides a direct, functional picture of the overall toxicity properties of the compound mixture present in the leachate samples (Matejczyk et al., 2011). Furthermore, it is well known that biological and chemical transformations in leachates can lead to the formation of toxic substances from relatively innocuous organic compounds, and that these are a potential risk to the environment and living organisms (Slack et al., 2005). In such cases, biological tests represent a useful complement to physical-chemical analyses, as they investigate the interactions between pollutants. Genotoxicity bioassays are tools for assessing the effect of numerous contaminants, or a combination of those, on the integrity of the genetic material, and appear to be more helpful in estimating such risks than a precise chemical characterization alone (Lah et al., 2008).

In conclusion, this chapter aimed at providing a summary of plant cytogenetic assays available to monitor the (geno)toxic potential of environmental pollutants, emphasizing the applications, methodologies and endpoints that can be assessed, and their correlations with other biological tests applied for the same purposes.

2. The Use of Higher Plants in Environmental Pollution Approaches

Plants are immobile organisms. Under pollution stress, roots are the plant's primary point of contact with contaminants. To some extent, plants could overcome environmental stress by developing efficient and specific physio-biochemical mechanisms (Sandalio et al., 2001). The influence of environmental mutagens on a plant depends not only on the type and dose of mutagen, exposure time and interaction with other factors, but also on plant species, genotype and developmental stage (Gichner, 2000). Moreover, plant response to mutagenic treatment can be considered at different levels of organization: from DNA, chromosome and genome to the whole organism (Maluszynska and Juchimiuk, 2005). Several plant tests are commonly used to biomonitor pollution levels and evaluate the effects of toxic and mutagenic agents present in the natural environment.

In order to prevent DNA modifications caused by environmental factors, it is important to understand the biological consequences of DNA damage, as well as the molecular tools that lead to repair or alterations of the genetic material. Numerous genotoxicity assay systems have been developed to identify DNA-reactive compounds. The available data show that plant bioassays are useful for detection of genotoxic contamination in the environment, and in the establishment of controlling systems. Plant systems are able to detect a wide range of genetic damages, including gene mutation and chromosome aberrations (Maluszynska and Juchimiuk, 2005). In this regard, several higher plant bioassays for screening and monitoring environmental mutagens have been established in recent years (Constantine and Owens, 1982; Grant, 1982), for instance using *Allium cepa*, *Hordeum vulgare*, *Arabidopsis thaliana*, *Glycine max*, *Vicia faba* and *Zea mays*, among other suitable bioindicators of genetic toxicity caused by environmental pollutants (Liu et al., 2005). From these, *A. cepa* appears as the most efficient and often applied test for the purpose of assessing chromosome damage, owing to the species' large chromosomes, amenable to the study of chromosome aberrations in somatic cells (Grant et al., 1994, 1998; Caritá and Marin-Morales, 2008). Other reported plants such as *Z. mays* and *Tradescantia* are used in the study of gene mutations and chromosome aberrations (Grant, 1994), while *V. faba* is currently used in micronucleus assays to evaluate the genotoxic potential of contaminated materials (Foltête et al., 2011).

Genetic assays involving higher plants have the following advantages: (i) relatively low cost and easy handling; (ii) size of plant chromosomes, showing good correlation with other bio-testing systems; (iii) short generation times; (iv) wide range of environmental conditions, pH and temperature under which assays can be conducted; (v) evaluation of genotoxicity from single chemicals to complex mixtures; (vi) wide range of genetic endpoints, such as cytological alterations and gene mutations in whole plants, leaves, embryos and pollen; (vii) detection of intermediate mutagenic metabolites (promutagens), when combined with microbial assays; (viii) high sensitivity (i.e. few false negatives) in predicting carcinogenicity of tested agents; (ix) existence of many hundreds of genetic loci that can be monitored; and (x) relatively brief training of technicians to conduct plant assays (Grant, 1994).

3. Plants Cytogenetic Assays to Assess Environmental Contamination

As previously mentioned, plant bioassays are amenable for measuring the genotoxicity of environmental pollutants, including aqueous leachates originated from MSW. Several plant species have been used as bioindicators, and a variety of tests, including cytogenetic protocols, have been employed (Juchimiuk and Maluszynska, 2005). These tests are commonly used for biomonitoring the extent of pollution and evaluating the toxicity and mutagenicity of environmental contaminants on vegetal organisms (Liu et al., 2005; Matsumoto et al., 2006). In this sense, it is relevant to consider that roots are usually the primary point of contact of plants with contaminants (Fiskesjö, 1988; Sandalio et al., 2001; Liu et al., 2005).

Many plant species find a wide range of applications as indicators in cytogenetic procedures. In the same context, tests in plants are relatively inexpensive and simple to perform. Owing to the size of their chromosomes, higher plants are suitable for cytological analyses, and have shown good correlation with other bio-testing systems (Fiskesjö, 1985). Plant roots are extremely useful for biological testing. Because root tips are often first to be exposed to chemicals naturally dispersed in soil or water (Fiskesjö, 1988), their observation constitutes a rapid and sensitive method for environmental monitoring.

Cytogenotoxicity of environmental pollutants assessed through plant species is based on analysis of the cell cycle, thus identifying the nature of the studied compound as genotoxic, mutagenic or carcinogenic (Leme and Marin-Morales, 2009). Cell cycle analysis considers the percentage of dividing cells, the type and frequency of chromosome aberrations in mitotic cells, and the frequency of micronuclei and other nuclear alterations occurring in interphasic cells. Further, the following parameters are observed: (i) mitotic index (MI), calculated as the fraction of dividing cells in the total of observed cells; (ii) percentage of chromosome aberrations (CAs), expressed in number of aberrations – c-metaphases, stickiness, bridges and laggards – divided by the total number of cells; and (iii) nuclear aberrations (NAs), determined by the frequency of condensed nuclei, lobated nuclei, and poly- and micronucleated cells in interphase (Andrade et al., 2008; Leme et al., 2008).

Considering these parameters, CAs define the genotoxicity of the substance or compound, whereas MI and some NAs are employed to determine cytotoxicity and analyze micronuclei in order to verify mutagenicity (Leme and Marin-Morales, 2009). MI is reliable for determining presence of cytotoxic compounds in the environment, and is a suitable test for biomonitoring pollution levels (Smaka-Kincl et al., 1996; Fernandes et al., 2007). Inhibition of mitotic index can be attributed to effects of environment chemicals on DNA and protein synthesis in a biological system (Chauhan et al., 1998). The mechanisms of action of these chemicals, leading to occurrence of CAs and NCs, as well as the type of these alterations, are discussed below.

The mentioned plant bioassays also detect three different genetic endpoints: the formation of chromosome aberrations, the induction of micronuclei, and the induction of point mutations. These studies may also consider different cell types (mitotic and meiotic plant cells) (Andrade-Vieira et al., 2011).

3.1. Chromosome Aberration (CA)

Genotoxic agents cause DNA damage, which is either subsequently repaired or leads to DNA alterations. CAs are the consequence of DNA double-strand breakage that stays unrepaired or is improperly repaired. Broken chromosome ends lacking telomeres become "sticky", and may fuse with other broken chromosome ends. The result of these chromosomal rearrangements are acentric fragments; dicentric bridges, observed in mitotic cells of the first cell cycle after mutagenic treatment; or micronuclei in the interphasic cells of the following cell cycle (Campos et al., 2008). Moreover, the mechanism of action of the substance or compound on the genetic material can be classified by CA assays: clastogenic, when involving chromosome breakages, or aneugenic, when leading to laggard chromosomes and/or affecting the spindle (Leme and Marin-Morales, 2009).

Sticky chromosomes are an aneugenic type of CA, representing a change in chromosome structure that leads to loss of characteristics typical of normal chromosome condensation, resulting in formation of agglomerates (Babich et al., 1997). The sticky appearance is due to action of the toxic compounds on the physical-chemical structure of the DNA and/or protein, leading to complex formations involving DNA phosphate groups, DNA condensation or the formation of inter- and intra-chromatid links (El-Ghamery et al., 2003). This change is irreversible and generally leads to cell death.

Another aneugenic CA frequently observed as a consequence of environmental pollutants is the c-metaphase. This type of change is thus named based on the observation of chromosomes around the equator of the cell, as also seen in cells treated with colchicine. The observation of c-metaphasic cells indicates that the toxic agent acts on the spindle fibers, preventing continuation of the mitotic cycle (Leme and Marin-Morales, 2009). As a result, the cell cycle is interrupted at metaphase, and the chromosomes are seen condensed, with very well defined centromere scattered inside the cell (Fiskesjö, 1985).

In addition, influence of chemicals on the spindle fibers and consequent abnormality in microtubule assembly induce multipolar anaphases and lagging chromosomes in metaphases or anaphases. Multipolar anaphases reflect the disturbances in microtubule dynamics (Haroun and Shehri, 2001). Lagging chromosomes are probably induced by interference in the normal chromatid separation, which leads to an unequal number of chromosomes in the daughter nuclei, and subsequently to the formation of daughter cells with unequally sized or irregularly shaped nuclei at interphase (El-Ghamery et al., 2003). In some cases, the lagging chromosomes are surrounded by a nuclear membrane, forming micronuclei.

In the context of clastogenic alterations, it is possible to observe anaphase/telophase bridges and chromosome fragments in plant cytogenetic tests. Bridges can be attributed to chromosome breaks, stickiness or breakage and reunion of broken ends (El-Ghamery et al., 2003). Bridges can also result from terminal deletion or loss of telomeres, following the fusion of broken chromosome ends, leading to the formation of dicentric chromosomes that attach to both spindle poles. Consequently, chromosomes are pulled simultaneously to both poles of the microtubule spindle during chromosome segregation (Hoffelder et al., 2004). Chromosome breaks lead to the formation of acentric chromosome fragments. The presence of chromosome fragments is thus an indication of chromosome breaks, being a consequence of anaphase/telophase bridges (Sharma and Sen, 2002; Singh, 2003).

3.2. Micronucleus (MN)

The micronucleus test was developed in parallel to CA assays, as micronuclei originate from the acentric fragments or lagging chromosomes that fail to incorporate into the daughter nuclei during telophase of the mitotic cells (Fiskesjö, 1985; Yi and Meng, 2003). As a consequence, they appear as an extranuclear body of chromatin, resulting from chromosome breakage or aneuploidy, respectively reflecting both clastogenic and aneugenic actions of a substance or compound (Leme and Marin-Morales, 2009). Ultimately, cell death may occur due to deletion of primary genes (Yi and Meng, 2003).

Additionally, MN frequency is a good indicator of the cytogenetic effects of tested chemicals. It is a sensitive biomarker, emphasizing the greater relevance of genotoxic biomarkers compared with the classical endpoints – CAs (Foltête et al., 2011). Besides, the

MN test can be conducted in mitotic cells, e.g. in *Allium* or *Vicia* MN assays, or in meiotic cells, e.g. in the Trad-MN assays described below.

3.3. *Tradescantia* Bioassays

Tradescantia bioassays are among the few higher plant systems frequently used for environmental genotoxicity detection (Ma et al., 1994, 1996; Gong et al., 2003).

Two genotoxic assays based on *Tradescantia* are available: i) the *Tradescantia* staminal hair assay (Trad-SHM), and ii) the *Tradescantia* micronucleus assay (Trad-MN) (Grant, 1999). Trad-SHM (Stamen Hair Mutation) is an assay based on the phenotypical observation that reflects the mutation effects on the genome. It is a somatic mutation bioassay in which expression of the heterozygous dominant blue character of the stamen hair cell is prevented, resulting in the appearance of the recessive pink color (Underbrink et al., 1973). It was determined that the pink pigmentation depended on a pair of alleles from a single locus, with allele B (blue) dominant over allele b (pink) (Rodrigues et al., 1997). Trad-MN is a genotoxic assay based on counting the MN found in pollen grain mother cells in the tetrad phase. MN formation results from whole or fragmented chromosomes that were lost, failing to incorporate into the nucleus of the main daughter cell during cell division, remaining in the cytoplasm of interphasic cells (Rodrigues et al., 1997; Feretti et al., 2008). This aberration can cause cell death due to deletion of primary genes (Yi and Meng, 2003), consequently leading to pollen non-viability. MN induction reflects clastogenic and/or aneugenic events. MN formation as a consequence of fragments is a clastogenic effect, whereas their formation from lagging chromosomes represents an aneugenic effect (Leme et al., 2008).

These tests are very useful tools for screening mutagenic potential in the environment, and are considered more efficient and sensitive than CA assays in mitotic cells (Ma et al., 1994; Knasmüller et al., 2003; Meireles et al., 2009).

4. Advantages of Using Allium Test as Indicator of Environmental Pollution

Among the plant species cited here, *A. cepa* is the most used as test system to evaluate the mechanism of action of environmental pollutants, by analysis of CA, MN and NA, as well as by Comet assay (Leme and Marin-Morales, 2009; Yildiz et al., 2009). In this sense, *A. cepa* roots are regarded as one of the pioneering materials to detect the effects of chemicals on plant chromosomes in cytogenetic studies. Furthermore, the *A. cepa* test, developed by Levan in 1938, has been used by the Environmental Protection Agency (US EPA) to evaluate toxicity and genotoxicity of industrial effluents (Fiskesjö, 1993; Grant, 1994; Mitteregger Júnior, 2007). The *A. cepa* test is a simple, sensitive and rapid bioassay, widely used as standard for biomonitoring environmental contaminants, toxicity of chemical compounds and evaluation of potential anticancer properties (Kurás and Malinowska, 1978; Podbielkowska et al., 1981, 1995; Kupidłowska et al., 1994; Keightley et al., 1996; Majewska et al., 2003;) using various genotoxicity parameters (Kurás et al., 2006). In addition, *A. cepa* has a number of very convenient characteristics: being available all year round; containing rather homogenous meristematic cells; exhibiting only 16 chromosomes, which are very long and well visible; getting stained easily; and allowing easy detection of chromosome aberrations. Considering all this, the test is a fast and inexpensive method, allowing investigation of universal mechanisms involved in plant response to treatment of root tip cells with a mixture of toxic wastes (DeMarini, 1991; Fiskesjö, 1993).

Apart from its sensitivity, this method also presents high correlation with other test systems. These features are essential for an accurate assessment of environmental risks, as well as for successful extrapolation of test results, obtained in exposed organisms, to other species. Regarding these characteristics, Fiskesjö (1985) showed that the *A. cepa* test presents sensitivity to similar that of some algae and human lymphocyte test systems. Rank and Nielsen (1994) showed a correlation of 82% between *A. cepa* test and carcinogenicity assays in rodents; they also demonstrated that the former was more sensitive than Ames or microscreen tests. Furthermore, studies of sensitivity amongst higher plants have also shown that *A. cepa* is more sensitive than some other species, such as *Vicia faba* (Ma et al., 1995; Migid et al., 2007). Feretti et al. (2008) showed that the CA test in *A. cepa* was more sensitive than the MN test in *Tradescantia* in analyzing the genotoxicity of chlorite and chlorate ions. The authors concluded that using different test systems is important for a more comprehensive assessment of the different action aspects of an agent, this way providing a complete overview of its hazards. *A. cepa* test system also has a good correlation with mammals, as reported by Chauhan et al. (1999), who compared the *A. cepa* test with rodent bone marrow and human lymphocytes tests in a study of cytogenetic effects of pyrethroid insecticides. Based on these results, the authors suggested validating the *A. cepa* test as an alternative for mammal test systems, to monitor the genotoxic potential of environment chemicals. Regarding the evaluation of landfill leachates, Bortolotto et al. (2009) recently demonstrated that the bioassay using *A. cepa* is more sensitive than using *Daphnia similis* and *Artemia salina*, and as efficient as using *Geophagus brasiliensis* in Comet assay to demonstrate the genotoxic effects of leachates. Furthermore, because tests to detect risks for the human health may not be carried out in humans, for ethical reasons, and considering the urgent need for an accurate evaluation of environmental risks, plant bioassays, for instance the *A. cepa* test, appear as a reliable alternative for screening environmental contamination and extrapolating data to other target organisms, such as humans (Leme and Marin-Morales, 2009).

Conclusion

A number of environmental pollution issues were identified here, especially in regard of MSW and production of leachate at landfills, as well as lack of information about leachate physical-chemical properties and the necessity of assessing its risks to humans and the environment as a whole. In view of these problems, it is concluded that the plant cytogenetic bioassays provided in this chapter are fast, sensitive and reliable tools to provide answers to these issues, by enabling the determination of genotoxic potential of contaminants present in landfill leachates. By allowing assessment of several genetic endpoints, plant cytogenetic bioassays make it possible to evaluate the mechanisms of action of contaminants/chemicals present in landfill leachates on the DNA of exposed organisms. This represents an efficient method for screening environmental contamination, and provides insights into the risks to human health and other target organisms.

References

Andrade, L.F. et al. Cytogenetic alterations induced by SPL (spent potliners) in meristematic cells of plant bioassays. *Ecotoxicol. Environ. Saf.* 71, 706–710, 2008.

Andrade-Vieira, L. F. et al. Genotoxicity of SPL (Spent Pot Lining) as measured by the Tradescantia micronucleus (Trad-MCN) and stamen hair mutation (Trad-SHM) bioassays. *Ecotoxicology and Environmental Safety.* , v.74, p.2065 - 2069, 2011.

Babich, H et a. The Allium test—a simple, eukaryote genotoxicity assay. *Am. Biol. Teach.* 59, 580–583, 1997.

Bakare, A.A.et al. Effect of stimulated leachate on chromosomes and mitosis in roots of Allium cepa L.. *Environ. Biol.* 21, 251–260, 2000.

Bakare, A.A. et al. An in vivo evaluation of induction of abnormal sperm morphology in mice by landfill leachates. *Mutat.Res.*582, 28–34, 2005

Bardena, D. et al. A combined approach to investigate the toxicity of an industrial landfill's leachate: Chemical analyses, risk assessment and in vitro assays. *Environmental Research*, v. 111, p. 603–613, 2011.

Benfenati, E. et al. A protocol to evaluate organic compounds present in a landfill. In: Velinni, A.A.(Ed.),*Landfill Research Trends,* Nova Science, pp.141–166, 2007.

Benfenati, E. et al. Comparative studies of the leachate of an industrial landfill by gas chromatography–mass spectrometry, liquid chromatography–nuclear magnetic resonance and liquid chromatography– mass spectrometry. *J Chromatogr.* A 831, 243–256, 1999.

Bernard, C. et al. Estimation of the hazard of landfills through toxicity testing of leachates: 2. Comparison of physico-chemical characteristics of landfill leachates with their toxicity determined with a battery of tests. *Chemosphere*, v. 35, p. 2783–2796, 1997.

Bortolotto, T. et al. Evaluation of the toxic and genotoxic potential of landfill leachates using bioassays. Environmental Toxicology and Pharmacology, v. 28, p. 288–293, 2009.

Campos, J. M. S. et al. Chapter 5: Genetic Toxicology and Environmental Mutagenesis in Allelopathic Interactions. In: Plant Bioassays ed.Houston : Studium Press, 2008, p. 1-344.

Caritá, R. and Marin-Morales, M.A. Induction of chromosome aberrations in the *Allium cepa* test system caused by the exposure of seeds to industrial effluents contaminated with azo dyes. *Chemosphere*, v. 72, p. 722-725, 2008.

Chauhan, L.K.S. et al. Cytogenetic effects of cypermethrin and fenvalerate on the root meristem cells of Allium cepa, *Environ. Exp. Bot.* 42, 181–189, 1999.

Chauhan, L.K.S. et al. Diuron induced cytological and ultrastrucutural alterations in the root meristem cells of *Allium cepa. Pesticide Biochemistry Physiology*, v. 62, p. 152-163, 1998.

Christensen, T.H. et al. Attenuation of landfill leachate pollutants in aquifers. *Crit. Rev. Environ. Sci. Technol.* 24, 119–202, 1994.

Christensen, T.H. et al. Biochemistry of landfill leachate plums. *Appl. Geochem.*16, 659–718, 2001.

Constantin, M.J. and OWENS, E.T. Introduction and perspective of plant genetic and cytogenetic assays. *Mutation Research,* v. 99, p. 37-49, 1982.

Demarini, D.M. Environmental mutagens/complex mixtures. In: LI, A.P.; HEFLICH, R.H. (Ed.). *Genetic toxicology*. Boca Raton: CRC Press, 1991. p. 290.

El-Fadel, M. et al. Environmental impacts of solid waste and filling. *J. Environ. Manage*. 50, 1–25, 1997.

El-Ghamery, A.A. et al. Evaluation of cytological effects of Zn2+ in relation to germination and root growth of *Nigella sativa* L. and *Triticum aestivum* L. *Mutation Research*, v. 537, p. 29-41, 2003.

Feretti, D. et al. Evaluation of chlorite and chlorate genotoxicity using plant bioassays and in vitro DNA damage tests. *Water Res*. 42, 4075 – 4082, 2008.

Fernandes, T.C.C. et al. Mechanism of micronuclei formation in polyploidizated cells of *Allium cepa* exposed to trifluralin herbicide. *Pesticide Biochemistry and Physiology*, v. 88, p. 252-259, 2007.

Fiskesjö, G. The *Allium* test as a standard in environmental monitoring. *Hereditas*, v. 102, p. 99-112, 1985.

Fiskesjö, G. The *Allium* test in wastewater monitoring. *Environmental Toxicology and Water Quality*, v. 8, p. 291-298, 1993.

Fiskesjö, G. The *Allium* test: an alternative in environmental studies: the relative toxicity of metal ions. *Mutation Research*, v. 197, p. 243-260, 1988.

Foltête, A-S. et al. Improvement of Vicia-micronucleus test for assessment of soil quality: A proposal for international standardization. *Chemosphere*, 85, 1624–1629, 2011.

Gichner, T. et al. Maleic hydrazide induces genotoxic effects but no damage detectable by the Comet assay in tobacco and field beans. *Mutagenesis*, v. 15, p. 385-389, 2000.

Gong, P. et al. Genotoxicity of 2,4- and 2,6-dinitrotoluene as measured by the *Tradescantia* micronucleus (Trad-MCN) bioassay. *Mutat. Res*. 538, 13-18, 2003.

Grant, W F. and Owens, E.T. Chromosome aberration assays in Crepis for the study of environmental mutagens. *Mutation Research*, v. 410, p. 291-307, 1998.

Grant, W.F. Cytogenetic studies of agricultural chemicals in plants. In: Fleck, R.A.; Hollaender, A. (Ed.). Genetic toxicology: an agricultural perspective. New York: Plenum Press, 1982. p. 353-378

Grant, W.F. The present status of higher plant bioassay for detection of environmental mutagens. *Mutation Research*, v. 310, p. 175-185, 1994.

Grant, W.F. Higher plant assays for the detection of chromosomal aberrations and gene mutations-a brief historical background on their use for screening and monitoring environmental chemicals. *Mutat. Res*. 426, 107–112, 1999.

Haroun, S.A. and Shehri, M.A. Cytogenetic effects of Calotropis procera extract on *Vicia faba* L. *Cytologia*, v. 66, p. 373-378, 2001.

Hoffelder, D.R. et al. Resolution of anaphase bridges in cancer cells. *Chromosoma* 112: 389-397, 2004.

Jones, D.L. et al. Phytoremediation of landfill leachate. *Waste Manage*. 26, 825–837, 2006.

Keightley, A.M.et al. Coumarin and its 4- and 7- substituted derivatives as retardants of mitoses in Allium root promeristem. *International Journal of Pharmacognosy*, v. 34, n. 2, p. 105-113, 1996.

Kjeldsen, P. et al. Present and long-term composition of MSW landfill leachate: a review. *Crit. Rev. Environ.Sci.Technol*.32,297–336, 2002.

Knasmueller, S. et al. Use of human cell lines in toxicology. *Toxicol.* 191, 15-16, 2003.

Kupidłowska, E. et al. The effect of coumarins on root elongation and ultrastructure of meristematic cell protoplasts. *Annals of Botany*, v. 73, p. 525-530, 1994.

Kurás, M. and Malinowska, A. Influence of 1-b-D arabinofuranosylcystoine on mitotic activity of apical meristem of onion (*Allium cepa* L.) roots. *Acta Societatis Botanicorum Poloniae*, v. 47, p.173-187, 1978.

Kurás, M. et al. Changes in chromosome structure, mitotic activity and nuclear DNA content from cells of *Allium* test induced by bark water extract of *Uncaria tomentosa* (Willd.) DC. *Journal of Ethnopharmacology*, v. 107, p. 211-221, 2006.

Kwasniewska, J. et al. An assessment of the genotoxic effects of landfill leachates using bacterial and plant tests. *Ecotoxicology and Environmental Safety*, v. 75, p. 55–62, 2012.

Lah, B., et al. Genotoxicity evaluation of water soil leachates by Ames test, comet assay, and preliminary Tradescantia micronucleus assay. *Environ. Monit. Assess.* 139, 107–118, 2008.

Leme, D.M. and Marin-Morales, M.A. *A. cepa* test in environmental monitoring: A review on its application. *Mutat. Res.* 682, 71-81, 2009.

Leme, D.M. et al. Action mechanisms of petroleum hydrocarbons present in waters impacted by an oil spill on the genetic material of Allium cepa root cells. *Aquat.Toxicol.* 88, 214–219, 2008.

Liu, W. et al.DNA changes in barley (*Hordeum vulgare*) seedlings induced by cadmium pollution using RAPD analysis. *Chemosphere*, v. 61, p. 158-167, 2005.

Ma, T. H. et al. *In situ* monitoring with the *Tradescantia* bioassays on the genotoxicity of gaseous emissions from a closed lanfill site and an incinerator. *Mutat. Res.*, 359, 39-52., 1996.

Ma, T.H. et al. *Tradescantia* micronucleus bioassay. *Mutat. Res.* 310, 221–230, 1994.

Ma, T.H. et al. The improved *Allium/Vicia* root tip micronucleus assay for clastogenicity of environmental pollutants. *Mutation Research,* v. 334, p. 185-195, 1995.

Majewska, A. et al. Antimitotic effect, G2/M accumulation, chromosomal and ultrastrukture changes in meristematic cells of *Allium cepa* L. root tips treated with the extract from *Rhodiola rosea* roots. *Caryologia*, v. 56, p. 337-351, 2003.

Maluszynska, J. and JUCHIMIUK, J. Plant genotoxicity: a molecular cytogenetic approach in plant bioassays. *Plant Genotoxicity*, v. 56, p. 177-184, 2005.

Matejczyk, M. et al. Estimation of the environmental risk posed by landfills using chemical, microbiological and ecotoxicological testing of leachates.*Chemosphere* 82, 1017–1023, 2011.

Meireles, J. et al. Genotoxic effects of vehicle traffic pollution as evaluated by micronuclei test in *Tradescantia* (Trad-MCN). *Mutat. Res.* 675, 46–50, 2009.

Migid, H.M.A. et al. Use of plant genotoxicity bioassay for the evaluation of efficiency of algal biofilters in bioremediation of toxic industrial effluent. *Ecotoxicology and Environmental Safety*, v. 66, p. 57-64, 2007.

Mitteregger Júnior, H. et al. Evaluation of genotoxicity and toxicity of water and sediment samples from a Brazilian stream influenced by tannery industries. *Chemosphere*, v. 67, p. 1211-1217, 2007.

Paxeus, N. Organic compounds in municipal landfill leachates. *Water Sci. Technol.* 42, 323–333, 2000.

Podbielkowska, M. et al. Effect of coumarin and its derivatives on mitosis and ultrastructure of meristematic cells. *International Journal of Pharmacognosy*, v. 33,p. 7-15, 1995.

Podbielkowska, M. et al. Influence of methotrexate on respiration and ultrastructure of meristematic cells from *Allium cepa* roots. *Acta Societatis Botanicorum Poloniae*, v. 50, p. 563-566, 1981.

Rank, J. and Nielsen, M.H. Evaluation of the *Allium* anaphase-telophase test in relation to genotoxicity screening of industrial wastewater. *Mutation Research*, v. 312, p. 17-24, 1994.

Renou, S. et al. Landfill leachate treatment: Review and opportunity. *Journal of Hazardous Materials*, v. 150, p.468–493, 2008.

Rodrigues, G.S. et al. *Tradescantia* bioassays as monitoring systems for environmental mutagenesis: A review. *Crit. Rev. Plant Sci.* 16, 325-359, 1997.

Sandalio, L.M. et al. Cadmium-induced changes in the growth and oxidative metabolism of pea plants. *Journal of Experimental Botany,* v. 52, p. 2115-2126, 2001.

Sang, N. et al. Landfill leachate affects metabolic responses of Zea mays L. seedlings. *Waste Management*, v. 30, p. 856–862, 2010.

Schwarzbauer, J. et al. Occurrence and alteration of organic contaminants in seepage and leakage from water from a waste deposit landfill. *Water Res.* 36, 2275–2287, 2002.

Sharma, A.; Sen, S. Chromosome botany. Enfield: Science Publishers, 2002.

Singh, R.J. Plant cytogenetics. Boca Raton: CRC Press, 2003.

Slack, R.J. et al. Household hazardous waste in municipal landfills: contaminants in leachate. *Sci. Total Environ.* 337, 119–137, 2005.

Smaka-Kincl, V. et al. The evaluation of waste, surface and ground water quality using the Allium test procedure. *Mutat. Res.* 368, 171–179, 1996.

Underbrink, A.G. et al. *Tradescantia* stamen hairs: a radiobiological test system applicable to chemical mutagenesis. In: Chemical Mutagens – Principles and Methods for their detection (Ed., Hollaender, A.), pp. 171-207. Plenum Press, New York, 1973.

Yi, H.L. and MENG, Z. Genotoxicity of hydrated sulfur dioxide on root tips of *Allium sativum* and *Vicia faba*. *Mutation Research*, v. 537, p. 109-114, 2003.

Yildiz, M. et al. Determination of genotoxic effects of copper sulphate and cobalt chloride in *Allium cepa* root cells by chromossome aberration and comet assays. *Chemosphere* 75, 934-938, 2009.

INDEX

A

B

C

D

F

H

I

J

K

L

M

N

O

P

Q

R

S

T

U

V

W

X

Y

Z